SOIL SALINITY MANAGEMENT IN AGRICULTURE

Technological Advances and Applications

Innovations in Agricultural and Biological Engineering

SOIL SALINITY MANAGEMENT IN AGRICULTURE

Technological Advances and Applications

Edited by
S. K. Gupta, PhD, CE
Megh R. Goyal, PhD, PE

Apple Academic Press Inc. | Apple Academic Press Inc.
3333 Mistwell Crescent | 9 Spinnaker Way
Oakville, ON L6L 0A2 Canada | Waretown, NJ 08758 USA

©2017 by Apple Academic Press, Inc.

First issued in paperback 2021

Exclusive worldwide distribution by CRC Press, a member of Taylor & Francis Group
No claim to original U.S. Government works

ISBN 13: 978-1-77-463039-6 (pbk)
ISBN 13: 978-1-77-188443-3 (hbk)

Library and Archives Canada Cataloguing in Publication

Soil salinity management in agriculture : technological advances and applications / edited by S. K. Gupta, PhD, CE, Megh R. Goyal, PhD, PE.
(Innovations in agricultural and biological engineering)
Includes bibliographical references and index.
Issued in print and electronic formats.
ISBN 978-1-77188-443-3 (hardcover).--ISBN 978-1-315-36599-2 (PDF)
1. Soil salinization. 2. Soils, Salts in. 3. Soil management. 4. Reclamation of land. I. Gupta, S. K. (Suresh Kumar), 1949-, author, editor II. Goyal, Megh Raj, editor III. Series: Innovations in agricultural and biological engineering

| S595.S64 2017 | 631.4'16 | C2016-908100-1 | C2016-908101-X |

Library of Congress Cataloging-in-Publication Data

Names: Gupta, S. K. (Suresh Kumar), 1949- editor. | Goyal, Megh Raj, editor.
Title: Soil salinity management in agriculture : technological advances and applications / editors: S.K. Gupta, Megh R. Goyal.
Description: Waretown, NJ : Apple Academic Press, 2017. | Includes bibliographical references and index.
Identifiers: LCCN 2016056054 (print) | LCCN 2016058331 (ebook) (print) | LCCN 2016058332 (ebook) | ISBN 9781771884433 (hardcover : alk. paper) | ISBN 9781315365992 (ebook)
Subjects: LCSH: Soils, Salts in. | Salinity.
Classification: LCC S595 .S642 2017 (print) | LCC S595 (ebook) | DDC 631.8/2--dc23
LC record available at https://lccn.loc.gov/2016056054

Apple Academic Press also publishes its books in a variety of electronic formats. Some content that appears in print may not be available in electronic format. For information about Apple Academic Press products, visit our website at **www.appleacademicpress.com** and the CRC Press website at **www.crcpress.com**

CONTENTS

LIST OF CONTRIBUTORS

Sanjay Arora, PhD
Senior Scientist, ICAR – Central Soil Salinity Research Institute, Regional Research Station, Lucknow – 226002, Uttar Pradesh, India. Mobile: +91-7376277190; E-mail: aroraicar@gmail.com

Shrikant Badole, PhD
Assistant Professor, Department of Soil Science and Agricultural Chemistry, College of Agriculture, Golegaon, VNMKV, Parbhani – 431705, Maharashtra, India. Mobile: +91-9604777431, E-mail: shrikantbadole358@gmail.com

A. Krishna Chaitanya
PhD Research Scholar, Department of Agricultural Chemistry and Soil Science, Bidhan Chandra Krishi Viswavidyalaya, Mohanpur, Nadia – 741252, West Bengal, India. Mobile: +91-8697414173, E-mail: krch3737@outlook.com

K. V. Chaitanya, PhD
Associate Professor, Department of Biotechnology, GITAM University, Visakhapatnam – 530045, Andhra Pradesh, India. Mobile: +91-8912840246, E-mail: viswanatha.chaitanya@gmail.com

Anil R. Chinchmalatpure, PhD
Principal Scientist and Head, ICAR – Central Soil Salinity Research Institute, Regional Research Station, Bharuch 392012, Gujarat, India. E-mail: rcanil2014@gmail.com; Mobile: +91-2642225673

Devendra Kumar Choudhary, PhD
Assistant Professor, Amity Institute of Microbial Technology (AIMT), Block E-3, 4th Floor, Amity University Campus, Gautam Buddha Nagar, Sector-125, Noida – 201313, UP, India, Mobile: +91-120-2431182, E-mail: dkchoudhary1@amity.edu, devmicro@rediffmail.com

J. C. Dagar, PhD
Ex Assistant Director General (ICAR), ICAR – Central Soil Salinity Research Institute, Karnal 132 001, Haryana, India, Mobile: +919416301661, E-mail: dagarjc@gmail.com.

Yogita Deshmukh
PhD Research Scholar, Agronomy and Soil Science Division, CSIR-Central Institute of Medicinal and Aromatic Plants, Lucknow – 226015, India, Mobile: +91-7408857399, E-mail: yogita.deshmukh2@gmail.com

Dipankar Ghorai, PhD
Subject Matter Specialist (I/C), Krishi Vigyan Kendra, ICAR – Central Research Institute of Jute and Allied Fibers (ICAR), Burdwan – 713403, West Bengal, India, Mobile: +91-9433122515, E-mail: dipankarghoraikvk@gmail.com

Megh R. Goyal, PhD, PE
Retired Professor in Agricultural and Biomedical Engineering, University of Puerto Rico – Mayaguez Campus; and Senior Technical Editor-in-Chief in Agriculture Sciences and Biomedical Engineering, Apple Academic Press Inc., PO Box 86, Rincon – PR – 00677, USA. E-mail: goyalmegh@gmail.com

Arbind Kumar Gupta
PhD Research Scholar, Department of Agricultural Chemistry and Soil Science, Bidhan Chandra Krishi Viswavidyalaya, Mohanpur, Nadia – 741252, West Bengal, India. Mobile: +91-7044112393, E-mail: arbind4gupta@gmail.com

S. K. Gupta, PhD
INAE Distinguished Professor, ICAR – Central Soil Salinity Research Institute (Zarifa Farm), Karnal – 132001, Haryana, India. Mobile: +91-9416081613; E-mail: drskg1949@yahoo.com

Madan Kumar Jha, PhD
Professor, AgFE Department, IIT Kharagpur, Kharagpur – 721302, West Bengal, India. Tel: +91-3222-283116(O), E-mail: madan@agfe.iitkgp.ernet.in

Amrita Kasotia, PhD
UGC-RGNF Senior Research Fellow, Department of Science, Faculty of Arts, Science and Commerce, Mody University of Science and Technology, Lakshmangarh, Sikar – 332311, Rajasthan, India, Mobile: +91-1573-225001, E-mail: amritakasotia@gmail.com

Puja Khare, PhD
Scientist, Agronomy and Soil Science Division, CSIR-Central Institute of Medicinal and Aromatic Plants, Lucknow – 226015, India, Mobile: +918004923033, E-mail: kharepuja@rediffmail.com

Sk. Khasim Beebi, PhD
Associate Professor, Department of Biotechnology, GITAM University Visakhapatnam – 530045, Andhra Pradesh, India, Mobile, +91 8912840246, E-mail: khasimb@yahoo.com

Deepesh Machiwal, PhD
Senior Scientist, Central Arid Zone Research Institute (CAZRI), Regional Research Station, Kukma – 370105, Bhuj, Gujarat, India; Tel.: +91-2832–271238(O); E-mail: dmachiwal@rediffmail.com

Biplab Pal
PhD Research Scholar, Department of Agricultural Chemistry and Soil Science, Bidhan Chandra Krishi Viswavidyalaya, Mohanpur, Nadia – 741252, West Bengal, India. Mobile: +91-9474371998, E-mail: biplab.psb@gmail.com

Ch. Ramakrishna, PhD
Professor, Department of Environmental Sciences, GITAM Institute of Sciences, GITAM University, Visakhapatnam – 530045, Andhra Pradesh, India Mobile: +91-8912840451, E-mail:chrk2020@gmail.com

G. V. Ramana, PhD
Senior Research Fellow, Department of Biotechnology, GITAM Institute of Technology, GITAM University, Visakhapatnam – 530045, Andhra Pradesh, India. Mobile: +91-8912840246, E-mail: eruditeramana@gmail.com

H. S. Sen, PhD
Former Director, ICAR – Central Research Institute for Jute and Allied Fibers, Barrackpore – 700120, West Bengal, India. Present address: 2/74 Naktala, Kolkata – 700047, West Bengal, India, Mobile: +91-987418962, E-mail: hssen.india@gmail.com

D. K. Sharma, PhD
Director, ICAR – Central Soil Salinity Research Institute, Karnal – 132001, Haryana, India. Tel.: +91-184-2290501, E-mail: dk.sharma@icar.gov.in

K. S. Shylaraj, PhD
Professor, Plant Breeding and Genetics, Rice Research Station, Kerala Agricultural University, Vyttila, Kochi – 682019, Kerala, India. Mobile: +91-9846789150; E-mail: shylarajks@gmail.com

Anshuman Singh, PhD
Scientist (Horticulture), ICAR – Central Soil Salinity Research Institute, Karnal – 132001, Haryana, India. Tel.: +91-184-2290501, E-mail: anshumaniari@gmail.com

Y. P. Singh, PhD
Principal Scientist, ICAR – Central Soil Salinity Research Institute, Regional Research Station, Lucknow – 226005, Uttar Pradesh, India, Mobile: +91-7309563010, E-mail: ypsing_5@yahoo.co.in

A. K. Sreelatha, PhD
Assistant Professor, Soil Science and Agricultural Chemistry, Rice Research Station, Kerala Agricultural University, Vyttila, Kochi – 682019, India, Mobile: +91-9446328761, E-mail: sreelatha.ak@kau.in; aksreelathavp@gmail.com

Ajit Varma, PhD
Distinguished Scientist & Professor of Eminence, Amity Institute of Microbial Technology (AIMT), Block E-3, 4th Floor, Amity University Campus, Gautam Buddha Nagar, Sector-125, Noida – 201313, UP, India, Tel.: +91-120-2431182, E-mail: ajitvarma@amity.edu

N. P. S. Yaduvanshi, PhD
Principal Scientist, Division of Soil Science and Agricultural Chemistry, Indian Agricultural Research Institute, New Delhi – 110012, India. Mobile: +91-9582380066, E-mail: npsyaduvanshi@gmail.com

LIST OF ABBREVIATIONS

ABA	abscisic acid
ACC	aminocyclopropane-1-carboxylic acid
*acd*S	ACC deaminase structural gene
ACO	ACC oxidase
ACS	ACC synthase
AESR	agro-ecological Sub Regions
ALA	5-aminolevulinic acid
AMT	ACC-N-malonyl transferase
ANN	artificial neural network
ANOVA	analysis of variance
APX	ascorbate peroxidase
ATAF	arabdopsis transcription activator factor
B:C ratio	benefit:cost ratio
BAW	best available water
BS	basic slag
bZIP	basic leucine zipper
CA	correspondence analysis
CAM	Crassulacean acid metabolism
CAT	catalase
CBF	C-repeat binding factors
CBLs	calmodulin and calcineurin B-like proteins
CCA	canonical correlation analysis
CCE	calcium carbonate equivalent
CCSHAU	Chaudhary Charan Singh Haryana Agricultural University
CDPKs	Ca^{2+}-dependent protein kinases
CDPKs	calcium dependant protein kinases
CEC	cation exchange capacity
Chl	chlorophyll
COR	cold-regulated genes
CRRI	Central Rice Research Institute
CSSRI	Central Soil Salinity Research Institute
CUC	cup-shaped cotyledons

DA	discriminant analysis
DAP	diammonium phosphate
DAPG	2,4-diacetylphlorogucinol
DAS	days after sowing
DNA	deoxyribonucleic acid
DREB	dehydration responsive element binding
DREB2	dehydration responsive element-binding factor
DTPA	diethyline-triamine-penta-acetic acid
EC	electrical conductivity
EC_e	electrical conductivity of saturation extract
EC_{iw}	electrical conductivity of irrigation water
EC_t	threshold electrical conductivity
EL	electrolyte leakage
EMS	ethyl methanesulfonate
ePGPR	extracellular-PGPR
epm	equivalents per million
EPS	exopolysaccharide
ESP	exchangeable sodium percentage
ETC	electron transport chain
EU	European Union
FAO	Food and Agricultural Organization
FAO-AGL	Food and Agricultural Organization-The Land and Water Development Division
fw	fresh weight
FYM	farmyard manure
GACC	γ-glutamyl-ACC
GAs	Gibberellins
GDP	gross domestic product
GGT	γ-glutamyl-transpeptidase
GHGs	greenhouse gases
GI	geographical indication
GIS	geographical information system
GITAM	Gandhi Institute of Technology and Management
GOI	Government of India
GR	gypsum requirement
GWQI	ground water quality index

ha	hectare
HCA	hierarchical cluster analysis
HCN	hydrogen cyanide
HKT	high affinity K^+ transporter
HR	hypersensitive reaction
IAA	indole-3-acetic acid
IAM	indole-3-acetamide
IAN	indole-3-acetonitrile
ICAR	Indian Council of Agricultural Research
IFPRI	International Food Policy Research Institute
INR	Indian Rupees
IPCC	Intergovernmental Panel on Climate Change
iPGPR	intracellular-PGPR
IPyA	indole-3-pyruvate pathway
ISR	induced systemic resistance
IST	induced systemic tolerance
ITK	indian traditional knowledge
JA	jasmonic acid
JA-ACC	jasmonyl-ACC
JAR1	jasmonic acid resistance 1
K	potassium
Kcal	kilo calories
KSPB	Kerala State Planning Board
KVK	*Krishi Vigayan Kendra*
LC_{50}	50% of the lethal concentration
LD_{50}	50% of the lethal dose
LEA	late embryogenesis abundant
LEW	leaf epicuticular waxes
LF	leaching fraction
LR	leaching requirement
LR	lime requirement
M ha	million hectare
M-t	metric ton
MACC	1-malonyl-ACC
MAPKs	mitogen-activated protein kinases
MDA	malonaldehyde

MDHAR	monodehydroascorbate reductase
meq L^{-1}	milliequivalent per liter
mg L^{-1}	milligrams per liter
MIA	modified index of agreement
mM	molar mass
mmol L^{-1}	millimole per liter
MNNG	methyl-N'-nitro-N-nitrosoguanidine
MNSE	modified nash-sutcliffe efficiency
MOU	memorandum of understanding
MYB	myeloblastosis
MYC	myelocytomatosis
N	nitrogen
NAAS	National Academy of Agricultural Sciences
NABARD	National Bank for Agriculture and Rural Development
NAD	nicotinamide adenine dinucleotide
NAM	no apical meristem
NO	nitric oxide
NPK	nitrogen, phosphorus and potash
NT	no tillage
NV	neutralizing value
OC	organic carbon
P	phosphorus
PAU	Punjab Agricultural University
PC(s)	principle component(s)
PCA	phenazine-1-carboxylic acid
PCA	principal component analysis
PFCs	polyfluorinated compounds
PG	phospho-gypsum
PGPB	plant growth promoting bacteria
PGPR	plant growth promoting rhizobacteria
pH$_s$	soil reaction of saturation extract
Pi	inorganic-phosphate
PLD	phospholipid
PLDA	*Pokkali* Land Development Agency
PLS	partial least squares
PM	poultry manure

POD	peroxidase
ppm	parts per million
ppt	parts per thousand
PSB	phosphate solubilizing bacteria
PUFA	polyunsaturated fatty acids
PVS	participatory varietal selection
Px	peroxidase
QTL	quantitative trait locus
QTLs	quantitative trait locii
R	correlation coefficient
R&D	Research and Development
R^2	coefficient of determination
RDF	recommended dose of fertilizers
RMSE	root mean square error
ROS	reactive oxygen species
RS	remote sensing
RSAC	Remote Sensing Application Centre
RSC	residual sodium carbonate
RSC_{iw}	residual sodium carbonate of irrigation water
RWC	relative water content
SA	salicylic acid
SAM	S-adenosyl-methionine
SAR	sodium adsorption ratio
SAR	systemic acquired resistance
SAR_{iw}	sodium adsorption ratio of irrigation water
SAS	salt affected soils
SERB	Science and Engineering Research Board
SIC	site implementation committee
SIP	site implementation plan
SMHE	*Salvia miltiorrhiza* hydrophilic extract
SOD	superoxide dismutase
SOS	salt overly sensitive
SPM	sulphitation press mud
SSD	subsurface drainage
SW	saline water
TAM	trypamine

TDS	total dissolved solids
TSO	tryptophan side-chain oxidase
TSS	total soluble salts
UNFCCC	UN Framework Convention on Climate Change
UNICEF	United Nations Children's Fund
UPBSN	UP *Bhumi Sudhar Nigam*
US	United States
USA	United States of America
USAID	United State Agency for International Development
USDA	United States Department of Agriculture
USSL	United States Salinity Laboratory
UV	ultraviolet
WHO	World Health Organization
WQI	water quality index
WRI	World Resources Institute
WUC	water uptake capacity
WUG	water users groups
ZF-HD	zinc-finger homeodomain

FOREWORD BY GURBACHAN SINGH

Salinity is a global problem extending across all the continents in more than 100 countries of the world. As per an assessment of the FAO – Land and Plant Nutrition Management Service, more than 6% of the world's land is affected by either salinity or sodicity. It is a major threat to irrigated agriculture as 20% of the irrigated lands are afflicted with water logging and/or soil salinity, commonly designated as twin problems. The primary and secondary salinization is adversely impacting on-farm agricultural production that gets upscaled to regional and national levels having implication on food security, environmental health, and economic welfare. With most of the good quality lands already committed, these otherwise desolate looking lands could provide an opportunity to increase food production by restoring their productivity. It could go a long way in meeting the foreseen 57% increase in global food production by the year 2050 to feed the growing world population at the current levels of per capita food supply. Besides, the data generated globally has proved that costs of prevention or land reclamation are much less than the costs to the society if land degradation continues.

The prevention and/or reclamation of salt affected soils require team effort by groups of experts from various disciplines such as soil science, agronomy, plant physiology, hydrology, agro-forestry, plant breeding, genetic engineering, computer sciences, water sciences, environmental sciences including modelers in various disciplines. My own experience show that a changed land use to manage and live with salts is a low cost alternate to land reclamation and may be the only option in cases where amelioration is impractical or too expensive. Translation of such integrated scientific interventions and technologies to field practices must be supported by pertinent policies, well-designed salinity management plans and supportive institutions. Besides, stakeholder's participation being crucial, skilling human resource, capacity development of farmers and relying on local resources and indigenous knowledge would play an important role in any such endeavor.

I am happy to note that technical solutions, policy tools, and the integrated catchment management approaches adopted by various Governments have made significant headway in addressing the salinization of land and water resources. However, the integrated approach amongst the various stakeholders is either lacking or people from one discipline are unaware of what is being attempted in the other disciplines.

I am happy to note that editors of the current volume have compiled comprehensive chapters contributed by experts across many disciplines. This volume covers many important issues in the field of soil salinity management in agriculture. These chapters have been categorized in well-defined sections. I congratulate the duo for their efforts in bringing out this publication covering a very topical subject that might have serious economic and social consequences, if left unattended.

I am sure that the publications of this nature would help in sensitizing the world communities in taking effective and integrated remedial actions as per local needs so as to eradicate the menace of land and water salinity. I believe that this publication will be quite handy to researchers, policy planners, students and field practitioners alike and must be in their shelves for ready reference.

Gurbachan Singh, PhD
Chairman, Agricultural Scientists Recruitment
Board, Formerly Agricultural Commissioner,
Government of India, Assistant Director General
(Agronomy and Agro-Forestry), ICAR, Krishi
Anusandhan Bhavan – 1, New Delhi, India.
Former Director, Central Soil Salinity Research
Institute, Karnal, India.

PREFACE 1 BY S. K. GUPTA

Soil salinity, an incipient problem, comes to the fore only when some damage has already been done. Besides adverse impacts on agricultural productivity, the consequences of soil salinization can be quite damaging. Civilizations in southern Mesopotamia and in several parts of the Tigris–Euphrates valley were wiped out in the past because of water logging and soil salinity. Damages in the Aral Sea Basin are the living example of how things can go wrong with improper land and water management practices. There are other basins in Australia, China, India, Pakistan, and the United States that are grappling with salt related land and water degradation.

Soil and water salinity adversely affect on-farm, regional and national interests having a major impact on food security, land, and water quality and environment resulting in serious economic and social problems in rural and urban communities. There are wide variations in the figures of the extent of the area affected by soil salinity from one assessment to another yet the sizable area in different continents especially in irrigated lands is causing concerns in more than 100 countries around the globe. With many intersecting challenges such as climate change, deforestation, and fresh water shortages resulting in the exploitation of saline water in agriculture, soil salinity is posing unprecedented dangers to the sustainability of agricultural systems and the environment. Learning from what went wrong coupled with scientific innovations, understanding of the natural and anthropogenic causes, and prevention and reclamation of salt affected lands has hugely improved. Catastrophic losses experienced in the past can easily be avoided if the current knowledge on the subject gets translated into scientifically proven field practices. Advancements in researches as well as application of research results in this multi-disciplinary arena require an inter-disciplinary approach drawing knowledge and experiences of the experts from soil science, agronomy, engineering, hydrology, other water sciences, plant sciences, environmental sciences, modelers, and computer sciences. Equally crucial would be the stakeholder participation in the field applications of these technologies.

The current volume *Soil Salinity Management in Agriculture: Technological Advances and Applications* under *Innovations in Agricultural and Biological Engineering* has been designed keeping all these issues in view to provide readers a comprehensive picture of the saline environment and plant interactions. The chapters in the current volume have been written by experts in their respective sphere, and cover major issues related to salinity management in agriculture. The editors for the sake of convenience have grouped these chapters into three sections namely:

I.   Emerging Trends and Technologies in Salinity Management
II. Mechanisms of Salt Tolerance
III. Soil Salinity Management in Crop Production

Part I with four chapters dwells upon emerging trends and technologies in salinity management in soils or water covering arid, semi-arid and coastal eco-regions. While the first chapter by Sharma and Singh deals with the emerging trends of salinity research in India. The included technologies and practices have the replication potential in many developing and developed countries as many cost-cutting techniques including use of industrial wastes in land reclamation programs are included. The second chapter specifically deals with similar issues related to use of saline water in agriculture. It advocates living with salts which may be the only option when reclamation is either expensive or impractical to adopt. The coastal ecosystem have the high potentiality of production of a large number of goods and services valued at about US$ 12–14 trillion annually, and is confronted with high risk of soil salinization due to the multiple issues including the climate change—is the theme of discussion by Sen and Ghorai. In recent times, ground water salinity and its pollution has assumed serious overtones around the globe because of its use across many economic sectors. Unlike surface waters, to pinpoint the sources of contamination of ground water could be a major challenge for the ground water management and planning. The chapter by Machiwal and Jha briefly reviews the past technologies and describes the recent advances made in this vital arena of salinity management in agriculture.

Plants when exposed to salt stress activate their salt tolerance mechanisms through stress sensing and signaling so as to regulate the plant salinity stress response. For example, some halophytes respond to salinity by taking up sodium and chloride at high rates and then accumulating these ions in their leaves. Many glycophytes respond to stress by salt exclusion particularly through low rates of net transport of sodium or chloride, or both from root to shoot. Plant adaptations to salinity have been categorized into three types, namely osmotic stress tolerance, Na^+ or Cl^- exclusion, and the tolerance of plant tissues to accumulated Na^+ or Cl^-.

Part II deals with these and similar issues in various plants types. The chapter by Gururaja Rao et al. lists some of the technological interventions to green the barren saline Vertisols using *Salvadora persica*, a facultative halophyte, a potential source for seed oil, and some forage grasses. Four grasses namely *Dichanthium annulatum* and *Leptochloa fusca* for saline water-logged soils, and *Aeluropus lagopoides* and *Eragrostis* species for saline water-irrigated lands have been included. Ramana et al. argue that not much is known about the abiotic stress responses of *Plectranthus* species grown for their ethno botanical use as ornamental, medicinal and economic plants. This chapter looks at the impacts on morphology, physiology and biochemical processes in the six genotypes of *Plectranthus* species to arrive at their suitability for cultivation in saline environment. Desmukh and Khare in their chapter argue that as per the estimates of World Health Organization (WHO), nearly 80% of the world populations rely on medicinal herbs, the requirement of which can be met only through commercial cultivation of medicinal plants. Since land is already a constraint to meet the food demand, it is the otherwise problem soils and/or harsh climatic conditions under which these plants will have to be cultivated. As such, their chapter reviews the current status of studies conducted to assess the effect of salinity stress on growth parameters and metabolites of medicinal plants.

Reclamation and sustainable crop production on reclaimed lands has been the major thrust area of research in salinity management in agriculture. Part III deals with this theme and related issues includes five chapters wherein application of emerging technologies and stakeholders participation in sustainable land reclamation programs is highlighted. Biological reclamation of salt affected lands, a low cost eco-friendly approach, is emerging as a new thrust area of research where plants and microbes are increasingly used to reclaim these lands. A comprehensive review of bacterial-mediated amelioration processes in plants under salt stress has been made by Kasotia and his colleagues. Singh argues that the lack of success of many breeding programs in developing commercially success-ful salt tolerant crops is due to limited evaluation of genetic material in idealized conditions that does not represent the actual field situations. Sus-tainable land reclamation using salt tolerant cultivars developed through farmers' participatory approach is the subject matter of discussion by him. Pokkali rice cultivation has passed down from one generation to another generation from more than 3000 years; and this has been discussed in the framework of symbiotic nature of rice and shrimp cultivation. This strat-egy helps to manage lands experiencing multiple stresses due to water logging, soil salinity, and irrigation water salinity, a common feature of the coastal ecosystems.

One of the important limiting factors for optimal use of land resources for higher productivity especially in sub-humid and humid regions is soil acidity—acid soils occupying about 3.95 billion ha globally. The chapter by Chaitanya et al. presents comprehensive technological interventions to manage such lands including the acid sulfate soils. Yaduvanshi, in the last chapter of this volume, has emphasized that limited nutrient availability in salt affected the soils. As a result of host of unfavorable physico-chemical conditions, the production potential of reclaimed lands as well as impacts the sustainability of the program. He emphasizes that appropriate prescription of macro and micro-nutrients, in right quantities, at the right time and place, from the right source, and in the right combination could play a major role in sustaining the reclamation benefits.

I believe that the salinization of the land and water is now a very seri-ous threat to the health and utility of soil, vegetation, rivers and ground water around the world. Considering the looming food security crisis of the 2050, no let-up can be allowed in our efforts to understand salt toler-ance mechanisms, develop new technologies and apply the existing ones for prevention, living with salts and/or sustainable land reclamation pro-grams can be allowed. This volume may give the readers a good feel of the efforts being made in these respective areas.

I take this opportunity to thank all the contributors for sparing their valuable time and painstaking efforts made in preparing timely submission and updating their edited manuscripts in a time bound manner to make timely publication of this volume possible. I believe that the current volume will serve as a good repository of latest information and will be highly useful to researchers, policy planners, teachers, students especially the post graduate students, development agencies and other who are interested in salinity management for resolving global food security issues and its relationship with the environment. Finally, I thank the editorial staffs who have been involved in this project.

—S. K. Gupta, PhD
Lead Editor

PREFACE 2 BY MEGH R. GOYAL

According to *https://en.wikipedia.org/wiki/Soil_salinity*, the soil salinity is the salt content in the soil; and the process of increasing the salt content is known as salinization, which can be caused by natural processes (mineral weathering or by the gradual withdrawal of an ocean) or through artificial processes such as irrigation. The ions responsible for salinization are: Na^+, K^+, Ca^{2+}, Mg^{2+}, and Cl^-. As the Na^+ (sodium) predominates, soils can become *sodic*. Sodic soils present particular challenges because they tend to have very poor structure, which limits or prevents water infiltration and drainage. Over long periods of time, as soil minerals weather and release salts, these salts are flushed or leached out of the soil by drainage water in areas with sufficient precipitation. In addition to mineral weathering, salts are also deposited via dust and precipitation. In dry regions, salts may accumulate leading to naturally saline soils. Human practices can increase the salinity of soils by the addition of salts in irrigation water. Proper irrigation management can prevent salt accumulation by providing adequate drainage water to leach added salts from the soil. Disrupting drainage patterns that provide leaching can also result in salt accumulations. Salinity in drylands can occur when the water table is between two to three meters from the surface of the soil. The salts from the ground water are raised by capillary action to the surface of the soil. This occurs when ground water is saline (which is true in many areas), and is favored by land use practices allowing more rainwater to enter the aquifer than it could accommodate. Salinity from irrigation can occur over time wherever irrigation occurs, since almost all water (even natural rainfall) contains some dissolved salts. When the plants use the water, the salts are left behind in the soil and eventually begin to accumulate. Since soil salinity makes it more difficult for plants to absorb soil moisture, these salts must be leached out of the plant root zone by applying additional water. This water in excess of plant needs is called the leaching fraction. Salinization from irrigation water is also greatly increased by poor drainage and use of saline water for irrigating agricultural crops. The consequences of salinity are: detrimental effects on plant growth and yield; damage to infrastructure (roads, bricks, corrosion of pipes and cables); reduction of water quality for users, sedimentation problems; soil erosion ultimately, when crops are too strongly affected by the amounts of salts. Salinity is an important land degradation problem. Soil salinity can be reduced by leaching soluble salts out of soil with excess irrigation water. Soil salinity control involves water table control and flushing in combination with tile drainage or another form of subsurface drainage. A comprehensive treatment of soil salinity is available from the United Nations Food and Agriculture Organization. High levels of soil salinity can be tolerated if salt-tolerant plants are grown. Sensitive crops lose their vigor already in slightly saline soils, most crops are negatively affected by (moderately) saline soils, and only salinity resistant crops thrive in severely saline soils.

According to "R. Brinkman (1980), Saline and Sodic Soils. In: Land reclamation and Water Management, pp. 62–68. International Institute for Land Reclamation and Improvement (ILRI), Wageningen, The Netherlands," the salinized areas (in million ha) are: 69.5 in Africa; 53.1 in Near and Middle East; 19.5 in Asia and Far East; 59.4 in Latin America; 84.7 in Australia; 16.0 in North America; and 20.7 in Europe.

One can download free the LeachMod software for simulating leaching of saline irrigated soil from *http://www.waterlog.info/leachmod.htm*. LeachMod is designed to simulate the depth of the water table and the soil salinity in irrigated areas with a time step as selected by the user (from 1 day to 1 year). The program uses small time steps in its calculations for a better accuracy. In case of a leaching experiment with measured soil salinities, LeachMod can automatically optimize the leaching efficiency by minimizing the sum of the squares of the differences between measured and simulated salinities. The root zone can consist of 1, 2, or 3 layers. LeachMod allows the introduction of a subsurface drainage system in a transition zone between root zone and aquifer, and subsequently it determines the drain discharge. When the irrigation/rainfall is scarce and the water table is shallow, LeachMod will calculate the capillary rise and reduce the potential evapotranspiration to an actual evapotranspiration. LeachMod can also take into account upward seepage from the aquifer or downward flow into it. The latter flow is also called natural subsurface drainage. This model is somewhat similar to SaltCalc. On one hand, the water management options are fewer (e.g., re-use of drainage or well water for irrigation do not feature here), but the model is more modern in the sense that the variable input for each time step is given in a table so that the calculations over all the time steps are done at one step. Moreover, by inserting the observed values of soil salinity in the data table, the model optimizes the leaching efficiency of the soil automatically. On 11 July 2015, LeachMod was updated to include more rigorous data checks.

I know what the cooperating authors have emphasized in their chapter for this book volume. I am a staunch supporter of preserving our natural resources. Importance of wise use of our natural resources has been taken up seriously by Universities, Institutes/Centers, Government Agencies and Non-Government Agencies. I conclude that the agencies and departments in soil salinity management have contributed to the ocean of knowledge.

This book also contributes to the ocean of knowledge on soil salinity management. Agricultural and Biological Engineers (ABEs) with expertise in this area work to better understand the complex mechanics of soil salinity. ABEs are experts in agricultural hydrology principles, such as controlling drainage, and they implement ways to control soil erosion and study the environmental effects of sediment on stream quality.

The mission of this book volume is to serve as a reference manual for graduate and undergraduate students of agricultural, biological and civil engineering; horticulture, soil science, crop science, and agronomy. I hope that it will be a valuable reference for professionals who work with soil salinity management; for professional training institutes, technicals agricultural centers, irrigation centers, agricultural extension service, and other agencies that work with micro irrigation programs. I cannot guarantee the information in this book series will be enough for all situations.

After my first textbook, *Drip/Trickle or Micro Irrigation Management* by Apple Academic Press Inc., and response from international readers, I was motivated to bring out for the world community a ten-volume series on *Research Advances in Sustainable Micro Irrigation*. The website *http://www.appleacademicpress.com* gives details on these ten book volumes. I have already published five book volumes under book series, *"Innovations and Challenges in Micro Irrigation."*

At the 49th annual meeting of the Indian Society of Agricultural Engineers at Punjab Agricultural University (PAU) during February 22–25 of 2015, a group of ABEs and FEs convinced me that there is a dire need to publish book volumes on focus areas of agricultural and biological engineering (ABE). This is how the idea was born for new book series titled *Innovations in Agricultural and Biological Engineering*. This book, *Soil Salinity Management in Agriculture: Technological Advances and Applications*, is the ninth volume under this book series.

My classmate and longtime colleague, Dr. S. K. Gupta, joins me as a Lead Editor of this volume. In the last 45 years, Dr. Gupta holds exceptional professional qualities with his expertise in Soil Salinity Management. In addition, Dr. Gupta is a research scientist and distinguished Professor at ICAR – Central Soil Salinity Research Institute (Zarifa Farm), Karnal – 132001, Haryana, India. His contribution to the contents and quality of this book has been invaluable.

I would like to thank the editorial staff, Sandy Jones Sickels, Vice President, and Ashish Kumar, Publisher and President at Apple Academic Press, Inc., for making every effort to publish the book when the diminishing water and food resources are a major issue worldwide. Special thanks are due to the AAP Production Staff for the quality production of this book.

I request the readers to offer your constructive suggestions that may help to improve the next edition. The reader can order a copy of this book from *http://appleacademicpress.com*.

I express my deep admiration to my family and colleagues for their understanding and collaboration during the preparation of this book volume. As an educator, there is a piece of advice to one and all in the world: *Permit that our almighty God, our Creator, provider of all and excellent teacher, feed our life with Healthy Food Products and His Grace—and Get married to your profession.*

—*Megh R. Goyal, PhD, PE*
Senior Editor-in-Chief

WARNING/DISCLAIMER

READ CAREFULLY

The goal of this book volume on *Soil Salinity Management in Agriculture: Technological Advances and Applications* is to guide the world engineering community on how to manage soil salinity stress for economical crop production. The reader must be aware that the dedication, commitment, honesty, and sincerity are most important factors in a dynamic manner for complete success. It is not a one-time reading of this compendium. Read and follow every time, it is needed.

The editors, the contributing authors, the publisher and the printer have made every effort to make this book as complete and as accurate as possible. However, there still may be grammatical errors or mistakes in the content or typography. Therefore, the contents in this book should be considered as a general guide and not a complete solution to address any specific situation in soil salinity. For example, one type of solution does not fit all case studies of soil salinity and to all crops.

The editors, the contributing authors, the publisher and the printer shall have neither liability nor responsibility to any person, any organization or entity with respect to any loss or damage caused, or alleged to have caused, directly or indirectly, by information or advice contained in this book. Therefore, the purchaser/reader must assume full responsibility for the use of the book or the information therein.

The mentioning of commercial brands and trade names are only for technical purposes. It does not mean that a particular product is endorsed over to another product or equipment not mentioned.

All weblinks that are mentioned in this book were active on December 31, 2016. The editors, the contributing authors, the publisher and the printing company shall have neither liability nor responsibility, if any of the weblinks is inactive at the time of reading of this book.

BOOK ENDORSEMENT

The volume *Soil Salinity Management in Agriculture* presents a holistic picture of reclamation and management of salt affected soils, should cater to the needs of those interested in soil/water salinity–plant-environment interactions.

—*Alok K. Sikka, PhD*
Deputy Director General (Natural Resources Management),
Indian Council of Agricultural Research, Krishi Anusandhan Bhavan II,
New Delhi, India

OTHER BOOKS ON AGRICULTURAL AND BIOLOGICAL ENGINEERING BY APPLE ACADEMIC PRESS, INC.

Management of Drip/Trickle or Micro Irrigation
Megh R. Goyal, PhD, PE, Senior Editor-in-Chief

Evapotranspiration: Principles and Applications for Water Management
Megh R. Goyal, PhD, PE, and Eric W. Harmsen, Editors

Book Series: Research Advances in Sustainable Micro Irrigation
Senior Editor-in-Chief: Megh R. Goyal, PhD, PE
  Volume 1: Sustainable Micro Irrigation: Principles and Practices
  Volume 2: Sustainable Practices in Surface and Subsurface Micro
             Irrigation
  Volume 3: Sustainable Micro Irrigation Management for Trees and Vines
  Volume 4: Management, Performance, and Applications of Micro
             Irrigation Systems
  Volume 5: Applications of Furrow and Micro Irrigation in Arid and
             Semi-Arid Regions
  Volume 6: Best Management Practices for Drip Irrigated Crops
  Volume 7: Closed Circuit Micro Irrigation Design: Theory and Appli-
             cations
  Volume 8: Wastewater Management for Irrigation: Principles and
             Practices
  Volume 9: Water and Fertigation Management in Micro Irrigation
  Volume 10: Innovation in Micro Irrigation Technology

Book Series: Innovations and Challenges in Micro Irrigation
Senior Editor-in-Chief: Megh R. Goyal, PhD, PE
  Volume 1: Principles and Management of Clogging in Micro Irrigation
  Volume 2: Sustainable Micro Irrigation Design Systems for Agricultural
             Crops: Methods and Practices

Volume 3: Performance Evaluation of Micro Irrigation Management: Principles and Practices

Volume 4: Potential Use of Solar Energy and Emerging Technologies in Micro Irrigation

Volume 5: Micro Irrigation Management: Technological Advances and Their Applications

Volume 6: Micro Irrigation Engineering for Horticultural Crops: Policy Options, Scheduling, and Design

Volume 7: Micro Irrigation Scheduling and Practices

Volume 8: Engineering Interventions in Sustainable Trickle Irrigation

Book Series: Innovations in Agricultural and Biological Engineering
Senior Editor-in-Chief: Megh R. Goyal, PhD, PE

- Dairy Engineering: Advanced Technologies and their Applications
- Developing Technologies in Food Science: Status, Applications, and Challenges
- Emerging Technologies in Agricultural Engineering
- Engineering Interventions in Agricultural Processing
- Engineering Practices for Agricultural Production and Water Conservation: An Interdisciplinary Approach
- Flood Assessment: Modeling and Parameterization
- Food Engineering: Modeling, Emerging issues and Applications.
- Food Process Engineering: Emerging Trends in Research and Their Applications
- Food Technology: Applied Research and Production Techniques
- Modeling Methods and Practices in Soil and Water Engineering
- Processing Technologies for Milk and Milk Products: Methods, Applications, and Energy Usage
- Soil and Water Engineering: Principles and Applications of Modeling
- Soil Salinity Management in Agriculture: Technological Advances and Applications
- Technological Interventions in Management of Irrigated Agriculture
- Technological Interventions in the Processing of Fruits and Vegetables
- Engineering Interventions in Foods and Plants
- Technological Interventions in Dairy Science: Innovative Approaches in Processing, Preservation, and Analysis of Milk Products
- Novel Dairy Processing Technologies: Techniques, Management, and Energy Conservation

ABOUT THE EDITOR

S. K. Gupta, PhD, CE, received his BSc (1970) and MSc (1976) degrees in Agricultural Engineering from Punjab Agricultural University, Ludhiana, India; and PhD in 1984 from Jawahar Lal Nehru Technological University, Hyderabad, India. Since 1971, he has worked at Central Soil Salinity Research Institute, Karnal, India, in various capacities, such as Principal Scientist; Head, Division of Drainage and Water Management; Head Indo-Dutch Network Project; and Project Coordinator, Use of Saline Water in Agriculture. At present, he is working as INAE Distinguished Professor at the same place after superannuating from the Central Soil Salinity Research Institute, Karnal, India. He has published 13 books. His area of expertise is irrigation and drainage engineering. He is chief editor of the *Journal of Water Management*, published by the Indian Society of Water Management (ISWM), New Delhi, India. Readers may contact him at: *drskg1949@yahoo.com*

ABOUT THE SENIOR EDITOR-IN-CHIEF

Megh R. Goyal, PhD, PE, is, at present, a Retired Professor in Agricultural and Biomedical Engineering from the General Engineering Department in the College of Engineering at University of Puerto Rico–Mayaguez Campus; and Senior Acquisitions Editor and Senior Technical Editor-in-Chief in Agricultural and Biomedical Engineering for Apple Academic Press Inc.

Dr. Goyal received his BSc degree in Engineering in 1971 from Punjab Agricultural University, Ludhiana, India; his MSc degree in 1977; and PhD degree in 1979 from the Ohio State University, Columbus; and his Master of Divinity degree in 2001 from Puerto Rico Evangelical Seminary, Hato Rey, Puerto Rico, USA.

Since 1971 he has worked as Soil Conservation Inspector (1971); Research Assistant at Haryana Agricultural University (1972–1975) and the Ohio State University (1975–1979); Research Agricultural Engineer/Professor at Department of Agricultural Engineering of UPRM (1979–1997); and Professor in Agricultural and Biomedical Engineering at General Engineering Department of UPRM (1997–2012). He spent one-year sabbatical leave in 2002–2003 at Biomedical Engineering Department, Florida International University, Miami, USA.

He was first agricultural engineer to receive the professional license in Agricultural Engineering in 1986 from College of Engineers and Surveyors of Puerto Rico. On September 16, 2005, he was proclaimed as "Father of Irrigation Engineering in Puerto Rico for the twentieth century" by the ASABE, Puerto Rico Section, for his pioneer work on micro irrigation, evapotranspiration, agroclimatology, and soil and water engineering. During his professional career of 45 years, he has received awards such as: Scientist of the Year, Blue Ribbon Extension Award, Research Paper Award, Nolan Mitchell Young Extension Worker Award, Agricultural Engineer of the Year, Citations by Mayors of Juana Diaz and Ponce, Membership Grand Prize for ASAE Campaign, Felix Castro Rodriguez Academic Excellence, Rashtrya Ratan Award and Bharat Excellence Award and Gold Medal, Domingo Marrero Navarro Prize, Adopted son of Moca, Irrigation Protagonist of UPRM, Man of Drip Irrigation by Mayor of Municipalities of Mayaguez/Caguas/Ponce and Senate/Secretary of Agriculture of ELA, Puerto Rico.

He has authored more than 200 journal articles and textbooks, including *Elements of Agroclimatology* (Spanish) by UNISARC, Colombia; two *Bibliographies on Drip Irrigation.* His published books with Apple Academic Press include *Management of Drip/Trickle or Micro Irrigation and Evapotranspiration: Principles and Applications for Water Management,* among others. Dr. Goyal is the senior editor of three book series for AAP: Research Advances in Sustainable Micro Irrigation, Innovations and Challenges in Micro Irrigation, and Innovations in Agricultural and Biological Engineering. Readers may contact him at: *goyalmegh@gmail.com.*

EDITORIAL

Apple Academic Press Inc., (AAP) will be publishing various book volumes on the focus areas under book series titled *Innovations in Agricultural and Biological Engineering*. Over a span of 8 to 10 years, Apple Academic Press Inc., will publish subsequent volumes in the specialty areas defined by American Society of Agricultural and Biological Engineers (ABEs) (http://asabe.org).

The mission of this series is to provide knowledge and techniques for ABEs. The series aims to offer high-quality reference and academic content in Agricultural and Biological Engineering (ABE) that is accessible to academicians, researchers, scientists, university faculty, and university-level students and professionals around the world. The following material has been edited/modified and reproduced below [From: "Megh R. Goyal (2006). Agricultural and Biomedical Engineering: Scope and Opportunities. Paper Edu_47 Presentation at the Fourth LACCEI International Latin American and Caribbean Conference for Engineering and Technology (LACCEI' 2006): Breaking Frontiers and Barriers in Engineering: Education and Research by LACCEI University of Puerto Rico – Mayaguez Campus, Mayaguez, Puerto Rico, June 21–23"]:

WHAT IS AGRICULTURAL AND BIOLOGICAL ENGINEERING (ABE)?

"Agricultural Engineering (AE) involves application of engineering to production, processing, preservation and handling of food, fiber, and shelter. It also includes transfer of technology for the development and welfare of rural communities," according to http://isae.in. "ABE is the discipline of engineering that applies engineering principles and the fundamental concepts of biology to agricultural and biological systems and tools, for the safe, efficient and environmentally sensitive production, processing, and management of agricultural, biological, food, and natural resources systems," according to http://asabe.org. "AE is the branch of engineering involved with the design of farm machinery, with soil management, land development, and mechanization and automation of livestock farming, and with the efficient planting, harvesting, storage, and processing of farm commodities," the definition by *http://dictionary.reference.com/browse/agricultural+engineering.*

"AE incorporates many science disciplines and technology practices to the efficient production and processing of food, feed, fiber and fuels. It involves disciplines like mechanical engineering (agricultural machinery and automated machine systems), soil science (crop nutrient and fertilization, etc.), environmental sciences (drainage and irrigation), plant biology (seeding and plant growth management), animal science (farm animals and housing) etc.," as indicated by *http://www.ABE.ncsu.edu/academic/agricultural-engineering.php.*

According to *https://en.wikipedia.org/wiki/Biological_engineering,* "Biological Engineering (BE) is a science-based discipline that applies concepts and methods of biology to solve real-world problems related to the life sciences or the application thereof. In this context, while traditional engineering applies physical and mathematical sciences to analyze, design and manufacture inanimate tools, structures and processes, biological engineering uses biology to study and advance applications of living systems."

SPECIALTY AREAS OF ABE

Agricultural and Biological Engineers (ABEs) ensure that the world has the necessities of life including safe and plentiful food, clean air and water, renewable fuel and energy, safe working conditions, and a healthy environment by employing knowledge and expertise of sciences, both pure and applied, and engineering principles. Biological engineering applies engineering practices to problems and opportunities presented by living things and the natural environment in agriculture. BA engineers understand the interrelationships between technology and living systems, have available a wide variety of employment options. The *http://asabe.org* indicates that "ABE embraces a variety of following specialty areas." As new technology and information emerge, specialty areas are created, and many overlap with one or more other areas.

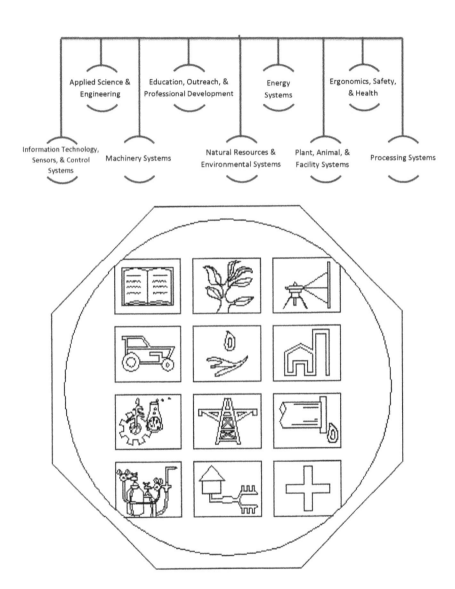

1. **Aqua Cultural Engineering:** ABEs help design farm systems for raising fish and shellfish, as well as ornamental and bait fish. They specialize in water quality, biotechnology, machinery, natural resources, feeding and ventilation systems, and sanitation. They seek ways to reduce pollution from aqua cultural discharges, to

reduce excess water use, and to improve farm systems. They also work with aquatic animal harvesting, sorting, and processing.

2. **Biological Engineering** applies engineering practices to problems and opportunities presented by living things and the natural environment.

3. **Energy:** ABEs identify and develop viable energy sources—biomass, methane, and vegetable oil, to name a few—and to make these and other systems cleaner and more efficient. These specialists also develop energy conservation strategies to reduce costs and protect the environment, and they design traditional and alternative energy systems to meet the needs of agricultural operations.

4. **Farm Machinery and Power Engineering:** ABEs in this specialty focus on designing advanced equipment, making it more efficient and less demanding of our natural resources. They develop equipment for food processing, highly precise crop spraying, agricultural commodity and waste transport, and turf and landscape maintenance, as well as equipment for such specialized tasks as removing seaweed from beaches. This is in addition to the tractors, tillage equipment, irrigation equipment, and harvest equipment that have done so much to reduce the drudgery of farming.

5. **Food and Process Engineering:** Food and process engineers combine design expertise with manufacturing methods to develop economical and responsible processing solutions for industry. Also food and process engineers look for ways to reduce waste by devising alternatives for treatment, disposal and utilization.

6. **Forest Engineering:** ABEs apply engineering to solve natural resource and environment problems in forest production systems and related manufacturing industries. Engineering skills and expertise are needed to address problems related to equipment design and manufacturing, forest access systems design and construction; machine-soil interaction and erosion control; forest operations analysis and improvement; decision modeling; and wood product design and manufacturing.

7. **Information and Electrical Technologies Engineering** is one of the most versatile areas of the ABE specialty areas, because it is applied to virtually all the others, from machinery design to soil

testing to food quality and safety control. Geographic information systems, global positioning systems, machine instrumentation and controls, electromagnetics, bioinformatics, biorobotics, machine vision, sensors, spectroscopy: These are some of the exciting information and electrical technologies being used today and being developed for the future.

8. **Natural Resources:** ABEs with environmental expertise work to better understand the complex mechanics of these resources, so that they can be used efficiently and without degradation. ABEs determine crop water requirements and design irrigation systems. They are experts in agricultural hydrology principles, such as controlling drainage, and they implement ways to control soil erosion and study the environmental effects of sediment on stream quality. Natural resources engineers design, build, operate and maintain water control structures for reservoirs, floodways and channels. They also work on water treatment systems, wetlands protection, and other water issues.

9. **Nursery and Greenhouse Engineering:** In many ways, nursery and greenhouse operations are microcosms of large-scale production agriculture, with many similar needs—irrigation, mechanization, disease and pest control, and nutrient application. However, other engineering needs also present themselves in nursery and greenhouse operations: equipment for transplantation; control systems for temperature, humidity, and ventilation; and plant biology issues, such as hydroponics, tissue culture, and seedling propagation methods. And sometimes the challenges are extraterrestrial: ABEs at NASA are designing greenhouse systems to support a manned expedition to Mars!

10. **Safety and Health:** ABEs analyze health and injury data, the use and possible misuse of machines, and equipment compliance with standards and regulation. They constantly look for ways in which the safety of equipment, materials and agricultural practices can be improved and for ways in which safety and health issues can be communicated to the public.

11. **Structures and Environment:** ABEs with expertise in structures and environment design animal housing, storage structures, and

greenhouses, with ventilation systems, temperature and humidity controls, and structural strength appropriate for their climate and purpose. They also devise better practices and systems for storing, recovering, reusing, and transporting waste products.

CAREER IN AGRICULTURAL AND BIOLOGICAL ENGINEERING

One will find that university ABE programs have many names, such as biological systems engineering, bioresources engineering, environmental engineering, forest engineering, or food and process engineering. Whatever the title, the typical curriculum begins with courses in writing, social sciences, and economics, along with mathematics (calculus and statistics), chemistry, physics, and biology. Student gains a fundamental knowledge of the life sciences and how biological systems interact with their environment. One also takes engineering courses, such as thermodynamics, mechanics, instrumentation and controls, electronics and electrical circuits, and engineering design. Then student adds courses related to particular interests, perhaps including mechanization, soil and water resource management, food and process engineering, industrial microbiology, biological engineering or pest management. As seniors, engineering students work in a team to design, build, and test new processes or products.

For more information on this series, readers may contact:

Ashish Kumar, Publisher and President
Sandy Sickels, Vice President
Apple Academic Press, Inc.,
Fax: 866-222-9549
E-mail: ashish@appleacademicpress.com
http://www.appleacademicpress.com/
publishwithus.php

Megh R. Goyal, PhD, PE
Senior Editor-in-Chief
Innovations in Agricultural and Biological Engineering
E-mail: goyalmegh@gmail.com

PART I

SALINITY MANAGEMENT: EMERGING TECHNOLOGIES AND TRENDS

CHAPTER 1

EMERGING TRENDS IN SALINITY RESEARCH: AN INDIAN PERSPECTIVE

D. K. SHARMA and ANSHUMAN SINGH

CONTENTS

1.1 INTRODUCTION

High salt concentration in the root zone severely limits agricultural productivity in most of the irrigated regions of the world [53]. The widespread menace of salinity can be gauged by the fact that globally over 1100 million hectare (M ha) lands are salt affected [75]. Although saline and sodic soils are predominant in arid and semi-arid climates, they also occur in humid and coastal regions [1, 53]. The salt affected lands in rainfed (arid and semi-arid) regions account for about 70% of the global area under cereals [55] and this fact points to the urgent need for technological interventions to sustain and enhance the agriculturally important assets in these resource scarce regions which seem to be extremely vulnerable to compounded impacts of land degradation, water shortages and climate variability. The rapid and alarming rate of expansion of salt affected soils (SAS) is a pointer to the looming threat for the world food security [21]. Besides, diverse forms of land degradation [5, 22] could prove disastrous as vast tracts of fertile soils would become barren and unproductive. Conversely, arresting the soil and environmental processes responsible for deterioration in soil health coupled with technological interventions to sustain and enhance productivity of this precious natural resource would significantly improve the quality of human life [10, 23].

Of late, many intersecting challenges such as climate change, deforestation and fresh water shortages have posed unprecedented dangers to the sustainability of agricultural systems in many parts of the world. Extreme climate events have increased in frequency in the past few decades often causing severe harm to agro-biodiversity [28], soil health [36] and irrigation water availability [39]. Massive deforestation could inflict huge damages to productive soil and water resources [20]. Currently, global (fresh) water resources are overstretched [73] and climate change, aquifer salinization, increasing inter-sector competition for good quality water use and water pollution [58] are likely to worsen the situation such that fresh water would increasingly become difficult to access and use, if not rare. The adverse impacts of most of these changes could be very severe in saline environments necessitating a thorough scrutiny of the existing technologies for salinity management as many of them may no longer be effective to address these diverse constraints.

This chapter presents the current trends in salinity research, constraints in field applications of available technologies and future research agenda consistent with current global trends for sustainable agriculture in salt affected soils from a developing country perspective.

1.2 SALINITY AS A FORM OF LAND DEGRADATION

Saline and sodic soils occur under almost all climatic conditions. Nonetheless, they are predominant in arid and semi-arid regions having dry soils, scanty rainfall, nutrient deficiencies and high evapotranspiration losses [53]. The arid soils are usually rich in salts of geogenic origin, which have a tendency to move upward and accumulate in surface soil under continuous irrigation. In fact, introduction of irrigation in these regions substantially alters the geo-hydrological properties of the soils often resulting in profuse salt movements from the lower to upper horizons [66]. The contemporary global assessments on salinity menace [21, 49, 68, 75, 76] hint at socio-economic pitfalls and hidden environmental costs of what may be referred to as a complex inter-relationship among diverse soil, environmental and anthropogenic processes causing massive deterioration in soil health and environmental quality. Available evidences reveal that along with soil erosion and acidity, salinity and sodicity stresses together represent a formidable challenge to global agriculture and life supporting natural resources spanning, *inter alia*, biodiversity, fertile soils and fresh water [19, 26, 49, 53].

According to a recent study, Middle East, Australia and North Africa together constitute nearly half (45%) of the global area under saline and sodic soils. The South Asian region comprising of India accounts for about 5% of the total world area under SAS. Interestingly, it could be a matter of relief to the land managers and salinity researchers as bulk (85%) of these degraded lands are only slightly (65%) to moderately (20%) salt affected and accordingly may not pose substantial hindrances in reclamation as existing technical know-how, with slight modifications, could enable their sustainable management. The real challenge, however, would lie in preventing desalinization of reclaimed soils and in ameliorating the remainder (15%) extremely affected soils [75]. The current estimated salt affected area in India (6.73 M ha) is expected to substantially increase in ensuing decades [57, 58] if appropriate technological interventions and policy measures are not initiated in time. The state wise distribution of saline and sodic soils in India (Table 1.1) indicates that states of Gujarat (2.2 M ha), Uttar Pradesh (1.37 M ha), Maharashtra (0.61 M ha), West Bengal (0.44 M ha) and Rajasthan (0.38 M ha) together account for almost 75% of SAS in the country. What is more worrisome is the fact that majority of saline tracts are also often underlain with marginal quality saline and sodic ground water compounding the problems [63]. Under such a situation, substantial modifications in future reclamation programs would be necessary for salinity management in a holistic manner. Given the increasing competition for fresh water use among different sectors of economy,

TABLE 1.1   State Wise Distribution of Salt Affected Soils in India (ha)

State	Saline soils	Sodic soils	Total
Andhra Pradesh	77,598	196,609	274,207
Andaman & Nicobar Islands	77,000	0	77,000
Bihar	47,301	105,852	153,153
Gujarat	1,680,570	541,430	2,222,000
Haryana	49,157	183,399	232,556
Karnataka	1893	148,136	150,029
Kerala	20,000	0	20,000
Madhya Pradesh	0	139,720	139,720
Maharashtra	184,089	422,670	606,759
Odisha	147,138	0	147,138
Punjab	0	151,717	151,717
Rajasthan	195,571	179,371	374,942
Tamil Nadu	13,231	354,784	368,015
Uttar Pradesh	21,989	1,346,971	1,368,960
West Bengal	441,272	0	441,272
Total	**2,956,809**	**3,770,659**	**6,727,468**

Source: NRSA and Associates 1996. Mapping of Salt Affected Soils in India, 1:250,000 map sheets, Legend NRSA. Hyderabad (Published under https://creativecommons.org/licenses/by/4.0/).

dangers of climate change induced abrupt water shortages, rapidly diminishing fresh water reserves and massive water pollution, the successful agricultural production will, to a great extent, depend on the use of marginal quality saline water and drainage effluents in soil reclamation and crop irrigation in times to come [58].

All soils invariably contain soluble salts, but certain soil and environmental conditions coupled with faulty crop management practices induce rapid salinity development [50]. Depending on soil reaction (pH_s) and electrical conductivity (EC_e) of saturation extract, exchangeable sodium percentage (ESP) and the sodium adsorption ratio (SAR), three distinct categories of SAS are saline, sodic and saline-sodic soils [63]. Considering the commonly adopted management practices, ICAR-Central Soil Salinity Research Institute (CSSRI) has suggested that saline-sodic soils could either be treated as saline or sodic depending upon the need for application of amendments. In spite of favorable physical environment and good water permeability, plant growth in saline soils is hampered initially by osmotic stress and subsequently by specific ion toxicities [43]. While osmotic stress is due to restricted water flux to the roots [27], disruption of ion homeostasis and failure of cellular transport channels to prevent excessive uptake of Na^+ ions are responsible for the nutritional imbalances and toxicities [31]. Excessive cellular Na^+ concentrations not only drastically limit K^+ absorption but also prove fatal to key metabolic reactions such as photosynthesis and protein synthesis [27]. Besides alkaline reaction and high exchangeable sodium percentage (ESP) which causes nutrient imbalances, plant growth in sodic soils is also hampered due to restricted water and air movements, low water holding capacity and poor seedling emergence [44].

1.3 SALINITY RESEARCH IN INDIA: ACHIEVEMENTS AND LESSONS

The establishment of CSSRI in 1969 to conduct basic and applied researches in salt stress management culminated into the development and wide adoption of many promising technologies. The major technological breakthroughs made in the last 45 years include development of approximation soil and water salinity maps of India, gypsum-based reclamation package for sodic soils, sub-surface and bio-drainage techniques for ameliorating waterlogged saline lands, release of several salt tolerant cultivars in rice, wheat and mustard and different technologies for coastal saline soils and difficult to reclaim black Vertisols. Gradual improvements have played a key role in increasing and sustaining the popularity of these technologies among the farmers of diverse agro-ecological regions. Considering the specific requirements of post-reclamation phase, research experiments have been started to optimize different resource conservation technologies, *viz.*, zero till wheat, direct seeded rice, sprinkler irrigation, mulching with crop residues for sustaining the health and productivity of reclaimed and semi-reclaimed soils. Widespread menace of secondary salinization, particularly in north-western region and Indo-Gangetic plains, has proved catalytic for prioritized research on land modification models and saline aquaculture. Many tree-based alternate land use systems, salt tolerant high value horticultural crops, low cost microbial crop growth enhancers and an integrated farming system model for small landholders have also been developed [57, 58, 63]. In spite of these credible achievements in productive management of SAS, salinity research is facing new challenges, which have necessitated a paradigm shift in future, agenda and priorities.

1.3.1 ENHANCING THE EFFICIENCY OF SUBSURFACE DRAINAGE

After years of gradual improvements in design and materials, subsurface drainage (SSD) technology to reclaim the waterlogged saline lands has come of age [24, 25]. With modest beginnings in north-western India, the practical utility of this technology has been successfully demonstrated in other salinity affected regions of the country and about 0.1 M ha waterlogged saline lands have been reclaimed till date. The reclaimed soils exhibit significant improvements in crop yield and cropping intensity. In spite of tangible socio-economic gains in terms of higher incomes and employment opportunities to the farmers and landless laborers, large scale implementation and maintenance of SSD projects are hampered by higher establishment costs, operational difficulties, lack of community participation and the problems encountered in disposal of drainage effluents [24, 25, 58]. On one hand, insufficient monetary resources restrict the implementation of the technology and on the other virtual non-existence of drainage societies and lack of collective responsibility results in lower than anticipated improvements at sites where SSD projects have been initiated. This state of affairs highlights the need to evolve effective institutional arrangements to ensure active community participation in SSD projects [54].

Formation of Farmers' Drainage Societies in Haryana state, entry of several contracting firms to undertake drainage activities on turnkey basis and reduced cost of drainage investigations as a result of emerging drainage guidelines (Table 1.2) are some pointers to the positive developments in this field in India [24, 25]. The socio-economic constraints apart, disposal of saline drainage effluents presents another big problem especially in landlocked regions. To overcome the environmental problems

TABLE 1.2 Guidelines for Drainage Design Under Various Agro-Climatic Conditions in India

Drainage coefficient (mm d⁻¹)			Drain depth (m)			Drain spacing (m)	
Climatic conditions	Range	Optimum value	Outlet conditions	Drain depth	Optimum depth (m)	Soil texture	Drain spacing
Arid	1–2	1	Gravity	0.9–1.2	1.1	Light	100–150
Semi-arid	1–3	2	Pumped	1.2–1.8	1.5	Medium	50–100
Sub-humid	2–5	3				Heavy (Vertisols)	30–50

Source: Gupta [25].

due to localized disposal, use of evaporation ponds is suggested which again faces the problems of higher establishment costs and specific design requirements [71]. Under these circumstances, irrigation with saline drainage water in cyclic and/or blended mode with fresh water is an attractive option. While available results point to ample prospects for the profitable reuse of saline drainage water in salt leaching and irrigation of crops [14, 25, 41, 59, 60], they also hint for future refinements in developing appropriate management practices such as pre-sowing irrigation with fresh water and use of salt tolerant cultivars [59, 60] for enhanced acceptability at farmers' fields.

1.3.2 BIO-DRAINAGE, LAND SHAPING AND SALINE AQUACULTURE

The slow penetration of SSD technology has generated interest in exploiting rapidly transpiring, salt tolerant trees to prevent salinity build-up in canal commands. It is essentially a preventive measure based on bio-energy driven removal of excess soil water [29]. The results obtained so far validate the usefulness of this intervention which requires nominal one-time investment without any major recurring expenses [7, 52]. Salt tolerant trees (e.g., *Eucalyptus*, *Populus tremula* and bamboo) have given encouraging results when raised early to prevent secondary salinization in irrigation commands due to faulty on-farm water management [63]. There is a growing realization that appropriate land modification measures could substantially enhance the worth of this technology [57].

Land leveling and shaping, canal lining, restoration of abandoned water bodies and artificial ground water recharge structures are some of the intrinsic components of sustainable irrigation development projects. Policy flaws in implementation of these measures and neglect on the part of end users often cause congestion of natural drains resulting in water logging and salinity build-up with adverse consequences for agricultural productivity and soil health [3]. Evidence is mounting that difficult to reclaim waterlogged lands may be put to productive use by simple, low cost earth maneuvering measures such as farm ponds, deep furrow-high ridge planting systems and integrated rice-cum-fish culture [12, 40]. While economically viable land shaping technologies have become successful in coastal saline tracts of the country [40], research experiments are under-way to demonstrate their potential for rehabilitation of waterlogged soils in other salinity affected regions.

Saline aquaculture is being increasingly advocated as a practical solution to harness the productivity of waterlogged saline lands. This is consistent with current global trends as commercial land-based aquaculture using saline ground water has gained momentum in countries like Australia, Israel and USA [2]. Although research on shrimp and fish farming in saline environments is in a nascent stage in India, available evidences indicate that it could play an important role in improving farmers' livelihoods and food security [8, 35, 51] provided that improved management practices and safety regulations are in place to sustainably harness the regional endowments in the best possible way [35]. This concern emanates from the experience of neighboring Bangladesh where a sudden transition from rice to shrimp culture, devoid of scientific and environmental considerations, has jeopardized the health and productivity of traditional rice agro-ecosystems [48].

1.3.3   SEARCH FOR NEW AMENDMENTS FOR SODIC SOILS

Traditionally, sodic soils are reclaimed by replacing the sodium by calcium on exchange sites using chemical amendments and the subsequent removal of replaced sodium by leaching. In India, gypsum-based technology has ensured crop production in vast areas (~1.8 M ha) covered by sodic soils. The sustained future applications of this technology, however, face stiff price, supply and quality constraints which are hampering gypsum availability and use by the farmers. To allay these fears, prioritized research on alternative amendments has been started and the initial results are encouraging (Table 1.3). It has been shown that sodic soils treated with industrial by-products such as fly-ash [34, 69], press-mud [78, 79] and distillery spent wash (DSW) [33, 65] exhibit significant improvements in physico-chemical properties and crop yields. Clearly, these amendments have potential to replace gypsum in land reclamation programs with degree of replacement depending on factors such as characteristics of experimental soil (e.g., soil pH and ESP), prevailing climatic conditions and crops grown. The ameliorative efficacy of these amendments is often significantly enhanced by supplemental application of organic manures and crop residues.

TABLE 1.3 Efficacy of Different Alternative Amendments in Sodic Soils

Amendment	Results	Ref.
Fly-ash	Combined use of fly-ash (3%) and gypsum (60% of GR) gave highest grain and straw yields in rice and wheat and caused significant reductions in pH, ESP and SAR of the experimental soil over sole applications of fly-ash.	[34]
	Fly-ash (10 and 20 Mg ha^{-1}) treated sodic soils produced significantly higher rice and wheat yields and recorded substantial decrease in pH, TSS and ESP.	[69]
Press-mud	Press-mud application and wheat residue incorporation were found best to sustain rice-wheat yields in soils irrigated with high RSC (8.5 me L^{-1}) sodic water.	[78]
	Press-mud (10 Mg ha^{-1}) along with FYM (10 Mg ha^{-1}), gypsum (5 Mg ha^{-1}) and recommended fertilizer dose significantly enhanced rice and wheat yields under continuous sodic irrigation.	[79]
DSW	Application of 50% distillery effluent along with bio-amendments was best in improving the properties of sodic soil and in improved germination and seedling growth of pearl millet.	[33]

As deteriorated physical environment and poor hydraulic conductivity are major constraints to plant growth in sodic soils, polymers have been used as ameliorants and soil conditioners to improve aggregate stability [67], water infiltration rates [15, 74] and reduce run-off induced soil erosion [37]. It appears, however, that polymers used in these experiments were effective at only low to moderate ESP levels. As these are essentially soil column studies, conclusive evidence cannot be drawn regarding whether these polymers will be equally effective under field conditions characterized by challenges other than high ESP. In future, therefore, investigations will be essential to identify low-cost, natural polymers as amendments as well as experiments to determine their dosage, timings of field application and relative efficacy as compared to popular amendments (e.g., gypsum) being used and alternative amendments (e.g., press-mud and fly-ash) under evaluation. Recently, the prospects of using nano scale materials-variously described as nanomaterials, nanoparticles, engineered nanoparticles, manufactured nanoparticles and nano-enhanced materials-as ameliorants to restore the productivity of degraded soils has received wide attention [38, 47, 50, 81–83].

For example, a relatively abundant mineral zeolite is well known for favorable effects on soil cation exchange capacity and hydraulic conductivity. Accordingly, it has been evaluated for its soil ameliorating properties in calcareous [77] and sodic soils [38, 46]. Fine-grained calcareous loess soil treated with zeolite exhibited improved water infiltration and higher moisture retention capacity, which significantly reduced the surface runoff and soil erosion [77]. Decrease in runoff rate and soil loss in sodic soils mixed with Ca-zeolite was presumably due to reduced clay dispersion, improvement in soil aggregation and the subsequent increase in soil hydraulic conductivity [38]. These results imply that zeolite can potentially regulate water supply to crops under drought conditions and can be used as a potential ameliorant for enhancing water use efficiency in dry land agriculture. Formation of pedogenic calcium carbonate is the major chemical reaction responsible for chemical degradation and impaired hydraulic properties of sodic soils. In spite of higher ESP values (>15), presence of gypsum and Ca-zeolites prevented increase in soil pH and improved the hydraulic properties in degraded Vertisols [46]. They are also implicated in salinity alleviation as application of natural zeolites caused removal of considerable amounts of salts in poultry litter [72]. As these are essentially preliminary findings, future experiments are warranted to study the behavior of nanoparticles and similar compounds in soils, their interactions with other components of soil system and effects on soil microbes and plants so as to ascertain that their future use is environmentally viable.

1.3.4　DEVELOPING MULTIPLE STRESS TOLERANT CROPS

Cultivation of salt tolerant crops is seen as a cost-effective and environmental-friendly strategy to overcome the problems of high water footprint of salt leaching and use of cost prohibitive chemical amendments in reclamation of saline and sodic soils [57, 58]. It has been demonstrated that degraded SAS put under salt tolerant trees and crops exhibit improved physical, chemical and biological environment due to gradual increments in organic carbon and nutrient levels, improved hydraulic conductivity and higher microbial activity and a concurrent reduction in ESP and soluble salts [6, 42, 45]. While improved water infiltration results in decreased salt content [45], organic matter addition through leaf litter enhances the microbial activity in soils under tree plantations [32]. A number of viable candidate species have been identified and agronomic practices standardized for raising plantations in salt affected community lands lying unproductive. For example, *Prosopis juliflora, Acacia nilotica, Casuarina equisetifolia, Tamarix articulata* and *Leptochloa fusca* are promising species for reclaiming sodic soils [64].

Besides soil reclamation, agro-forestry plantations which act as strong carbon sinks are also valuable for alleviating fuel wood and forage shortages [61]. Tree-crop combinations based on indigenous fruits (*Aegle marmelos, Emblica officinallis* and *Carissa congesta*) as main components and field crops (cluster bean and barley) as subsidiary components give good results even with the use of moderate (EC_{iw} ~6 dS m^{-1}) to high (~10 dS m^{-1}) salinity waters [9]. Medicinal plants (*Plantago ovate, Aloe barbadensis* and *Andrographis paniculata*) are amenable to saline irrigation without any appreciable reductions in biomass [70].

To provide viable and cost-effective solutions to the resource poor farmers, salt tolerant varieties have been developed in rice, wheat and mustard (Table 1.4) which perform and yield well with the applications

TABLE 1.4 Salt Tolerance Threshold and Yield Potential of Crop Varieties Developed by ICAR-CSSRI

Crop	Variety	Salient features
Rice	CSR10	Its grain yield is about 3 t ha^{-1} in SAS (pH$_2$ up to 9.9 and EC$_e$ up to 9.0 dS m^{-1}).
	CSR13	It gives about 3 t ha^{-1} yield in saline (EC$_e$ up to 9 dS m^{-1}) and sodic soils (pH$_2$ up to 9.9).
	CSR23	It yields fairly well (~4 t ha^{-1}) in SAS (pH$_2$ up to 9.9 and EC$_e$ up to 10 dS m^{-1}).
	CSR27	It has high yield potential (~4 t ha^{-1}) and relatively early maturity (125 days) under salt stress (pH$_2$ up to 9.9 and EC$_e$ up to 10 dS m^{-1}).
	CSR30	Although it gives relatively low (2–2.5 t ha^{-1}) yield under high salinity (EC$_e$ up to 7 dS m^{-1}) and sodicity (pH$_2$ up to 9.5) stresses, it is a fine grained (Basmati) type and fetches premium price.
	CSR36	It has high yield potential (~4 t ha^{-1}) in salt-affected soils (pH$_2$ up to 9.9 and EC$_e$ up to 10 dS m^{-1}).
Wheat	KRL 1–4	A selection from Kharchia-65, it gives 2.5–3.5 t ha^{-1} grain yield in sodic (pH$_2$ up to 9.3) and saline (EC$_e$ up to 7 dS m^{-1}) soils.
	KRL 19	It produces 2.5–3.5 t ha^{-1} yield in sodic (pH$_2$ up to 9.3) and saline (EC$_e$ up to 7 dS m^{-1}) soils.
	KRL 210	Its yield potential is 5.5 t ha^{-1} in normal soils and 3–5 t ha^{-1} in salt affected soils (pH$_2$ up to 9.3 and EC$_e$ up to 6 dS m^{-1}) soils.
	KRL 213	It yields well (3.3 t ha^{-1}) under salt stress (pH$_2$ up to 9.2 and EC$_e$ up to 6 dS m^{-1}) and gives good performance in areas having brackish and saline (EC$_{iw}$ 10 dS m^{-1}; RSC 8–9 meq L^{-1}) ground water.
Mustard	CS 52	Its seed yield is about 1.5 t ha^{-1} in saline (EC$_e$ 6–8.5 dS m^{-1}) and sodic (pH$_2$ up to 9.3) soils.
	CS 54	It gives about 1.6 t ha^{-1} seed yield in SAS (EC$_e$ 6–9 dS m^{-1} and pH$_2$ up to 9.3).
	CS 56	Its seed yield is above 1.6 t ha^{-1} in saline (EC$_e$ 6–9 dS m^{-1}) and sodic (pH$_2$ up to 9.3) soils.

of small amounts of chemical amendments [57, 58]. There is a growing realization, however, that exclusive focus on breeding for salt tolerance would no longer work and that development of multiple stress tolerant crop genotypes must be prioritized by integrating molecular and genomics tools with conventional breeding approaches.

1.3.5   EXPLOITING MICROBIAL BIOREMEDIATION APPROACH

In recent years, emphasis has gradually shifted to exploit the potential of alternative technologies for sustaining crop growth and yields in saline soils. The objective is to employ economically realistic and practically feasible solutions for salt stress mitigation in an environmental-friendly way. In this regard, concerted research efforts in past decades have led to the identification and development of many promising soil inhabiting microorganisms known to alleviate salt stress in plants. These soil microflora, collectively referred to as plant growth promoting rhizobacteria (PGPR), act by diverse mechanisms such as up-regulation of endogenous plant growth promoting hormones, production of volatile organic compounds and extracellular enzymes, and improved availability of nutrients for enhanced tolerance to abiotic stresses [56].

Given the fact that such soil inhabiting micro-organisms exhibit considerable salt tolerance and have potential to promote plant growth in saline and sodic soils [4], experiments have been conducted to isolate and utilize effective strains in salinity management in both field and high value horticultural crops. For example, endophytic bacteria induced sodicity tolerance in polyembryonic mango rootstocks (GPL-1 and ML-2) which was presumably due to higher activity of extracellular enzymes such as amylase, protease, cellulase and lipase [30].

The underlying physiological mechanisms through which these microorganisms alleviate salt stress in plants include higher uptake of K^+ ions, improvement in water absorption and leaf water relations, stability of chlorophyll pigments and increase in photosynthesis, elevated levels of antioxidant enzymes and expression of genes involved in salt tolerance [56]. Although effective in alleviating salt stress in crops, use of microbial inoculants is limited due to higher costs and lack of technical know-how. To circumvent these constraints, a low-cost microbial bio-formulation 'CSR–BIO, ' based on a consortium of *Bacillus pumilus*, *Bacillus thuringenesis* and *Trichoderma harzianum* on dynamic media, has been developed. It acts as a soil conditioner and nutrient mobilizer and significantly increases the productivity of rice, banana, vegetables and gladiolus in sodic soils [11].

1.3.6   SUSTAINING THE PRODUCTIVITY OF RECLAIMED SOILS

To sustain the agricultural productivity of ameliorated saline lands, with specific focus on small and marginal farmers, crop diversification with resource use efficient high value crops, promotion of integrated farming and large scale adoption of resource conservation technologies are absolutely essential. It is well established that labor- and resource use-intensive integrated farming systems significantly enhance the income and employment opportunities for farm families, and ensure provision of home grown nutritious food. Bringing marginal lands under cultivation and productivity enhancement of small farms could address the twin challenges of relentless natural resource degradation and ever swelling food demands. While smallholder cultivators supported by a favorable policy environment played a crucial role in the success of Green Revolution, the future of smallholder-led agricultural development is likely to face constraints such as poor infrastructure and inefficient market linkages highlighting the need to increase investments in agricultural research and development (R&D) [17, 18].

In this backdrop, an integrated model consisting of diverse crop and animal components (field and horticultural crops, fishery, cattle, poultry and beekeeping) has been developed for obtaining high and regular incomes while ensuring higher resource use efficiencies. Experiments are in progress to demonstrate the efficacy of similar farming models in waterlogged sodic soils, saline Vertisols and coastal saline soils [57]. With adverse socio-economic and environmental impacts of intensive cropping during Green Revolution being well known, large scale adoption of resource conservation technologies such as zero tillage in wheat, direct seeded rice, residue incorporation and mulching, sprinkler irrigation- has become imperative for sustainable crop production in reclaimed saline and sodic soils [58].

1.4   LOOMING THREATS REQUIRE IMMEDIATE ATTENTION

Apart from previously mentioned challenges, alarming scale of secondary salinity in irrigated commands [13, 62], repeated instances of resodification and resalinization of amended soils [58], climate change induced alterations in growing conditions [16, 76] and severe water shortages [73] may create huge stumbling blocks in reclamation and management of salt-affected lands. It is likely that emerging challenges would drastically reduce the efficiency of many of the current salinity management practices and would thus necessitate a thorough revision of existing packages and practices to make them more robust and relevant for the future reclamation programs.

Massive secondary salinization in many parts of country in particular and globally in general has caused huge damage to biodiversity, productive soils and local cropping systems. In most of the cases, gradual reversion of reclaimed saline and sodic soils to the original state is largely due to faulty crop and water management practices [57, 58]. In developing countries like India, agriculture sector will be worst affected by abrupt climatic aberrations. Even slight changes in current temperature and rainfall patterns would cause huge production losses in arid zones [16], while projected sea level rise and the consequent increase in salt intrusion coupled with increased frequency of cyclonic storms would undermine the productivity of coastal agro-ecosystems [80]. As these problems are essentially anthropogenic in nature- largely due to faulty crop and water management practices—an integrated approach based on ground water recharge, improved on-farm water management, cultivation of less irrigation demanding crops, cultivation of stress resilient cultivars, integrated nutrient management and adoption of resource conservation technologies is suggested to minimize their severity and adverse impacts [58].

1.5 SUMMARY

Sustainable management of soil and water resources will be crucial for ensuring sufficient and secure access to food to a burgeoning population. Evidence that climate change would adversely affect crop production is worrisome as most of the developing countries are likely to be worst affected by the climate change induced food scarcity. The present global agriculture scenario is a pointer to the urgent need for bringing salt affected marginal lands under cultivation through frontier technology driven research and development programs so as to lessen the pressure on productive soils as well as to ensure gradual improvements in health and quality of these degraded lands though agronomic measures. Authors began their discussion with a general picture of salt affected environments and emerging constraints which are likely to exacerbate the challenges for sustainable agriculture. Then, they summarized the characteristics of and constraints to plant growth in saline and sodic soils. After a brief review of technological achievements, they highlighted the means and ways to address the critical gaps for sustainable and profitable cropping in salt-affected soils. They concluded by emphasizing the need to put these degraded resources under food production through technological interventions. While many salinity mitigation technologies require a thorough reworking to make them relevant to the changing needs, many new approaches offer ample opportunities to harness the productivity of saline environments without much effort.

KEYWORDS

- agricultural productivity
- alternative amendments
- amendments
- aquifer salinization
- arid climate
- bio-drainage
- climate change
- coastal regions
- crop varieties
- drainage
- food security
- gypsum
- irrigation management
- land degradation
- land shaping

- **polymers**
- **reclamation**
- **rice-wheat**
- **salinity**
- **salt affected soils**
- **salt stress**
- **secondary salinity**
- **sodic**
- **soil health**
- **stress tolerance**
- **sub-surface drainage**
- **sustainable management**
- **water resources**
- **water shortages**

REFERENCES

1. Abrol, I. P., Yadav, J. S. P., & Massoud, F. I. (1988). *Salt-affected Soils and Their Management*. FAO Soils Bulletin 39, Rome, Italy.
2. Allan, G. L., Fielder, D. S., Fitzsimmons, K. M., Applebaum, S. L., Raizada, S., Burnell, G., & Allan, G. (2009). Inland saline aquaculture. In: *New Technologies in Aquaculture: Improving Production Efficiency, Quality and Environmental Management*, 1119–1147.
3. Ambast, S. K., Gupta, S. K., & Singh, G. (2007). *Agricultural Land Drainage: Reclamation of Waterlogged Saline Lands*. Central Soil Salinity Research Insitute, Karnal, India, 231p.
4. Arora, S., Vanza, M. J., Mehta, R., Bhuva, C., & Patel, P. N. (2014). Halophilic microbes for bio-remediation of salt affected soils. *African Journal of Microbiology Research*, 8, 3070–3078.
5. Bai, Z. G., Dent, D. L., Olsson, L., & Schaepman, M. E. (2008). Proxy global assessment of land degradation. *Soil Use and Management*, 24, 223–234.
6. Bhojvaid, P. P., & Timmer, V. R. (1998). Soil dynamics in an age sequence of *Prosopis juliflora* planted for sodic soil restoration in India. *Forest Ecology & Management*, 106, 181–193.
7. Chhabra, R., & Thakur, N. P. (1998). Lysimeter study on the use of bio-drainage to control water logging and secondary salinization in (canal) irrigated arid/semi-arid environment. *Irrigation and Drainage Systems*, 12, 265–288.

8. CSSRI. 2013. *Annual Report, 2013–14*. Central Soil Salinity Research Institute, Karnal, India.

9. Dagar, J. C., Tomar, O. S., Minhas, P. S., Singh, G., & Ram, J. (2008). *Dry Land Bio-saline Agriculture- Hisar Experience*. Technical Bulletin No. 6/2008, Central Soil Salinity Research Institute, Karnal, 28p.

10. Daily, G. C. (1995). Restoring value to the world's degraded lands. *Science-AAAS-Weekly Paper Edition*, 269, 350–353.

11. Damodaran, T., Rai, R. B., Jha, S. K., Sharma, D. K., Mishra, V. K., Dhama, K., Singh, A. K., & Sah, V. (2013). Impact of social factors in adoption of CSR BIO-A cost effective, eco-friendly bio-growth enhancer for sustainable crop production. *South Asian Journal of Experimental Biology*, 3, 158–165.

12. Datta, K. K. (2003). Technological options for poor quality saline water management in agriculture: scope & prospects. In: *Drainage for a Secure Environment and Food Supply*. Proceedings of 9th ICID International Drainage Workshop, Utrecht, Netherlands, 10–13 September 2003. International Institute for Land Reclamation and Improvement (ILRI).

13. Datta, K. K., & De Jong, C. (2002). Adverse effect of water logging and soil salinity on crop and land productivity in northwest region of Haryana, India. *Agricultural Water Management*, 57, 223–238.

14. Datta, K. K., Sharma, V. P., & Sharma, D. P. (1998). Estimation of a production function for wheat under saline conditions. *Agricultural Water Management*, 36, 85–94.

15. El-Morsy, E. A., Malik, M., & Letey, J. (1991). Polymer effects on the hydraulic conductivity of saline and sodic soil conditions. *Soil Science*, 151, 430–435.

16. Enfors, E. I., & Gordon, L. J. (2007). Analyzing resilience in dry land agro-ecosystems: a case study of the Makanya catchment in Tanzania over the past 50 years. *Land Degradation & Development*, 18, 680–696.

17. FAO, (2011). '*State of the World's Land and Water Resources for Food and Agriculture* (SOLAW)'. Food and Agriculture Organization of the United Nations, Rome, Italy.

18. FAO, WFP, and IFAD. (2012). *The State of Food Insecurity in the World 2012. Economic Growth is Necessary but Not Sufficient to Accelerate Reduction of Hunger and Malnutrition.* Rome, FAO.

19. Fitzpatrick, R. W. (2002). Land degradation processes. *ACIAR Monograph Series*, 84, 119–129.

20. Geist, H. J., & Lambin, E. F. (2002). Proximate causes and underlying driving forces of tropical deforestation: Tropical forests are disappearing as the result of many pressures, both local and regional, acting in various combinations in different geographical locations. *Bio-Science*, 52, 143–150.

21. Ghassemi, F., Jakeman, A. J., & Nix, H. A. (1995). *Salinisation of Land and Water Resources: Human Causes, Extent, Management and Case Studies*. CAB International.

22. Gibbs, H. K., & Salmon, J. M. (2015). Mapping the world's degraded lands. *Applied Geography*, 57, 12–21.

23. Gisladottir, G., & Stocking, M. (2005). Land degradation control and its global environmental benefits. *Land Degradation & Development*, 16, 99–112.

24. Gupta, S. K. (2002). A century of subsurface drainage research in India. *Irrigation and Drainage Systems*, 16, 69–84.
25. Gupta, S. K. (2015). Reclamation and management of water-logged saline soils. *Agricultural Research Journal*, 52, 104–115.
26. Hajkowicz, S., & Young, M. (2005). Costing yield loss from acidity, sodicity and dry land salinity to Australian agriculture. *Land Degradation and Development*, 16, 417–433.
27. Hauser, F., & Horie, T. (2010). A conserved primary salt tolerance mechanism mediated by HKT transporters: a mechanism for sodium exclusion and maintenance of high K+/Na+ ratio in leaves during salinity stress. *Plant, Cell and Environment*, 33, 552–565.
28. Heller, N. E., & Zavaleta, E. S. (2009). Biodiversity management in the face of climate change: a review of 22 years of recommendations. *Biological Conservation*, 142, 14–32.
29. Heuperman, A. F., Kapoor, A. S., & Denecke, H. W. (2002). *Bio-drainage: Principles, Experiences and Applications*. Knowledge Synthesis Report No. 6. Food & Agriculture Organization, Rome.
30. Kannan, R., Damodaran, T., & Umamaheswari, S. (2015). Sodicity tolerant polyembryonic mango root stock plants: a putative role of endophytic bacteria. *African Journal of Biotechnology*, 14, 350–359.
31. Katiyar-Agarwal, S., Verslues, P., & Zhu, J. K. (2005). Mechanisms of salt tolerance in plants. In: *Plant Nutrition for Food Security, Human Health and Environmental Protection*, 44–45.
32. Kaur, B., Gupta, S. R., & Singh, G. (2000). Soil carbon, microbial activity and nitrogen availability in agro-forestry systems on moderately alkaline soils in northern India. *Applied Soil Ecology*, 15, 283–294.
33. Kaushik, A., Nisha, R., Jagjeeta, K., & Kaushik, C. P. (2005). Impact of long and short-term irrigation of a sodic soil with distillery effluent in combination with bio-amendments. *Bio-Resource Technology*, 96, 1860–1866.
34. Kumar, D., & Singh, B. (2003). The use of coal fly ash in sodic soil reclamation. *Land Degradation and Development*, 14, 285–299.
35. Kutty, M. N. (2005). Towards sustainable freshwater prawn aquaculture–lessons from shrimp farming, with special reference to India. *Aquaculture Research*, 36, 255–263.
36. Lal, R. (2004). Soil carbon sequestration to mitigate climate change. *Geoderma*, 123, 1–22.
37. Levy, G. J., Levin, J., & Shainberg, I. (1995). Polymer effects on runoff and soil erosion from sodic soils. *Irrigation Science*, 16, 9–14.
38. Liu, R., & Lal, R. (2012). Nano-enhanced materials for reclamation of mine lands and other degraded soils: a review. *Journal of Nanotechnology*, doi: 10.1155/2012/461468.
39. Mall, R. K., Singh, R., Gupta, A., Srinivasan, G., & Rathore, L. S. (2006). Impact of climate change on Indian agriculture: a review. *Climatic Change*, 78, 445–478.
40. Mandal, S., Sarangi, S. K., Burman, D., Bandyopadhyay, B. K., Maji, B., Mandal, U. K., & Sharma, D. K. (2013). Land shaping models for enhancing agricultural productivity in salt affected coastal areas of West Bengal—an economic analysis. *Indian Journal of Agricultural Economics*, 68, 389–401.

41. Minhas, P. S. (1996). Saline water management for irrigation in India. *Agricultural Water Management*, 30, 1–24.

42. Mishra, A., Sharma, S. D., & Khan, G. H. (2003). Improvement in physical and chemical properties of sodic soil by 3, 6 and 9 years old plantation of *Eucalyptus tereticornis*: bio-rejuvenation of sodic soil. *Forest Ecology & Management*, 184, 115–124.

43. Munns, R. (1993). Physiological processes limiting plant growth in saline soils: some dogmas and hypotheses. *Plant, Cell & Environment*, 16, 15–24.

44. Murtaza, G., Ghafoor, A., & Qadir, M. (2006). Irrigation and soil management strategies for using saline-sodic water in a cotton-wheat rotation. *Agricultural Water Management*, 81, 98–114.

45. Nosetto, M. D., Jobbágy, E. G., Tóth, T., & Di Bella, C. M. (2007). The effects of tree establishment on water and salt dynamics in naturally salt-affected grasslands. *Oecologia*, 152, 695–705.

46. Pal, D. K., Bhattacharyya, T., Ray, S. K., Chandran, P., Srivastava, P., Durge, S. L., & Bhuse, S. R. (2006). Significance of soil modifiers (Ca-zeolites and gypsum) in naturally degraded Vertisols of the Peninsular India in redefining the sodic soils. *Geoderma*, 136, 210–228.

47. Pan, B., & Xing, B. (2012). Applications and implications of manufactured nanoparticles in soils: a review. *European Journal of Soil Science*, 63, 437–456.

48. Paul, B. G., & Vogl, C. R. (2011). Impacts of shrimp farming in Bangladesh: Challenges and alternatives. *Ocean and Coastal Management*, 54, 201–211.

49. Pitman, M. G., & Läuchli, A. (2002). Global impact of salinity and agricultural ecosystems. In: *Salinity: Environment-Plants-Molecules*. Springer Netherlands. 3–20.

50. Prost, R., & Yaron, B. (2001). Use of modified clays for controlling soil environmental quality. *Soil Science*, 166, 880–895.

51. Purushothaman, C. S., Raizada, S., Sharma, V. K., Harikrishna, V., Venugopal, G., Agrahari, R. K., Rahaman, M., Hasan, J., & Kumar, A. (2014). Production of tiger shrimp (*Penaeus monodon*) in potassium supplemented inland saline sub-surface water. *Journal of Applied Aquaculture*, 26, 84–93.

52. Ram, J., Dagar, J. C., Lai, K., Singh, G., Toky, O. P., Tanwar, V. S., Dar, S. R., & Chauhan, M. K. (2011). Bio-drainage to combat water logging, increase farm productivity and sequester carbon in canal command areas of northwest India. *Current Science*, 100, 1673–1680.

53. Rengasamy, P. (2006). World salinization with emphasis on Australia. *Journal of Experimental Botany*, 57, 1017–1023.

54. Ritzema, H. P., Satyanarayana, T. V., Raman, S., & Boonstra, J. (2008). Subsurface drainage to combat water logging and salinity in irrigated lands in India: Lessons learned in farmers' fields. *Agricultural Water Management*, 95, 179–189.

55. Rosegrant, M., Cai, X., Cline, S., & Nakagawa, N. (2002). *The Role of Rainfed Agriculture in the Future of Global Food Production*. Environment and Production Technology Division, International Food Policy Research Institute.

56. Ruzzi, M., & Aroca, R. (2015). Plant growth-promoting rhizobacteria act as biostimulants in horticulture. *Scientia Horticulturae*, http://dx.doi.org/10.1016/j.scienta.2015.08.042.

57. Sharma, D. K., & Chaudhari, S. K. (2012). Agronomic research in salt affected soils of India: An overview. *Indian Journal of Agronomy*, 57, 175–185.
58. Sharma, D. K., & Singh, A. (2015). Salinity research in India: achievements, challenges and future prospects. *Water and Energy International*, 58, 35–45.
59. Sharma, D. P., & Rao, K. V. G. K. (1998). Strategy for long term use of saline drainage water for irrigation in semi-arid regions. *Soil and Tillage Research*, 48, 287–295.
60. Sharma, D. P., & Tyagi, N. K. (2004). On-farm management of saline drainage water in arid and semi-arid regions. *Irrigation and Drainage*, 53, 87–103.
61. Sharma, D. K., Chaudhari, S. K., & Singh, A. (2014). In salt-affected soils agroforestry is a promising option. *Indian Farming*, 63, 19–22.
62. Singh, A., Krause, P., Panda, S. N., & Flugel, W. A. (2010). Rising water table: A threat to sustainable agriculture in an irrigated semi-arid region of Haryana, India. *Agricultural Water Management*, 97, 1443–1451.
63. Singh, G. (2009). Salinity-related desertification and management strategies: Indian experience. *Land Degradation and Development*, 20, 367–385.
64. Singh, G., Singh, N. T., & Abrol, I. P. (1994). Agroforestry techniques for the rehabilitation of degraded salt-affected lands in India. *Land Degradation & Development*, 5, 223–242.
65. Singh, R., Singh, N. T., & Arora, Y. (1980). The use of spent wash for the reclamation of sodic soils. *Journal of the Indian Society of Soil Science*, 28, 38–41.
66. Smedema, L. K., & Shiati, K. (2002). Irrigation and salinity: a perspective review of the salinity hazards of irrigation development in the arid zone. *Irrigation and Drainage Systems*, 16, 161–174.
67. Sumner, M. E. (1993). Sodic soils-New perspectives. *Soil Research*, 31, 683–750.
68. Thomas, D. S. G., & Middleton, N. J. (1993). Salinization: new perspectives on a major desertification issue. *Journal of Arid Environments*, 24, 95–105.
69. Tiwari, K. N., Sharma, D. N., Sharma, V. K., & Dingar, S. M. (1992). Evaluation of fly ash and pyrite for sodic soil rehabilitation in Uttar Pradesh, India. *Arid Land Research and Management*, 6, 117–126.
70. Tomar, O. S., & Minhas, P. S. (2004). Performance of medicinal plant species under saline irrigation. *Indian Journal of Agronomy*, 49, 209–211.
71. Tripathi, V. K., Gupta, S., & Kumar, P. (2008). Performance evaluation of subsurface drainage system with the strategy to reuse and disposal of its effluent for arid region of India. *Journal of Agricultural Physics*, 8, 43–50.
72. Turan, N. G. (2008). The effects of natural zeolite on salinity level of poultry litter compost. *Bioresource Technology*, 99, 2097–2101.
73. Vörösmarty, C. J., Green, P., Salisbury, J., & Lammers, R. B. (2000). Global water resources: vulnerability from climate change and population growth. *Science*, 289, 284–288.
74. Wallace, A., Wallace, G. A., & Abouzamzam, A. M. (1986). Amelioration of sodic soils with polymers. *Soil Science*, 141, 359–362.
75. Wicke, B., Smeets, E., Dornburg, V., Vashev, B., Gaiser, T., Turkenburg, W., & Faaij, A. (2011). The global technical and economic potential of bioenergy from salt-affected soils. *Energy & Environmental Science*, 4, 2669–2681.
76. Williams, W. D. 1999. Salinization: A major threat to water resources in the arid and semi-arid regions of the world. *Lakes & Reservoirs: Research & Management*, 4, 85–91.

77. Xiubin, H., & Zhanbin, H. (2001). Zeolite application for enhancing water infiltration and retention in loess soil. *Resources Conservation and Recycling*, 34, 45–52.
78. Yaduvanshi, N. P. S., & Sharma, D. R. (2007). Use of wheat residue and manures to enhance nutrient availability and rice-wheat yields in sodic soil under sodic water irrigation. *Journal of the Indian Society of Soil Science*, 55, 330–334.
79. Yaduvanshi, N. P. S., & Swarup, A. (2005). Effect of continuous use of sodic irrigation water with and without gypsum, farmyard manure, pressmud and fertilizer on soil properties and yields of rice and wheat in a long term experiment. *Nutrient Cycling in Agroecosystems*, 73, 111–118.
80. Yeo, A. (1998). Predicting the interaction between the effects of salinity and climate change on crop plants. *Scientia Horticulturae*, 78, 159–174.
81. Zahow, M. F., & Amrhein, C. (1992). Reclamation of a saline sodic soil using synthetic polymers and gypsum. *Soil Science Society of America Journal*, 56, 1257–1260.
82. Zhang, L., & Fang, M. (2010). Nanomaterials in pollution trace detection and environmental improvement. *Nano Today*, 5, 128–142.
83. Zhang, W. X. (2003). Nanoscale iron particles for environmental remediation: an overview. *Journal of Nanoparticle Research*, 5, 323–332.

CHAPTER 2

LIVING WITH SALTS IN IRRIGATION WATER

S. K. GUPTA

CONTENTS

2.1   INTRODUCTION

The water cycle on Earth is essentially a closed system, always having the same amount of about 1.386 billion km³. Around 97.5% of it is salty while remaining 2.5% is fresh water (Figure 2.1). The oceans, storehouses of saline water, cover roughly 71% of the area of the Earth. Bulk of the saline water constituting about 96.5% of the total global water is found in Oceans, Seas and bays. Most of it has an average salinity of about 35,000 ppm, i.e., that of the sea water [25]. Nearly 1.74% of the resource is contained in ice caps, glaciers, and permanent snow, constituting about 68.7% of the fresh water resource [25]. Ground water contains about 1.69% of which 0.93 is saline while the remaining fresh water [25]. Small quantities of water are also found in the atmosphere and in living beings. A major fraction of the fresh water is locked up in ice and in the ground. Only about 1.2% of all freshwater or 0.3% of the whole water is surface water, which serves the humanity on this earth (Figure 2.1). It is found in lakes, rivers, reservoirs and underneath the ground at shallow/medium depths which can be tapped at an affordable cost. Only this amount is available on a sustainable basis as it is regularly replenished by rain and snowfall constituting about 110,000 km^3 [16]. Agricultural water use accounts for about 75% of total global consumption—mainly in crop irrigation—while industrial use accounts for about 20%, and the remaining 5% is used for domestic purposes.

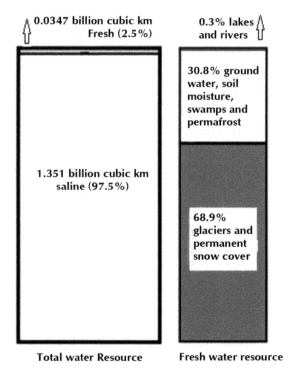

FIGURE 2.1   Distribution of global total and fresh water resource.

This chapter describes various technological options to avoid or blunt the adverse effects of saline irrigation. All the options are categorized into five groups. Options in the crop management group are selection of salt tolerant crops, crops varieties, application of saline water only at tolerant growth stages and improved cultural practices such as high seed rate to take care of likely mortality, seeding under conserved moisture and application of post germination irrigation, etc.

2.2 WHAT IS SALINE WATER?

Water that contains significant amounts of dissolved salts is designated with the generic term saline water. The total salt content of the irrigation water is represented as ppm or milligrams per liter (mg L^{-1}), millimole per liter (mmol L^{-1}) or electrical conductivity (EC) expressed in dS m^{-1} at 25°C. The purpose for which water is likely to be used determines the limits of salinity that makes it unfit for that particular use. In general categorization in Table 2.1 is used.

Saline water sources with salinity greater than 1000 parts per million (ppm) include ground water, reclaimed municipal effluent and agricultural drain flow. One of the concerns in using saline water for irrigation is soil salinization as noted in a number of publications [2, 4, 6, 8, 13, 17, 18, 24, 39]. But there is no doubt that long-term economic survival of humanity depends upon how we can overcome the problems caused by saline water and use it beneficially for the benefit of the mankind.

TABLE 2.1   Classification of Waters Based on EC

Water class	EC (dS m^{-1})	Type of water
Non saline	<0.7	Drinking/irrigation water
Slightly saline	0.7–2.0	Irrigation water
Moderately saline	2.0–10.0	Drainage/ground water
Highly saline	10.0–25.0	Drainage/ground water
Very highly saline	25.0–45.0	Very saline ground water
Brine	>45.0	Sea water

From irrigation point of view, all waters having EC_{iw} less than 2 dS m^{-1} are categorized as fresh (good) and others as saline. Several different guidelines are available guidelines to characterize saline water in different countries, even having large variations within a country depending upon location specific variables of soil, crops, kinds of salts, irrigation methods to be used and overall management inputs. Based on EC_{iw}, Sodium adsorption ratio (SAR_{iw}) and residual sodium Carbonate (RSC_{iw}), waters in India are characterized as good, saline, alkali and toxic as per classification criterion given in Table 2.2 that also makes distinction between the kinds of salts present in the water. The chemical composition of each kind of water characterized as per classification given in Table 2.2 is shown in Table 2.3.

2.2.1 EXTENT AND DISTRIBUTION

Ground water salinity is a widespread problem around the globe. Worldwide area with occurrences of saline and brackish ground water at shallow or intermediate depths approximates 24 million km^2 or about 16% of

TABLE 2.2   Classification of Saline Water

Water quality	EC_{iw} (dS m^{-1})	SAR_{iw} (mmolL^{-1})$^{1/2}$	RSC (meqL^{-1})
A. Good	< 2	< 10	< 2.5
B. Saline			
i. Marginally saline	2–4	< 10	< 2.5
ii. Saline	> 4	< 10	< 2.5
iii. High-SAR saline	> 4	> 10	< 2.5
C. Alkali water			
i. Marginally alkali	< 4	< 10	2.5–4.0
ii. Alkali	< 4	< 10	> 4.0
iii. High-SAR alkali	Variable	> 10	> 4.0
D. Toxic Water [22]	The toxic water has variable salinity, SAR and RSC but has excess of specific ions such as nitrate, boron, fluoride, chloride, sodium, silica or heavy metals such as selenium, cadmium, lead and arsenic, etc.		

Source: Ref. [34], Toxic waters [22].

TABLE 2.3 Chemical Composition of Good, Saline and Alkali Waters in Haryana

Parameter	Good	Marginally saline	Saline	High SAR saline	Marginally alkali	Alkali	High SAR alkali
EC (dS m⁻¹)	0.61	2.56	6.52	14.04	1.04	1.83	2.53
pH	8.60	8.30	8.3	8.55	8.05	8.20	8.50
Na (meq L⁻¹)	2.20	7.00	30.8	103.60	6.10	12.20	21.70
Ca (meq L⁻¹)	2.10	3.00	8.3	22.64	1.50	2.70	2.10
Mg (meq/L)	1.70	9.70	19.7	12.60	3.30	3.10	1.70
Cl (meq L⁻¹)	2.20	6.00	33.00	102.00	3.60	3.80	6.80
SO₄ (meq L⁻¹)	0.90	3.10	25.40	36.75	0.50	1.70	9.80
CO₃+HCO₃ (meq L⁻¹)	4.20	13.20	6.00	2.40	7.60	13.00	10.60
SAR (meq L⁻¹)¹ᐟ²	1.60	2.89	8.24	24.80	3.90	7.20	15.72
RSC (meq L⁻¹)	0.40	1.50	Nil	Nil	2.80	7.20	6.80
Mg/Ca ratio	0.81	3.23	2.37	0.56	2.20	1.15	0.81
Cl/SO₄ ratio	2.44	1.94	1.30	2.78	7.20	2.24	0.69
Village	Allipur	Barota	Bindrala	Durjanpur	Alipur	Amunpur	Kabulpur
Block	Nissing	Nissing	Assandh	B-Khera	Nissing	Nissing	Assandh
District	Karnal	Karnal	Karnal	Bhiwani	Karnal	Karnal	Karnal

the total Earth area [44]. It impacts about 1.1 billion people that live in areas with some level of significant ground water salinity. It causes health problems, decreases agricultural yields, turns fertile agricultural lands in barren field, reduces profits and jeopardizes livelihoods, increases costs of infrastructure maintenance and of industrial processes and impacts environment and eco-systems [44]. Notwithstanding the large spatial variations, high salinity ground waters are mostly encountered in arid regions.

Central Soil Salinity Research Institute (CSSRI), Karnal undertook a study to compile the information, what so ever was available, and prepared a map of ground water quality map of India for irrigation on a 1:6 million scale [20]. The map has four legends namely, good water, saline water, high-SAR saline water and alkali water as per classification reported in Table 2.3. The total area underlain with the saline ground water ($EC_{iw} > 4$ dS m^{-1}) is approximated as 0.19 million km^2 (Table 2.4) with annual replenishable recharge of 11,765 million m^3 yr^{-1} [9]. It is certain that the problem is more widespread than indicated by this value as the areas having salinity in the range of 2–4 dS m^{-1} have not been accounted.

Moreover, alkali waters are often encountered in area with low salinity. Another equally important source of ground water is the drainage water

TABLE 2.4   Estimated Area (M ha) Underlain with Saline Ground Water [41]

State	Total area of the state (M ha)	Area underlain with saline ground water (M ha)
Delhi	0.14	0.01 (7.14)
Gujarat	19.6	2.43 (12.39)
Haryana	4.4	1.14 (25.90)
Karnataka	19.1	0.88 (4.60)
Punjab	5.0	0.30 (6.00)
Rajasthan	34.2	14.10 (41.22)
Tamil Nadu	13.0	0.33 (2.53)
Uttar Pradesh	2.9	0.13 (4.48)
Total (India)	125.0	19.3 (15.44)

Note: Figures in brackets express percentage. M ha = million hectare

from agricultural lands and constitutes a significant source of irrigation such as in California, USA and Egypt. In the Imperial Valley alone annual discharge of drainage water to the Salton Sea amounts to about 1.5 million acre-feet with an average salt concentration of about 3,500 mg L^{-1}. Drainage reuse is being practiced in the Lower Egypt since 1970. At present, drainage reuse is widely practiced in Delta region providing about 4.0 BCM yr^{-1} of drainage water to be mixed with the fresh water of main canals. The government has an ambitious plan to expand drainage reuse to reach 8.0 BCM/year so as to improve the Nile water use efficiency and to expand the cultivated area. It will still leave no less than 8.0 BCM yr^{-1} that can be discharged to the sea, a minimum amount necessary to keep the salt balance of the Delta region [1].

2.2.2 *LIVING WITH SALTS IN SALINE WATER*

A number of potential measures have been proposed to mitigate the problems and adverse effects often associated with use of saline ground water for irrigation. In some measures it is attempted to reclaim the water or the irrigated lands so that cultivation on saline water irrigated land is sustainable. The other groups of measures accept the high ground water salinity but attempts to blunt/minimize the salinity impacts on the soils/crops [44]. The most widespread problem with saline water irrigation worldwide is soil salinization that adversely impacts the crop yields. The extent of irrigated lands affected by salinity vary from a low of 20–30 M ha in a total of 260 million ha of irrigated lands in the world [14] to 34 M ha [33] to 45 M ha out of the 230 million ha of irrigated lands [15]. The problems have arisen either due to development of water logging or use of saline ground water for irrigation but in any case these are associated with saline ground water. It is assessed that Pakistan (7.0 M ha), China (6.7 M ha), United States (4.9 M ha) and India (3.3 M ha) contributes to about half of the salinized irrigated lands [33]. This chapter presents some of the management technologies that can be adopted to live with the salts in the irrigation water or the irrigated lands. The monsoon climatic conditions in India and many other countries provide opportunities to adopt low cost options to live with the salts.

2.3   SOIL SALINIZATION OF IRRIGATED LANDS

Soils, which are otherwise salt free, are rendered unproductive by irrigation water containing excess amounts of salts. Several factors control the extent of salinity build-up in the soil profile. The five most important factors are: (a) quality of irrigation water, (b) number of irrigations, (c) soil texture, (d) leaching fraction, and (e) rainfall. Usually, steady state equation based on leaching fraction criterion is used to assess the build-up such that

$$EC_e = EC_i * X \tag{1}$$

where, EC_e is the EC of the soil solution saturation extract, EC_i is the EC of the irrigation water and X is the salt concentration factor (SCF), which depends upon the leaching fraction (LF).

The value of X varies from 3.2 for a LF of 0.05 to 0.7 for a LF of 0.6. In practice values of X vary widely from about 0.5 to >2.0 in India and 0.71 to 2.63 in Iran [8, 32, 43]. If LF and the water uptake pattern by the crop are known, one can get the steady state salt profile that is likely to develop with the use of a given quality of irrigation water (Figure 2.2). It shows the likely salinity profiles resulting from irrigation water salinity of 4 dS m⁻¹, a LF of 0.2 and two water uptake patterns of 40:30:20:10 and 60:25:10:05.

The leaching requirement (LR) that helps to restrict the soil salinity build-up to a pre-decided level can be calculated with the help of the following equation:

$$LR = EC_{iw}/(5EC - EC_{iw}) \tag{2}$$

where, EC is either EC_e at threshold or at an acceptable yield level below threshold, and LR is the fraction of the irrigation water that must be passed through the root zone to control soil salinity at a level not exceeding EC.

The distinction between LR and LF must be made. While the former is the required fraction, later is the actual leaching fraction attained under given field situations.

In monsoon climatic conditions, such steady state conditions never develop. Rather a dynamic equilibrium is maintained over the years through the salinization and desalinization processes (Figure 2.3). Following three processes are clearly reflected in this figure:

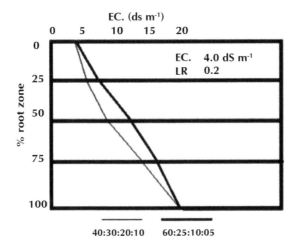

FIGURE 2.2   Steady state root zone salinity (ECe) likely to develop with water of 4 dS m⁻¹.

- Period of salinity build-up (Zone A) during November to April till equilibrium salinity is reached
- Period of salt redistribution (Zone B) within the root zone from April to mid June or till the onset of monsoon. During this period, soil salinity is usually highest in the surface layer (0–15 cm).
- Period of desalinization (Zone C) during monsoon season when rainfall is in excess over the potential evapotranspiration.

Under non-steady state conditions, the build-up besides the EC_{iw} also depends upon number of irrigations (Table 2.5). It also depends upon rainfall received during the monsoon (decides initial salinity) and that received during the cropping season and method of irrigation [21]. The soil salinity increases with increase in irrigation water salinity and number of irrigation as per regression Eq. (3) [21] and there is not much change after 4–5 irrigation (n = 4–5). Apparently, soil salinization can be controlled by reducing the number of irrigations with saline water. The soil salinity decreases with amount of rainfall received during the crop growing season. It is attributed to leaching of salts as well as to reduced numbers of irrigations required to fulfill the crop water requirement (Table 2.6).

$$EC_e = -2.26 + 0.904\ EC_{iw} + 1.235\ n \tag{3}$$

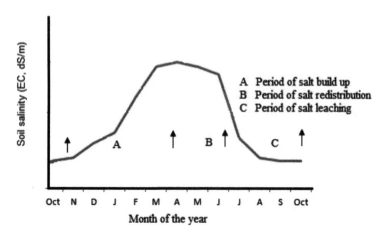

FIGURE 2.3   Salinization–desalinization cycle in a saline water irrigated land.

TABLE 2.5   Relationship between ECe and ECiw at Agra

Number of irrigations	Regression equations
2	$EC_e = 1.539 + 0.662\ EC_{iw}$
3	$EC_e = 2.143 + 0.903\ EC_{iw}$
4	$EC_e = 2.067 + 1.000\ EC_{iw}$
5	$EC_e = 0.623 + 1.212\ EC_{iw}$

TABLE 2.6   Regression Equation Showing Build-Up of Salinity As Influenced By Rainfall At Agra

Crop	Rainfall (mm)	No. of irrigations	Regression equations
Wheat	17.1	4	$EC_e = 2.437 + 1.291\ EC_{iw}$
Wheat	113.1	3	$EC_e = 1.759 + 0.693\ EC_{iw}$
Mustard	17.1	3	$EC_e = 2.758 + 1.070\ EC_{iw}$
Mustard	113.1	3	$EC_e = 1.795 + 0.499\ EC_{iw}$

For higher accuracy, more sophisticated models can be used to assess the salinity build-up in the soils, although their use is often limited because of lack of data required for their application. Irrespective of methodology of assessing soil salinity build-up, it is apparent that cultivating a salt tolerant crop requiring less numbers of irrigations will help to control the salinity build-up.

2.4   MANAGEMENT STRATEGIES FOR SALINE WATER

Increasing scarcity of irrigation water, need to manage water logging and soil salinity due to rising water table in irrigation commands and need to make efficient use of large reserves of saline ground water, it is of utmost importance to use in saline ground water for irrigation in a judicious manner. It should have minimum adverse impact on crop productivity and/or its quality and should not damage the soil resource for its sustainable use. Any management option for the judicious management must consider the interacting influence of the amount and type of salts in water and soils, nature of crop to be grown, type of soil to be irrigated, irrigation methods and related management options, nutrient applications and frequency and intensity of rainfall.

2.4.1   CROP MANAGEMENT

Plants differ widely in salt tolerance from sensitive glycophytes, whose normal growth is inhibited at low concentration of salts to the most resistant halophytes, which grow profusely on saline habitats. Though most crop plants are glycophytes, there is rather a wide range of salt tolerance amongst them, which should be exploited while using saline water. Beets and date palms, having halophytic ancestors, can be irrigated with saline water without much loss in yield. The crop selection must be based upon relative crop salt tolerance described and reported by many workers. Most data sets on crop tolerance are based on tolerance of crops to ECeof the soil rather than that of EC of irrigation water [6, 7, 24, 28, 30, 31]. As such, principles of soil salinization with saline water need to be applied to use these tables for crop selection (Table 2.7).

Ayers and Westcot [7] derived salt tolerance of crops on the basis of EC_{iw} assuming that SCF, X of Eq. (1) is 1.5 for a LF of about 0.15–0.20 and a crop uptake pattern of 40:30:20:10. Data set for few selected crops is reproduced in Table 2.9. Readers may refer the original publication for a complete record. Many other works have also appeared since then wherein crop tolerance has been expressed in terms of irrigation water salinity [19, 24, 35]. The later data are based on actual field experiments conducted

TABLE 2.7 Relative Crop Salinity Tolerance Rating

Relative crop salinity tolerance	Soil salinity (EC_e) at which yield loss begins (dS m^{-1})	Irrigation water salinity (EC_{iw}) at which yield loss begins (dS m^{-1})
Sensitive	<1.3	<1.0
Moderately sensitive	1.3–3	1.0–2.0
Moderately tolerant	3–6	2.0–4.0
Tolerant	6–10	4.0–8.0
Unsuitable unless reduced yield acceptable	>10	>8.0

Source: Mass [28], last column added by the author based on Ayers and Westcot [7] reasoning.

TABLE 2.8 Irrigation Water Salinity Limits for Selected Crops for 90% Yield Levels

	EC_{iw} for expected 90% of the optimum yield (dS m^{-1})			
Crops	Ayers and Westcot [7]	Lindsay [26]#	Minhas et al. [35]	
Barley	6.7	6.7	7.2–13.0	Silty loam to sandy loam
Maize	1.7	1.7	2.2–3.7	Clay loam to silty clay loam
Potato	1.7	1.7	2.1	Sandy loam
Rice	2.6	1.3	2.2–2.3	Silty clay loam to sandy loam
Sorghum	5.0	NR	2.6–7.0	Silty clay loam to sandy loam
Tomato	2.3	1.9	2.4	Sandy
Wheat	4.9	4.9	3.4 – 9.1	Clay to sandy loam*

For moderate to slow draining soils, Increase by 50% for well drained and decrease by 50% for medium to heavy clays; *Soil type for reported valued by Minhas et al. [35].

under the aegis of All India Coordinated Research Project (AICRP) on Management of Salt Affected Soils and Use of Saline Water in India (Table 2.8). Minor differences are visible although it emerges that under the dynamics of salinization-desalinization it should be possible to use relative higher salinity water for crop production without impairing the crop yields or the soil resource. Since a crop tolerant to salinity might be sensitive or semi-tolerant to alkali conditions, selection of crops is made to suit the irrigation water quality, i.e., saline, alkali/sodic or toxic (Table 2.2).

Moreover, a cropping sequence must be selected such that all the crops in the sequence are able to tolerate the characteristics of the irrigation water.

A general recommendations has emerged that semi-tolerant to tolerant crops with low water requirement should be grown while those requiring liberal water should be avoided. To quote an example, mustard is a crop which is salt tolerant and requires only one to two irrigations besides a pre plant irrigation. Experiments at Sampla indicated that highly saline drainage water can be used for post-plant irrigation to mustard without any substantial loss in its yield. In fact, grain yield was more when irrigated with 8 dS m^{-1} salinity water than with canal water (Table 2.9). It seems that the added salts at this level had a slight boosting effect on yield.

Besides intra-genic variations in salt tolerance of crops, wide inter-genic variation in the salt tolerance of crop cultivars has been documented. Among the 25 mustard cultivars, the highest germination index values were recorded for NRCD 509 and Varuna under control conditions with only minimal decline with EC_{iw} of 12 dS m^{-1}. Sensitive cultivars were SKM-425, JGM-03-02, CS-2100-5-6 and RH-0116 [40]. Usually there is a negative correlation between salt tolerance of varieties and their potential yields. For example, 'Damodar' in rice and 'Kharchia' in wheat are well documented for their salinity tolerance but have low yield potentials. Recent breeding efforts have changed this outlook as cultivars like 'CSR 30' for rice, 'HD-2560' of wheat, 'CS-52' of mustard and 'MESR-16' of cotton have been developed in India suggesting that it is possible to breed cultivars for high yield as well as high salt tolerances (Table 2.10).

Crops do not tolerate salinity equally at all stages of their growth (Table 2.11). Even a good salt tolerant crop sugar beet is sensitive to salts at germination stage although at later stages it shows quite a god tolerance. During the initial stages, the root zone being limited to surface few cm, most

TABLE 2.9   Effect of Drainage Water Salinity on Relative (%) Yield of Mustard

EC_{iw} (dS m^{-1})	1st year	2nd year	3rd year	Mean
0.5	100	100	100	100
8	100.7	120.1	116.7	112.5
15	94.1	90.4	88.0	90.8

TABLE 2.10 Promising Cultivars for Saline Environment

Crop	Varieties/Germplasm	
	India	Pakistan
Barley	Ratna, RL 345, RD 103, 137, K169	PK-30064, PK-30130, PK-30132, PK-30316
Cotton	DHY 286, CPD 404, G 17060, GA, MESR-16	NIAB-78, MNH-93
Mustard	CS 416, CS 330-1, Pusa Bold, CS 52, CS 54, CS 56	Gobi sarson
Pearl millet	MH 269, MH 331, MH 427, HHB 60	
Rice	CSR 23, CSR 36, CSR 30 (Basmati), TRY1, Vytilla 1, Canning 7	NIAB-6, IRO-6, KS-282
Safflower	HUS 305, A 1, Bhima	
Sorghum	SPV 475, 881, 678,669, CSH 11	Milo, JS-263, JS-1
Sugar beet*	Ramonsakya-06, Polyrava-E, Tribal, Maribo-resistapoly	Ramonsakya-06, Polyrava-E, Tribal, Maribo-resistapoly
Wheat	Raj 2325, Raj 2560, Raj 3077, WH 157, KRL 210, KRL 213, HD 2560	LU26S, Blue Silver, SARC-1 (well-drained), Blue Silver, SARC-3, Pb-85 (waterlogged)

Source: Anonymous [4]; Qureshi et al. [37]. *Varieties are tolerant to both alkali and saline environment.

salts from the evaporating soils concentrate there. As such, germination and early seedling establishment have been identified as the most crucial stages. Therefore, strategies to minimize the salinity of the seeding zone are needed to establish a good plant stand. The other critical periods for crops are when a phase change takes place say from vegetative to reproductive or heading and flowering to seed setting.

Conventional seeding of most crops is done when optimum moisture conditions for tillage and seedbed preparation are attained following a pre-sowing irrigation with good quality water. On the contrary, in saline water irrigated lands, seeds are exposed to soil water of higher salinity. Since the initial crop stand determines the final yield, cultural practices that help to establish good germination and crop stand are useful to live with the salts. Following agronomic manipulations help to achieve higher yields:

TABLE 2.11   Relative Tolerance of Crops at Different Growth Stages

Crop	Relative sensitivity
Agra, sandy loam soil	
Barley	Crown root initiation > pre-sowing > flowering/booting > jointing
Mustard	Pre-sowing* > flower initiation > secondary branching
Safflower	Pre-sowing = Rosette > flower initiation > main head opening
Wheat	Pre-sowing > flowering > milking > crown root initiation > jointing
Dharwad, silty clay loam, black soil	
Maize	Silking> tasseling > pre-sowing > knee height
Paddy	Transplanting > 50% flowering > end of tillering
Pigeon pea	Pre-sowing > flowering to pod development > flowering > seedling to flowering
Wheat	Pre-sowing > crown root initiation > milking > flowering > jointing
Bapatla, silty clay loam and loamy sand soil	
Black gram	Flowering to maturity > seeding > seeding to flowering
Groundnut	Germination to pegging > pegging to pod formation > pod formation to maturity
Onion	Transplanting to bud formation > bulb formation to bulb development > bulb development to maturity
Sunflower	Seeding > flower bud initiation > grain filling > germination to flowering

Source: Minhas and Gupta [34]. *Saline water used at the specified growth stage and non-saline water at all other stages.

- Apply a heavy pre-sowing irrigation with good quality water to push the salts down below the root zone. It will improve germination and early growth.

- Apply post-sowing irrigation to facilitate seed germination. Ensure that no soil crust is formed, as it might nullify the benefits of post sowing irrigation. In a field experiment, compared with the potential, the seed yield of Indian mustard in post-sowing irrigation following dry seeding could be sustained up to irrigation water salinity of 11 dS m^{-1}. Dry seeding and keeping surface-soil moist through sprinkler/post-sowing saline irrigation helps in better crop establishment.

- High concentration of salts not only reduces germination but causes mortality of young seedlings and reduces tillering and branching than a non-saline environment. Sharma et al. [40] found that germination of mustard reduced to 53% in a saline (12 dS m^{-1}) than a non-saline environment where it was 68% (distilled water). The average time of germination also increased from 1.59 days to 2.48 days. Therefore, a higher seed rate and/or closer spacing are used to maintain proper plant population. This can be achieved by using 20–25% higher seed rate over the recommended seed rate and narrowing the inter-row and/or intra-row spacing of row crops. In transplanted crops, increasing number of seedlings per hill or reducing row-to-row and/or plant-to-plant spacing to accommodate more seedlings helps in getting better yield.

- Practice rainwater conservation to leach down the salts.
- Since young seedlings are very sensitive to salts, older seedlings are recommended for transplanting.
- Saline water in general should not be used to grow summer crops during April to June.

2.4.2   SOIL MANAGEMENT

Saline water *per se* does not harm the crop unless salt accumulate in the root zone to the extent that the crop is unable to tolerate the accumulated salts. Accumulation of salts in soils depends upon the soil texture. Salt concentration factor X in Eq. (1) is lower in a light than in a fine textured soil. As a 'thumb rule' salt accumulation in saline water irrigated soils is nearly one half the salinity of irrigation water in coarse textured soils (loamy sand and sand), equal to that of irrigation water in medium textured sandy loam to loam soils and more than two times in fine textured soils (clay and clay loam). As such, saline water of higher EC can be applied to grow crops in coarse than fine textured soils. Crop production functions for wheat reveal that 50% reduction in yield of wheat occurs with saline waters of 17.5, 16.8, and 12.9 dS m^{-1} in loamy sand, sandy loam and silty clay loam. For the same texture groups these levels for mustard are 24.9, 18.3, and 14.7 dS m^{-1}, respectively.

As per guidelines prepared by CSSRI, Karnal, water having an EC$_{iw}$ of 12 dS m^{-1}can be used to grow tolerant and semi-tolerant crops in coarse textured soils, provided the annual rainfall is no less than 500 mm. But in fine textured soils, water with EC more than 2 dS m^{-1} would often create salinity problems. Even the leaching of salts through rainfall is governed to a great extent by soil texture. Analysis shows that to remove 80% of the salts accumulated during the period preceding monsoon would require 1.85, 0.95 and 0.76 cm of rainwater per cm soil depth in fine, medium and coarse textured soils [34]. It is attributed to low infiltration rates of fine textured soils (having high clay content) resulting in high runoff and evaporation of stagnated water. It reduces rainwater use efficiency to displace the salts. Adoption of measures such as tillage to open up the soil for better intake of rainwater and conservation of rainwater in soil via checking unproductive evaporation losses (soil/straw mulching) is useful during monsoon season.

Distribution of water and salts in soils vary with land form, as salts tend to move from wetter to drier areas. Thus, modifications in land configuration, modification in irrigation practice and seed placement at spots with lesser accumulation of salts is practiced to obtain higher yields. Seed bed design and seed placement should be so managed so as to minimize exposure of seeds and roots to salts particularly in the early sensitive stages of growth. Although, there are innumerable techniques to manage salts in this manner, yet for the purpose of this chapter it may suffice to illustrate this with few situations.

Planting seeds in the center of a single row raised bed will place the seed exactly in the area where salts concentrate (Figure 2.4A). To avoid such a situation, sowing of seeds should be done on the side of the ridges to avoid salts accumulating at the center of the ridge. In a double row raised planting bed, the two rows of seeds can be placed so that each is away from the accumulating salts at the center (Figure 2.4B). Growing seeds near the shoulders is even better alternative to avoid exposure to the salts provided crop can tolerate some stagnation (Figure 2.4C). For the larger seeded crops, even furrow seeding can be practiced with good response. Alternate furrow irrigation for single-row bed systems can help to raise the crop towards the wet furrow (Figure 2.4D, E). While adopting this practice, care should be taken that enough water is applied to wet all the way across the bed to prevent build-up of salts in the planted area. In this set-up, a problem may arise in this case if the other furrow is irrigated inadvertently

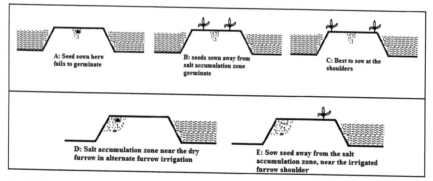

FIGURE 2.4 Seed placement in broad bed and furrow system irrigated in all furrows (top) and alternate furrow irrigation (bottom).

or the rainfall accumulates in both the furrows, disturbing the normal movement of salts in alternate furrow irrigation. Much better salinity control can be achieved by using sloping beds with seeds planted on the sloping side and the seed row placed just above the water line (Figure 2.5A, B).

2.4.3 IRRIGATION WATER MANAGEMENT

Irrigation with saline water improves productivity compared to limited irrigations with fresh water alone [10]. Experimental evidence has been generated to show that wheat productivity improved when saline water was used to supplement limited post plant irrigations at Agra (India) than without supplementation (Table 2.12).

On-farm irrigation management with saline water should adopt irrigation schedules that minimize irrigations, eliminate salinity build-up and

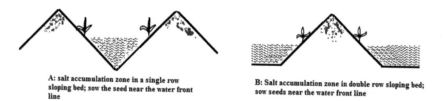

FIGURE 2.5 Sloping seedbeds commonly used for salinity and temperature control: single row (left, A) and double row (right, B).

TABLE 2.12   Effects of Supplement Irrigation with Medium and High Salinity Water on Wheat Yield

Year	Yield (t ha⁻¹)			
	One BAW	Two BAW	Three BAW	Mean
Only BAW without supplement	3.47	4.08	4.51	4.02
Supplemental and all irrigations with medium (6/8 dS m⁻¹) saline water				
	One BAW, remaining saline	Two BAW, remaining saline	Three BAW, remaining saline	All saline
2000–01	5.32	5.38	5.52	5.45
2001–02	4.95	4.92	5.37	5.05
2002–03	4.96	4.93	5.14	5.10
Average	5.08(46)*	5.08(26)*	5.34(18)*	5.20
Supplemental and all irrigations with highly (12 dS m⁻¹) saline water				
2000–01	5.06	5.26	5.44	4.64
2001–02	4.83	4.91	4.66	4.72
2002–03	4.77	4.88	4.94	4.69
Average	4.89(41)*	4.98(22)*	5.01(11)*	4.68

BAW is best available water;

*The values in parenthesis reveal increase in yield due to supplemental irrigations with saline water over average yield with fresh water without supplementation which were 3.47, 4.08 and 4.51 t ha⁻¹.

also ensures optimal crop production. Shallow depth high frequency irrigations are preferred in saline environment. Irrigations with saline water should be applied more frequently to reduce the cumulative water deficits (matric and osmotic) between the irrigation cycles. The distribution of water and salts in soils vary with the method of irrigation. Therefore, any irrigation method that creates and maintains favorable salt and water regimes in the root zone should be preferred. It would allow the plants to draw readily available water for their growth without any damage to the crop yield. It should be possible to achieve this with improved irrigation techniques. Even in surface irrigation methods, it could be ensured through proper land leveling, laser land leveling being preferred to achieve better results.

Furrow irrigation seems to be one of the effective methods of irrigation to provide salt free zone to the plants in the furrows since most salts either present in the soil or applied through irrigation would move towards the ridges of the furrows. Besides, by changing the size of the beds (broad bed and furrows) and/or the irrigation furrows (all furrows or alternate furrows) and placement of seeds, one can have several options as per the crop requirement discussed in Section 2.4.2.

The pressurized irrigation methods such as sprinkler and drip are more efficient as the quantity of water applied can be adequately controlled. Sprinkler irrigation is an ideal method to impose shallow depth high frequency regime. Leaching of soluble salts is more efficient in sprinkler irrigation as the water application rates can be controlled below the infiltration capacity of the soil. Sprinkler irrigation also helps in achieving uniform water application and salt leaching even under local topo-sequential differences in the field levels. The use of sprinkler irrigation helps in utilizing relatively more saline waters than surface irrigation methods. Yet these advantages gets subdued at higher salinity levels due to evaporation and consequent salt build-up in the water droplets falling on the leaves and harming the crop adversely [3]. Drip irrigation is even more beneficial as it also allows relatively more saline water use in crop production. The information generated so far clearly reveal the superiority of drip over the sprinkler method at all salinity levels for all the tested crops [23]. Even indigenous alternatives of drip irrigation such as pitcher irrigation and pipe irrigation allow the use of relatively more saline water for cultivating vegetables and horticultural crops [23].

Pre-sowing irrigation is given primarily to create optimal soil moisture conditions to facilitate tillage and seed bed preparation and to recharge the root zone with water for germination and to meet the ET needs of crops. In saline soils, pre-sowing irrigation helps to leach salts so as to allow a salt free zone for the seed germination and seedling establishment. It has been proved that relative yield of various crops with pre-sowing irrigation with fresh water followed by saline water throughout was always higher in comparison to when no fresh water was applied at pre-sowing irrigation (Table 2.13).

The conjunctive use of multi-source/multi-quality waters is defined as a strategy by which waters are used in a combined manner such that the

TABLE 2.13 Comparative Effect of Pre-Sowing Irrigation With Canal Water and Blended Saline Water on Wheat Yield (t ha^{-1})

EC_{iw} of blended water (dS m^{-1})	All irrigations with blended water	Only post-plant irrigation with blended water
0.5 (Canal water)	100.0	100.0
3.0	90.0	—
6.0	80.4	95.8
9.0	72.5	90.3
12.0	72.5	83.7
18.8	—	78.0

net output is more compared to the net output when each source/quality of water is used separately. In conjunctive use, fresh and saline water are used in a manner that salinity of the soil does not exceed a predefined level to grow a desired crop. Amongst the alternatives strategies shown in Figure 2.6, blending (on-farm, canal network and aquifer) and cyclic modes (in season sequential or switching modes) are more common in use.

When ground water is too saline for crop production, it is useful to dilute it with good quality water to reduce its salt content to a safe limit. Blending of saline and fresh waters is most promising proposition in areas where fresh water is available in sufficient quantities and on demand. Blending needs to be assessed on the basis of the EC, SAR, RSC or toxic constituents of the blended water such that most critical of them is brought below the pre-decided value through blending. If the salt concentration of the mix is less

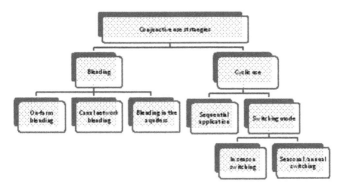

FIGURE 2.6 Conjunctive use options for fresh and saline waters.

than the threshold value of the crop to be grown, there should be no problem in the use of this option. In the other strategy, i.e., cyclic use, canal water is replaced with saline water in a pre-decided sequence/cycle. Saline water is substituted for good quality waters either when fresh water is unavailable and/or when the crop is at a more tolerant growth stage. The timing and the amount of water will depend upon the quality of water, the cropping pattern, the climate and the type of irrigation system. Cyclic mode is the most flexible technique for the conjunctive use of saline and fresh waters as both types of waters are independently available as against the mixing technique where only one quality of water is available following mixing [38, 39]. It has been argued that cyclic use is a better strategy than blending [34, 39].

The author had a critical look at the data sets on the relative yields of various crops under cyclic and blending modes and it emerged that cyclic mode is better than blending only if cycle begins with fresh water, treatment T3 being superior to treatment T4 (Table 2.14). If the cycle begins with saline water then the yield are usually less than blending. With a fresh look on the data sets, author concluded that if same amount of fresh and saline water is available at all times when irrigation is given to the crop, there are no yield differences between the cyclic use and blending modes of conjunctive use (Figure 2.7).

Irrespective of what has been stated, blending has proved to be the most promising alternative in Egypt where large-scale conjunctive use of drainage water is practiced [36]. Tanji et al. [42] calculated the flow-weighted concentration of water salinity and boron concentrations of the mix water when drainage water of 2.99 dS m^{-1} was mixed in canal water of 0.41 dS m^{-1}. The mixed supply has an EC of 2.19 as per conditions prevailing in Broadview Water District in 1976. Similar exercises are being conducted in India to understand the mix water salinity with various mixing ratios in Western Jamuna and Bhakra Canal Commands of Haryana to frame policies on conjunctive use. With calculated drainage volumes in 3 districts of Haryana, it has been shown that the salinity of mixed water will never exceed 1.7 dS m^{-1} if 30, 57 and 53 cusecs of discharge from vertical and horizontal drainage schemes having salinity of 6 dS m^{-1} is mixed in canal water. The discharge in the canal water varied from 14 to 84 m^3 s^{-1} depending upon the location of the blending site. The limited information for 28 m^3 s^{-1} flow is shown in Table 2.15, which shows that salinity remains less than 1.1 dS m^{-1}.

TABLE 2.14 Effect of Different Conjunctive Use Modes on Crop/Straw Yield of Wheat

Treatments		2001–02			2002–03		
		Grain yield (t ha⁻¹)	Straw yield (t ha⁻¹)	Weighted EC$_{iw}$ (dS m⁻¹)	Grain yield (t ha⁻¹)	Straw yield (t ha⁻¹)	Weighted EC$_{iw}$ (dS m⁻¹)
T1	BAW	3.15	5.63	8.81	3.22	5.19	9.76
T2	EC$_i$ 15	13.12	4.48	6.68	13.4	3.87	7.06
T3	1B:1S	8.14	5.44	9.05	8.33	5.03	9.56
T4	1S:1B	8.14	5.05	8.67	8.33	4.53	8.75
T5	1B:2S	9.80	5.18	8.52	10.02	4.17	8.28
T6	1S:2B	6.47	5.10	9.06	6.62	4.51	8.40
T7	2B:4S	9.80	5.49	9.95	10.02	4.99	9.52
T8	2S:4B	6.47	4.98	7.18	6.62	4.24	7.49
T9	3B:3S	8.14	5.59	9.85	8.33	5.06	9.76
T10	50B:50S	8.14	5.12	9.32	8.33	4.83	9.33
T11	30B:70S	10.10	4.91	8.25	10.4	4.31	8.49
T12	70B:30S	6.10	5.45	9.04	6.29	5.04	8.49
CD at 5%			0.38	0.57		0.46	0.57

BAW/B = Best available water (EC$_{iw}$ 3.6 dSm⁻¹); Saline/S = EC$_{iw}$ 15 dS m⁻¹.

T1 and T2 are control with all BAW and saline water, T3, T4, T5 and T6 are in season cyclic use sequential application modes (e.g., 1B:1S means 1 irrigation with BAW and 1 with saline water and cycle repeats), T7, T8 and T9 are in season cyclic switching modes (e.g., 2B:4S means BAW is switched to saline water following first two irrigations) and T10, T11 and T12 are blending modes (e.g., 50B and 50S means 50% of BAW and saline water are mixed). Weighted EC$_{iw}$ is calculated taking into account the rain water contribution as well.

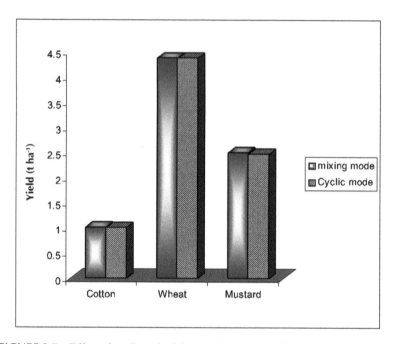

FIGURE 2.7   Effect of cyclic and mixing modes on crop yield.

TABLE 2.15   Effect of Blending Saline and Canal Water on Mixed Water Salinity, Haryana

Location (district)	Canal flow (cumecs)	EC canal water (dS m⁻¹)	Anticipated drainage flow (cumecs)	EC drainage water (dS m⁻¹)	EC mixed water (dS m⁻¹)
Jhajjar	28	0.4	1.48	6	0.68
Rohtak	28	0.4	1.60	6	0.70
Sonipat	28	0.4	0.84	6	0.56

Cumecs = $m^3\ s^{-1}$.

Another situation where blending has proved its usefulness is aquifer mixing of saline and fresh water where harvested fresh rain water is mixed in saline ground water through artificial ground water recharge technologies. The system has been tested and applied at a number of farmers' fields in Agra, Bhartapur and Mohindergarh regions of India with success. In this strategy, crops receive graded salinity water beginning with the least saline at pre-sowing or first irrigation followed by saline water of increasing salinity at subsequent irrigations. In general, highly saline water is brought down to 4–5 dS m^{-1} salinity at the time of first irrigation. By the time of fourth irrigation, it almost attains the original value of the ground water. The data of 9 such structures indicates such a trend for the years 2011–2014 (Table 2.16) [5]. This data also suggest that a variable mixing ratio strategy can be a good blending strategy.

The author modeled a variable mix ratio strategy wherein 1:0, 0.75:0.25, 0.5:0.5, 0.25:0.75 and 0:1 fresh: saline mix were assumed to have been applied for pre-sowing, 1st, 2nd, 3rd and 4th irrigation respectively to irrigate wheat crop. Using modeling capability of SWAP as validated by Verma et al. [45], the strategy was compared with the switching mode, where pre plant and first three irrigations were given with fresh water (Table 2.12), followed by one/two irrigations with saline water and a constant mixing ratio of 1:1 at all the stages. The relative transpiration in the variable mixing mode and the switching mode were the same while a decrease in the relative evapotranspiration was evident in the case of constant mixing ratio. Field results have also proved that former is superior to the constant mixing ratio mode. Notwithstanding the practical difficulties, it emerges that a variable mixing ratio mode performed better than the constant mixing mode. Overall, it emerges that conjunctive use is an art and can be manipulated to live with the salts. The exact design invariably is based upon crop characteristics, water qualities and the past experiences. Unintended conjunctive use is spreading fast in India and elsewhere both in the fresh and saline ground water regions. Although exact data are unavailable, the increasing number of private tube wells and the area irrigated by tube wells is a clear indicator of this phenomenon (Table 2.17).

TABLE 2.16 Effect of Blending Harvested Rainwater on Water Quality in Winter Crops (2011–2014)

Years	Initial EC of ground water (dS m^{-1})	EC$_{iw}$ at first irrigation (dS m^{-1})	EC$_{iw}$ at second irrigation (dS m^{-1})	EC$_{iw}$ at third irrigation (dS m^{-1})
2011–12	10.9–23.5	4.1–6.7	7.4–9.6	9.6–16.4
2012–13	10.9–23.5	7.7–10.7	9.3–13.4	10.3–17.2
2013–14	10.9–23.5	4.3–5.9	7.9–9.8	9.5–15.6

TABLE 2.17 Evolution of Unintended Conjunctive Use of Surface and Ground Water in Haryana, India

Year	Private irrigation tube wells (no.)	Net irrigated area (000, ha)	
		Canal	Tube well
1970–71	104,358	952	574
1980–81	332,027	1161	941
1990–91	497,571	1337	1248
2000–01	589,473	1476	1467
2010–11	753,357*	1236	1650

*2012–13.

2.4.4 NUTRIENT MANAGEMENT

Salt accumulation in saline water irrigated soils adversely impacts the nutrient availability to plants. It happens as a result of changes in the forms in which the nutrients are present in soils; increase in the losses as a result of leaching, denitrification in the case of nitrogen, precipitation in soils; interactive effects of cations and anions; and through the effects of complementary (non-nutrient) ions on nutrient uptake. Besides, saline soils are often poor in most essential plant nutrients owing to lack of vegetation and low organic matter content. Nitrogen deficiency is widespread in saline soils and a large fraction of the applied nitrogen is lost in gaseous forms under high soil salinity. It is useful to apply 25% more nitrogen than the soils irrigated with normal water. Availability of phosphorus increases up to a moderate level of salinity but thereafter it decreases. For chloride rich waters, additional dose of phosphorus is required if these soils tests low in phosphorus.

When saline water is used for irrigation, balanced use of essential nutrients is extremely useful to achieve optimum productivity. Bulky organic materials not only have the nutritive value, but also help to improve soil structure, which further influences leaching of salts to avoid/minimize accumulation in the root zone. The other advantages of integrated nutrient management include reduced volatilization losses, enhanced nitrogen-use efficiency and retention of nutrients in organic forms, which guards the plant nutrients from leaching and other losses.

2.4.5 RAINWATER MANAGEMENT

One of the fortunate situations with continental monsoon climate in many countries including India is the concentration of rains in a short span of 2–3 months (July–September). Under these conditions, rainfall has a profound effect on the soil salinization/desalinization cycle in the root zone. Water penetrating into soils during this period induces leaching of salts added through saline irrigation [17]. Besides, reduced numbers of irrigations due to high rainfall during the season result in lesser build-up of salts (Table 2.6). Large volume of experimental field data have clearly revealed that in most practical situations, saline water use even for prolonged periods, have not lead to excessive salt build-up in the root zone because of leaching due to rainfall. Salts, however, do accumulate during the winter season but gets leached out of the root zone during the following *kharif* season (Figure 2.8). Small differences can be noticed in the salinization-desalinization due to chemical constituents of water especially when the SAR of the water is high. Such differences must be kept in view while planning to live with salts.

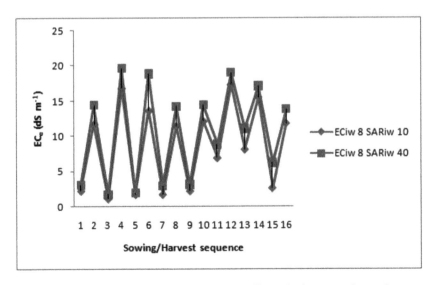

FIGURE 2.8   Soil salinity at the harvest of pearl millet and wheat crops in rotation.

Field studies reveal that the amount and frequency of rains, soil texture and on-farm land development determine the extent of salt leaching during the monsoon season. To remove 80% of salts present in the soil, it requires 1.85, 0.95 and 0.76 cm of rainwater per cm soil depth in fine, medium and coarse textured soils respectively. Thus, annual rainfall of 250 mm can leach 80% of accumulated salts from the surface 33 cm depth of coarse-textured soils. Gupta [21] showed that fields irrigated with saline water must be provided with strong dykes to conserve rainfall, leveled and tilled before the onset of monsoon so as to improve the rainwater use efficiency for salt leaching and to effectively rejuvenate the fields for the next crop-ping season.

In arid regions, where rainfall is relatively less, fallowing has been adopted as a strategy to live with the salts. Herein, fields are left fallow during winter if rainfall during the monsoon season is insufficient to leach down the salts. Thus, rainfall during the next monsoon season helps to rejuvenate the soil for the next crop. In arid parts of Rajasthan, India with scanty rainfall, cultivation every alternate year is adopted as a common practice. Dhir [13] evaluated this strategy and observed that the strategy is useful to live with the salts as economic returns are higher with the strat-egy than cropping every year.

2.5   OTHER STRATEGIES

In areas along the canal network and the coastal sandy tracts, a fresh water layer floating on saline water is often encountered. If the depth of this fresh water layer is quite deep, shallow tube wells are used to exploit this fresh water layer. If the layer is shallow or of medium depth, it is diffi-cult to exploit this layer using normal designs of tube wells. Upcoming of saline water layer results in mixing of saline and fresh water impairing the water quality. In order to skim the fresh water floating on the saline water, two innovative solutions have been developed. For the coastal sandy soils an improved technology (a replica of horizontal well) has been used to exploit the fresh water layer without disturbing the saline water (Figure 2.9). As soon as water level goes below the drain, there would be no flow so that saline water will not be disturbed.

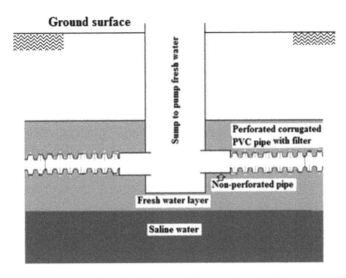

FIGURE 2.9   Schematic view of an improved Dorouv (Local Indian name for a conical structure) technology to skim fresh water.

The other strategy comprises of installing several shallow partially penetrating wells (multi-point wells) instead of one single well of large capacity. These wells are used together with one pumping unit to increase the discharge without upcoming of saline water, as this arrangement helps to distribute the upcoming over several wells. As a result, deterioration in water quality is avoided altogether or is minimal. In the later case, pumping schedules are designed with experience over one/two years. The pumping is stopped as soon as poor quality water begins to enter the tube wells. Instruments to monitor the water quality (EC) during pumping have also been designed to take knowledgeable decisions on deciding when to stop the pumps.

2.6   ALTERNATE USE OF SALINE WATER

Saline waters, which cannot be used for irrigation because they are deemed too salty, or there are other constraints in the use of saline water, alternate crops can be grown using such saline water. Many forage grasses, forest plants, aromatic and medicinal plants and even many fruit trees can be grown with saline water. Gooseberry (*Emblica officinalis*), *Karonda* (*Carissa carandus*), *Ber* (*Ziziphus mauritiana*) and *Bael* (*Aegle marmelos*) have been successfully grown with water salinity of 12 dS m^{-1} [11]. Coconut is also tolerant to saline water provided saline water is not allowed to stagnate for long periods. Andaman Tall and Katchal Tall varieties have been raised on raised beds with brackish water. Recently, commercial interest has been generated in mangrove palm (*Nypa fruticans*) for alcohol production. Many halophytes such as *Salvadora* and *Salicornia* have found favor because of their economic use [12]. Prawn culture is also possible with highly saline ground water in inland areas as has been demonstrated by Central Institute of Fisheries Education, Mumbai in Haryana. Without

TABLE 2.18 Irrigation Water Salinity Tolerance of Some Crops

Irrigation water salinity (dS m^{-1})	Kind of crops	Crops
>20	Forest and industrial plants	*Salvadora oleoides, Terminalia catappa, calophyllum inophyllum*, species of *Pandanus*
14	Medicinal	*Lepidium species**, *Withania somnifera* (Ashwagandha)
12	Under exploited species	*Cordia rothii, Azadirachta indica, Jatropha curcas, Adhatoda vasica, Ricinus communis,* and *Catharanthus roseus, Plantgo ovata* (Isabgol)
10	Aromatic	Vetiver, Palmarosa and Lemon grass
	Medicinal/flowers	Calendula (*Calendula officinalis*), Chrysanthemum (*Chrysanthemum indicum*) and German Chamomile (*Matricaria chamomilla*), Periwinkle (*Chatharanthus roseus*)*
8–10	Medicinal	*Egyptian henebene, Cassia angustifolia* (Indian senna)*
	Forestry	*Acacia nilotica, A. tortilis, A. farnesiana, Azadirechta indica, Eucalyptus tereticornis, Pithecellobium dulce, Prosopis juliflora, P. cineraria, Tamarix articulata,* and *Feronia limonia*
	Grasses	*Panicum laevifolium, P. antidotale, P. virgatum, P. maximum, Cenchris ciliaris, C. setigerus,* and *Brachiaria mutica*

Compiled from Dagar [11], *Lodha [27].

going into the details, a compilation of crops that could be irrigated with medium to high saline waters has been made (Table 2.18).

2.7   GUIDELINES FOR INTERPRETING WATER QUALITY FOR IRRIGATION

Based on the field experience and experimental evidences emerging from all over India, guidelines for use of saline water have been prepared by a team of experts drawn from AICRP Saline Water; CSSRI, Choudhary Charan Singh Haryana Agricultural University (CCSHAU) and Punjab Agricultural University (PAU) (Table 2.19).

Special Considerations for Table 2.18 are:

*   Use gypsum when saline water has SAR>20 and/or Mg/Ca ratio>3 or is rich in silica or causes surface stagnation of water during rainy season and crops grown are sensitive to water stagnation.

TABLE 2.19   Guidelines for Using Poor Quality Irrigation Water

Soil texture (% clay)	Crop tolerance	Upper limits of EC_{iw} (dS m^{-1}) in rainfall regions (mm)		
		< 350	350–550	550–750
Fine	S	1.0	1.0	1.5
(<30)	ST	1.5	2.0	3.0
	T	2.0	3.0	4.5
Moderately fine	S	1.5	2.0	2.5
(20–30)	ST	2.0	3.0	4.5
	T	4.0	6.0	8.0
Moderately coarse	S	2.0	2.5	3.0
(10–20)	ST	4.0	6.0	8.0
	T	6.0	8.0	10.0
Coarse	S	–	3.0	3.0
(< 10)	ST	6.0	7.5	9.0
	T	8.0	10.0	12.5

Saline waters (RSC < 2.5 meq L^{-1}).

S, ST and T denote sensitive, semi-tolerant and tolerant crops to salinity of irrigation water.

- In conjunctive use mode, canal water should be used at early growth stages including pre-sowing irrigation.
- If saline water is used for pre-sowing irrigation, 20% extra seed rate and a quick post-sowing irrigation will ensure better germination.
- In low rainfall areas, it would be appropriate to leave the fields fallow during the rainy season when SAR > 20 and waters of high salinity are used to irrigate the crops during the rabi season.
- Additional phosphorus fertilization is beneficial when Cl/SO_4 ratio is >2.0.
- For soils having (i) shallow water table (within 1.5 m during kharif), and (ii) hard sub-soil layer, the next lower EC_{iw} /alternate mode of irrigation (canal/saline) is applicable.
- Use of organic materials in saline environment enhances yields.
- Textural criteria should be applicable for all soil layers down to at least 1.5 m depth.

2.8   SUMMARY

Approximately 24 million km^2 area around the globe is underlain by saline or brackish ground water at shallow or intermediate depths. In India, areas having saline ground water with EC exceeding 4 dSm^{-1} is assessed at 0.19 million km^2. It may be an underestimate of the extent of the aquifers with poor quality water because alkali waters are often characterized by low salinity. Continuous application of saline water for irrigation causes soil salinization, the degree of salinization being controlled by several factors. Five most important ones are: (i) quality of irrigation water, (ii) number of irrigations, (iii) soil texture, (iv) leaching fraction, and (v) rainfall. In the monsoon climatic conditions, salinity remains in a dynamic equilibrium because of the salinization –desalinization cycle which is operative all through the year.

This chapter describes various technological options to avoid or blunt the adverse effects of saline irrigation. All the options are categorized into five groups. Options in the crop management group are selection of salt tolerant crops, crops varieties, application of saline water only at tolerant growth stages and improved cultural practices such as high seed rate to take care of likely mortality, seeding under conserved moisture and application of post germination irrigation etc. Under soil management group, it has emerged that relatively more saline water can be applied in light than heavy textured soils. As a 'thumb rule' salt accumulation in saline water irrigated soils is nearly one half the salinity of irrigation water in coarse textured soils (loamy sand and sand), equal to that of irrigation water in medium textured soils (sandy loam to loam) and more than two times in fine textured soils (clay and clay loam). Land forming can also be practiced to avoid salt accumulation zones.

Irrigation management is the key and includes a number of options such as high frequency low depth irrigation regime, pre-sowing irrigation with fresh water, switchover from surface to pressurized irrigation systems and the conjunctive use of saline and fresh water. It is brought out that sequential and switching modes, the two sub-groups of cyclic mode perform better than blending only if the cycle begins with fresh water. Application of 25% more nitrogen and integrated nutrient management also help to live with the salts.

Finally, the management of rainwater in the field through on-farm management and skipping of crops in low rainfall years especially in arid regions can help to extend the water supply in arid and semiarid regions. Besides, some skimming technologies to skim fresh water floating on saline water are included. In case water is unfit for cultivation or cannot be somehow be employed for crop cultivation, there is a great potential to use this for cultivation of halophytes, salt tolerant forage grasses, forest plants, aromatic and medicinal plants and even some fruit trees. It is concluded that the potential of vast saline water resource can be exploited through technologies that allows the crops to live with the salts, being much easier in monsoon climatic conditions.

KEYWORDS

- alternate use
- blending
- conjunctive use
- crop cultivars
- crop selection

- crop tolerance
- cyclic use
- desalinization
- Dorouv
- fallowing
- forage grasses
- forest plants
- furrow irrigation
- ground water
- irrigation water
- land form
- leaching fraction
- leaching requirement
- multi-point wells
- nutrient management
- pre-plant irrigation
- rainfall
- saline water
- salinization
- salt concentration factor
- skimming technologies
- supplemental irrigation
- switching mode
- water quality

REFERENCES

1. Abdel-Azim, R., & Allam M. N. (2005). Agricultural drainage water reuse in Egypt: strategic issues and mitigation measures. In: *Non-conventional Water Use: WASAMED Project. Bari* (Hamdy et al. Eds.). pp. 105–117.
2. Adelpour, A., Afzali, S. F., & Kamali, M. E. (2010). Evaluation of small projects on quantity and quality of water resources. Tenth Conference on Irrigation and Evaporation Reduction, Kerman, Iran.

3. Agarwal, M. C., & Khanna, S. S. (1983). *Efficient Soil and Water Management in Haryana*. CCS Haryana Agricultural University, Hisar Bulletin. 118p.

4. Anonymous (1998). *Annual Reports 1972–1998*. All India Coordinated Project on Management of Salt Affected Soils and Use of Saline Water in Agriculture, Central Soil Salinity Research Institute, Karnal-India.

5. Anonymous, (2012). *Biennial Report 2010–12*. Management of Salt Affected Soils and Use of Saline Water in Agriculture. Central Soil Salinity Research Institute, Karnal. 187 p.

6. Ayers, R. S., & Westcot, D. W. (1976). *Water Quality for Agriculture*. Irrigation and Drainage Paper, 29. FAO, Rome.

7. Ayers, R. S., & Westcot, D. W. (1985). *Water Quality for Agriculture*. Irrigation and Drainage Paper, 29, Revision-1. FAO, Rome.

8. Bhumbla, D. R. (1972). Water quality and use of saline water for crop production. *Symp. Soil and Water Management*, 1969. Hissar. 87–108.

9. CGWB, (1997). *Inland Ground water Salinity in India*. Government of India, Ministry of Water Resources, Central Ground Water Board, Faridabad, 62 p.

10. Chauhan, C. P. S., Singh, R. B., & Gupta, S. K. (2008). Supplemental irrigation of wheat with saline water. *Agricultural Water Management*, 95, 253–258.

11. Dagar, J. C. (2005). *Ecology, Management and Utilization of Halophytes*. Bulletin of the National Institute of Ecology, 15, 81–97.

12. Dagar, J. C., & Singh, N. T. (1994). Agro-forestry options in the reclamation of problem soils. In: *Trees and Tree Farming* (Thampan, P. K. Ed.). Peekay Tree Crops Development Foundation, Cochin (India): 65–103.

13. Dhir, R. P. (1977). Saline waters-their potentiality as a source of irrigation. In: *Desertification and its Control*. ICAR, New Delhi, 130–148.

14. FAO. (2000). *Crops and Drops: Making the Best Use of Water for Agriculture*. FAO, Rome, Italy (Advance edition).

15. Ghassemi F., Jakeman A. J., & Nix H. A. (1995). *Salinization of Land and Water Resources: Human Causes, Extent, Management and Case Studies*. UNSW Press, Sydney, Australia, and CAB International, Wallingford, UK.

16. Gleick, P. (2000). *The World's Water*. Island Press, New York.

17. Gupta, I. C., & Abichandani, C. T. (1970). Seasonal variations in the salt composition of some saline water irrigated soils of western Rajasthan. I. Effect of rainfall. *J. Indian Soc. Soil Sci*, 18, 429–435.

18. Gupta, I. C., & Gupta, S. K. (2001). *Use of Saline Water in Agriculture* (3rd revised edition). Scientific Publishers, Jodhpur. 299 p.

19. Gupta, I. C., & Yadav, J. S. P. (1986). Crop tolerance to saline irrigation waters. *J. Indian Soc. Soil Sci.*, 34, 279–286.

20. Gupta, R. K., Singh, N. T., & Sethi, M. (1994). *Ground Water Quality for Irrigation in India*. Tech. Bull No. 19. Central Soil Salinity Research Institute, Karnal. 13p.

21. Gupta, S. K. (1985). Leaching saline soils through rainfall. *J. Indian Soc. Soil Sci.*, 33, 128–136.

22. Gupta, S. K. (2011). Irrigation with saline and alkali waters: management strategies. In: *Salinity Management for Sustainable Agriculture in Canal Commands* (Dey and Gupta, Eds.). Central Soil Salinity Research Institute, Karnal. 69–81.

23. Gupta, S. K. (2015). Drip and indigenous alternatives for use of saline and alkali waters in India: Review. In: *Micro Irrigation Design Systems for Agricultural Crops: Methods and Practices, Volume 2* (Goyal and Panigrahi, Eds.). Oakville, ON, Canada: Apple Academic Press Inc.

24. Gupta, S. K., & Gupta, I. C. (2014). *Management of Saline and Waste Water in Agriculture*. Scientific Publishers, Jodhpur. 321 p.

25. Igor, Shiklomanov. (1993). World fresh water resources. In: *Water in Crisis: A Guide to the World's Fresh Water Resources*. (Gleick, P. Ed.). Oxford University Press, New York.

26. Lindsay, E. (2006). *Salinity Tolerance in Irrigated Crops*. NSW Department of Primary Industries, Australia (as available on Internet opened on 12.05.2015).

27. Lodha, Vandana. (2007). *Personal communication*. CSSRI, Karnal.

28. Maas, E. V. (1986). Salt tolerance of plants. *App. Agriculture Research*, 1, 12–26.

29. Mass, E. V. (1987). Salt tolerance of plants. In: *CRC Handbook of Plant Science in Agriculture*. (Cristie, B. R. Ed.). CRC Press Inc.

30. Mass, E. V. (1990). Crop salt tolerance. In: *Agricultural Salinity Assessment and Management Manual*. (Tanji, K. K. Ed.). ASCE, New York. 262–304.

31. Maas, E. V., & Hoffman, G. J. (1977). Crop salt tolerance: Current assessment. *J. Irrig. Drainage, Div. ASCE, 103 (IIR)*, 115–134.

32. Maskooni, E. K., Amiri, I., & Afzali, S. F. (2014). Soil salinity estimation with SCF and other traditional parameters. *Indian Journal of Fundamental and Applied Life Sciences*, 4, 506–510.

33. Mateo-Sagasta, J., & Burke, J. (2015). *Agriculture and Water Quality Interactions: a Global Overview. SOLAW Background Thematic Report – TR08* http://www.fao.org/fileadmin/templates/solaw/files/thematic_reports/TR_08.pdf (Opened on 09.05.2015).

34. Minhas, P. S., & Gupta, R. K. (1992). *Quality of Irrigation Water – Assessment and Management*. Indian Council of Agricultural Research, New Delhi, 123p.

35. Minhas, P. S., Sharma, O. P., & Patil, S. G. (1998). *25 years of Research on Management of Salt-Affected Soils and Use of Saline Water in Agriculture*. CSSRI, Karnal. 220 p.

36. MoWRI (Ministry of Water Resources and Irrigation) Arab Republic of Egypt. (2005). *Integrated Water Resources Management Plan*, http://en.wikipedia.org/wiki/Water_resources_management_in_modern_Egypt. Retrieved on April 30, 2015.

37. Qureshi, A. S., Iqbal, M., Anwar, N. A., Aslam, M., & Chaudhry, M. R. (1997). Benefits of shallow ground water. In: *Proc. of the Seminar on On-farm Salinity, Drainage and Reclamation*. (Chaudhry et al. Eds.). Publication No. 179. Lahore, Pakistan, IWASRI. Lahore.

38. Rhoades, J. D. (1974). Drainage for salinity control. In: *Drainage for Agriculture (J. van Schilfgaarde Ed.)*. Agronomy Monograph 17, American Society of Agronomy, Madison, Wisconsin.

39. Rhoades, J. D. (1984). Use of saline water for irrigation. *California Agriculture*, 41–43

40. Sharma, P., Sardana, V., & Banga, S. S. (2013). Salt tolerance of Indian mustard (*Brassica juncea*) at germination and early seedling growth. *Environmental and Experimental Biology*, 11, 39–46.

41. Sharma, S. K. (2000). Ground water management. In: *Conf. Proc. 3rd Water Asia 2000*. Interads Ltd. New Delhi.

42. Tanji, K. K., Iqbal, M. M., & Quek, A. F. (1977). Surface irrigation return flows vary. *California Agriculture*, 31, 30–31.

43. USDA. (1954). *Diagnosis and Improvement of Saline and Alkali Soils*. Agricultural Handbook 60, US Department of Agriculture, Washington DC.

44. Van Weert, F., & van der Gun, J. (2012). Saline and brackish ground water at shallow and intermediate depths: genesis and world-wide occurrence. IAH Congress. Niagara Falls, Canada.

45. Verma, A. K., Gupta, S. K., & Isaac, R. K. (2014). *Simulation Modeling for Saline water Use in Agriculture* (e-book). LAP Lambart Academic Publishing, Saarbrucken, Germany. 226 p.

CHAPTER 3

COASTAL ECOSYSTEMS: RISK FACTORS FOR DEVELOPMENT AND THREATS DUE TO CLIMATE CHANGE

H. S. SEN and DIPANKAR GHORAI

CONTENTS

3.1 INTRODUCTION

In one of its report by IFPRI [16], it has projected that globally about 1 billion people are 'absolutely' or 'ultra' poor living on less than US$1.00 a day and about 800 million people are hungry. Although, this level of poverty has decreased marginally overall from 28.6% of the population in 1990 to 18.0% in 2004, large changes in regional disparities in poverty level have been observed during the period, 1990–2004, from 39–47% in South Asia, 38–17% in East Asia and Pacific, 19–31% in Sub-Saharan Africa, and 4–5% in Latin America and Caribbean. Various risk factors have been identified to affect poverty in different time frames: Socio-technological constraints including macro-economic imbalances limiting agricultural and related productivity, rising food prices and resource scarcity, climate change and lack of environmental sustainability, lack of infrastructure, ethnic and other social crises, inappropriate mindset for the acceptance of improved technologies, health related issues, etc.

Of all the major ecosystems which factor in agricultural or food production, being at the very base of poverty alleviation program, coastal is probably the most important one because of its high potentiality and production of a large number of value added goods, yet confronted with high risk factors due to a multiple of issues, which is the theme of discussion in this chapter.

Current challenges and management of the coastal ecosystem have been overviewed in Wikipedia [42]. Different countries with coastal boundaries have varying proportion of the total area exposed to the sea, expressed as "Ratio of coastline area to total area of the country, km/km^2)." Some top-most countries [43] are: Tokelau (10,100), Federal States of Micronesia (8,706), Palau (3,316), Northern Mariana Islands (3,107), Maldives (2,147), Monaco (2,051), Marshall Islands (2,044), Cocos (Keeling) Islands (1,857), Gibraltar (1,846) Macau (1,464), Nauru (1,429), Kiribati (1,409), Saint Martin (1,093), Pitcairn Islands (1,085), Seychelles (1,079), and Christmas Island (1,030). India, Bangladesh and USA have low coastline/area ratio of 2, 4 and 2, respectively.

However, it is not the coast/ area ratio alone but also the total coastline, population density and anthropogenic factors, topography and related soil properties, protection measures undertaken, and natural disasters caused by the sea and through its interaction with climate and under-sea tectonic movement of the earth, that factor in not only to influence the agricultural production but also the nature and extent of vulnerability of the ecosystem per se in a country.

This chapter discusses high risk factors due to a multiple of issues related to coastal ecosystems and global climate change.

3.2 COASTAL ECOSYSTEMS: NEED FOR A THRUST

Coastal ecosystems have an economic value beyond their esthetic benefit supporting human lives and livelihoods. The combined global value of goods and services from coastal ecosystems is about US$12–14 trillion annually that is larger than the United States' Gross Domestic Product (GDP) in 2004 [26]. The problems of livelihood in these areas are compounded manifolds owing to a series of technological, administrative and socio-economic constraints.

A holistic look at the interaction matrix of factors, which are interdependent on each other, impacting on the coastal ecosystem is presented schematically in Figure 3.1. Unfortunately, at the global level, until very recently, not much serious and concerted attention has been paid for mitigating the problems for sustainable development in the coastal ecosystem. Attempt to improve agriculture in this ecosystem, the focal theme of this paper, though should be at the central stage from daily livelihood point of view, it is still at the back seat in majority of the areas. This is possibly because of the 'slow-poisoning kind of effect' of this sector, arising out of poor agricultural practice and/ or inability for the poor to pay for the commodities as a result of insufficient food production, that normally goes un-noticed among the poverty-stricken mass, vis-à-vis catastrophic effects with heavy toll on lives and properties due to climatic disasters. Exceptions are India, Bangladesh and possibly a few other countries paying concerted attention on the coastal ecosystem for improvement in agricultural front in particular.

3.3 COASTAL ECOSYSTEM: DEFINITIONS AND DISTRIBUTION

A coastal ecosystem includes estuaries and coastal waters and lands located at the lower end of drainage basins, where stream and river systems meet the sea and are mixed by tides. The ecosystem includes: Saline, brackish (mixed saline and fresh) and fresh waters, as well as coastlines and the adjacent lands. Coastal wetlands are commonly called as lagoons, salt marshes or tidelands [38]. According to World Resources Institute [51], coastal areas may be commonly defined as the interface or transition

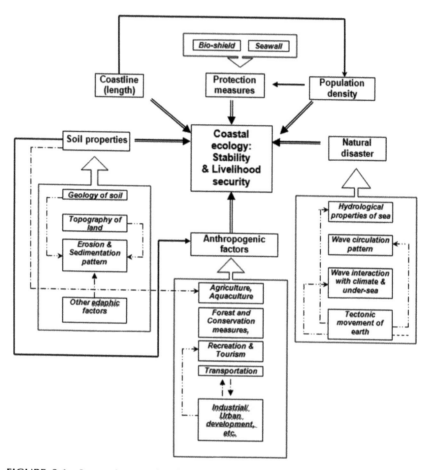

FIGURE 3.1 Interaction matrix of factors influencing stability and livelihood of the coastal ecosystem [32].

areas between land and sea, including large inland lakes. Coastal areas are diverse in function and form, dynamic, and do not lend themselves well to definition by strict spatial boundaries. Unlike watersheds, there are no exact natural boundaries that unambiguously delineate coastal areas at the global scale. The world coastline extends from 350,000–1,000,000 km in length, depending upon how finely the 'length' is resolved. More comprehensively, the coastal ecosystem has been defined as representing the transition from terrestrial to marine influences and vice versa [29]. It comprises not only shoreline ecosystems, but also the upland watersheds draining into coastal waters, and the near shore sub-littoral ecosystems influenced by land-based activities. Functionally, it is a broad interface between land and sea that is strongly influenced by both.

Soils in the coastal ecosystem along with their characteristics have been described comprehensively on a global scale but no attempt has been made to delineate the zones from inlands based on scientific criteria. Estimates in world have generally been arbitrarily done based on length of the coastline times a fixed distance landward, varying from 50 to 200 km as followed in different countries, from the shore assuming the zone representing coastal ecosystem different from that for inland part of the country. Velayutham et al. [40] for the first time described soil resources and their potentials for different Agro-ecological Sub Regions (AESR) in the coastal ecosystems of India showing total of 10.78 million hectare (M-ha) area under this ecosystem (including the islands). Soil in the coastal ecosystem *per se* does not have separate significance as far as its productivity is concerned unless it is considered in association with other relevant ecological factors describing the ecosystem owing to the latter's significant influence on threatening its very stability – a fact, unlike any other ecosystem. It should therefore be necessary, in priory, to delineate and characterize the coastal soils in each country based on sound scientific criteria, and alongside consider the relevant ecological factors which render the ecosystem concerned generally fragile in nature due to various risk factors, often complementing with each other, involved for planning for sustainable development with a holistic approach (Figure 3.1).

This topic in the current chapter is discussed in a series of two sections, the first of which is devoted to the various risk factors including climatic change threatening the stability, and the second one will discuss various issues impacting on productivity in agriculture and finally the integrated coastal area planning towards livelihood security in the ecosystem.

3.4   COASTAL ECOSYSTEMS: RISK FACTORS

Dirk et al. [7] estimated that the 51% of the world's coastal ecosystems appear to be at significant risk of degradation from development related activities. Europe, with 86% of the coastline at high or medium risk, and Asia, with 69% in these categories, is the region most threatened by degradation. Worldwide, nearly three-fourths of marine protected areas within 100 km of continents or major islands appear to be at risk. These were preliminary estimates and lack precision as indicated by WRI [51]. However, the data suggest already an alarming state towards destabilizing the ecosystem, notwithstanding that the estimate did not even take into consideration other important factors like agricultural and allied developments, deforestation, fishing, population density, and climatic disturbances with significant adverse contribution.

3.4.1 COMPONENTS OF COASTAL ECOSYSTEMS

The 'main' ecosystems of the coastal areas, besides taking into account about 50–100 km area landward to be designated as coastal plain and utilized mostly by agriculture and allied activities as well as for domicile and a few other occupational purposes, are reefs, salt marshes and the remaining continental shelves The global distribution of major classified into components, like estuaries, macrophyte communities, mangroves, coral reefs, salt marshes and the remaining continental shelves components of the coastal ecosystems. The global distribution of major components is shown in Figure 3.2.

3.4.2 SALT MARSH

According to Encyclopedia Britannica [10], salt marsh is an area of low, flat, poorly drained ground that is subject to daily or occasional flooding by salt water or brackish water and that is covered with a thick mat of grasses and such grass like plants as sedges and rushes. Salt marshes are common along low seacoasts, inside barrier bars and beaches, in estuaries, and on deltas and are also extensive in deserts and other arid regions that are subject to occasional overflow by water containing a high content of salts. Maritime salt marshes often extend many miles inland and are variably subject to tidal action; inland brackish marshes are found frequently on mineral substrates of alluvial and lacustrine origin.

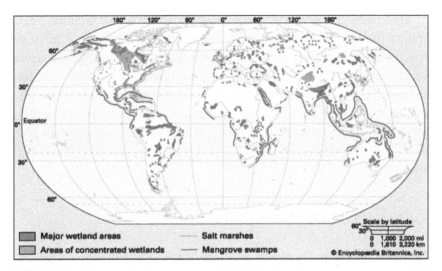

FIGURE 3.2   Major wetland areas and world distribution of mangrove swamps and salt marshes (Source: By courtesy of Encyclopaedia Britannica, Inc., copyright 1997; used with permission.[10])

According to Gedan et al. [11], salt marshes provide more ecosystem services to coastal populations than any other environment. Coastal wetlands all over the world have vanished or are threatened in spite of various international agreements and national policies. Losses due to human activities include effects of urbanization, development of tourism resort, industrial pollution, increase of inflow nutrients from the upstream reclaimed lands, changes in hydrology, conversion to aquaculture ponds and some drillings for gas exploitation. In addition, as a transition zone between land and sea, coastal wetlands are particularly vulnerable to sea-level rise caused by both oceanic thermal expansion and the melting of Artic and Antarctic glaciers as consequence of global warming [52]. According to them, the diversity of salt marsh plant species increases with increasing latitude. This contrasts with mangrove diversity, which is highest in the lower latitudes of the tropics. In Australia, when salt marshes and mangroves coexist, salt marshes are typically found at higher elevations where they are inundated less frequently than mangroves. However, this is not always true in an international context. When sea grass beds are found adjacent to salt marshes and mangroves, many material links and shared plant and animal communities can exist.

Oz Coasts [22] reported characterization of salt marsh sediments generally consisting of poorly sorted anoxic sandy silts and clays. Carbonate concentrations are generally low, and concentrations of organic material are generally high. As with salt flats, the sediments may have salinity levels that are much higher than that of seawater. These sediments are also usually anoxic and have large accumulations of iron sulfides. Disturbing these acid sulfate soils can cause sulfuric acid to drain into coastal waterways. Salt marshes are often associated with salt flats or exposed bare areas.

3.4.3   SALT FLATS

Salt flats, or saline supratidal mudflat facies, occur in dry evaporative environments (often in the tropics) that undergo infrequent tidal inundation. Sediments comprise poorly-sorted sandy silts and clays, including mineral deposits, such as gypsum and halite which form crusts. Salt flats tend to be low gradient, and mostly featureless, with a varying degree of algal colonization, and often with vertically accreting algal mats. They generally occur above mean high water spring, and experience infrequent inundation by king tides. The high salinity levels (surface and ground water) in these environments often preclude the growth of higher vegetation and biota (some infauna and epifauna may occur at lower elevations). Salt flats are habitats for birds, particularly during the wet season [22].

3.4.4   CORAL REEFS

Coral reefs are aragonite structures produced by living organisms. In most reefs the predominant organisms are colonial cnidarians that secrete an exoskeleton of calcium carbonate [44]. Coral reefs are estimated to cover 284,300 km^2, with the Indo-Pacific region (including the Red Sea, Indian Ocean, Southeast Asia and the Pacific) accounting for 91.9% of the total (Figure 3.3). Southeast Asia accounts for 32.3% of that figure, while the Pacific including Australia accounts for 40.8%. Atlantic and Caribbean coral reefs only account for 7.6% of the world total.

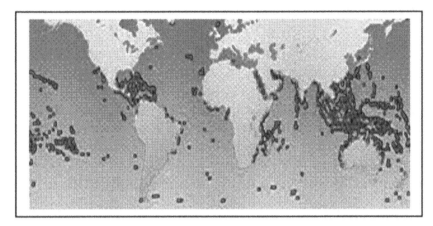

FIGURE 3.3   World distribution of coral reef [44].

Coral reefs are either restricted or absent from the west coast of the Americas, as well as the west coast of Africa. This is due primarily to upwelling and strong cold coastal currents that reduce water temperatures in these areas. Corals are also restricted from off the coastline of South Asia from Pakistan to Bangladesh. They are also restricted along the coast around north-eastern South America and Bangladesh due to the release of vast quantities of freshwater from the Amazon and the Ganges Rivers, respectively. Although corals are found in temperate and tropical waters, shallow water reefs are formed only in a zone extending at most from 30°N to 30°S of the equator.

Coral reefs cover < 0.5% of the ocean floor and 90% of the marine species are directly or indirectly dependent on them [26]. About 20% of coral reefs have been destroyed in the last few decades and an additional 20% or more are severely degraded, particularly in the Caribbean Sea and parts of Southeast Asia. Coral bleaching, which results from rising ocean temperatures caused by climate change is also increasing and further threatens this valuable resource.

A lagoon on the other hand is a body of comparatively shallow salt or brackish water separated from the deeper sea by a shallow or exposed sandbank, coral reef, or similar feature. Thus, the enclosed body of water behind a barrier reef or barrier islands or enclosed by an atoll (an island of coral that encircles a lagoon partially or completely) reef is called a lagoon [45].

3.4.5 MANGROVES SWAMPS

A mangrove is a plant and *mangal* is a plant community and habitat where mangroves thrive. Mangroves are found in tropical and sub-tropical tidal areas worldwide, like Africa, Americas (including Caribbean Islands), South America, Asia, Australasia, and Pacific Islands. The 15 countries having significant areas under mangrove swamp are shown in Figure 3.4. The areas are typically characterized by high degree of salinity and water logging due to tidal inundation. Areas under *mangals* include estuaries and marine shorelines. Plants develop physiological adaptations to overcome the problems of anoxia, high salinity and frequent tidal inundation. About 110 species have been identified as belonging to the *mangal*. Each species has its own capabilities and solutions to these problems. Small environmental variations within a *mangal* may lead to greatly differing methods of coping with the environment. Therefore, the mix of species at any location within the intertidal zone is partly determined by the tolerances of individual species to physical conditions, like tidal inundation and salinity, but may also be influenced by other factors such as predation of plant seedlings by crabs.

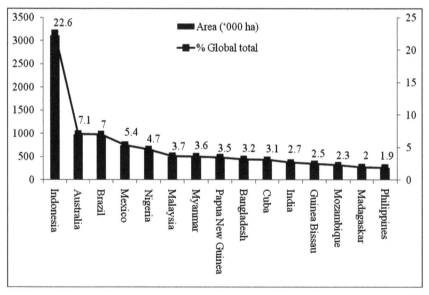

FIGURE 3.4 The 15 most mangrove-rich countries and their global share: Used with permission [14].

Mangroves protect the coast from erosion, surge storms (especially during hurricanes), and tsunamis. Their massive root system is efficient at dissipating wave energy. Likewise, they slow down tidal water enough that its sediment is deposited as the tide comes in and is not re-suspended when the tide leaves, except for fine particles. As a result, mangroves build their own environment. Because of the uniqueness of the mangrove eco-systems and their protection against erosion, they are often the object of conservation programs including national Biodiversity Action Plans [46].

Mangroves support unique ecosystems, especially on their intricate root systems (Figure 3.5). The mesh of mangrove roots produces a quiet marine region for many young organisms. In areas where roots are permanently submerged, they may host a wide variety of organisms, including algae, barnacles, oysters, sponges, and bryozoans, which all require a hard sub-stratum for anchoring while they filter feed. Shrimps and mud lobsters use the muddy bottom as their home. Mangrove crabs improve the nutritional quality of the *mangal* mud for other bottom feeders by mulching the man-grove leaves. In at least some cases, export of carbon fixed in mangroves is important in coastal food webs. The habitats also host several commercially

FIGURE 3.5   Above and below water view at the edge of the mangal [46].

important species of fish and crustaceans. In Vietnam, Thailand, the Philippines, and India, mangrove plantations are grown in coastal regions for the benefits they provide to coastal fisheries and other uses [46].

In the last 50 years, as much as 85% of the mangroves have been lost in Thailand, the Philippines, Pakistan, Panama and Mexico, globally the value being about 50%. An estimated 35% of mangroves have been removed due to shrimp and fish aquaculture, deforestation, and freshwater diversion. In Indonesia alone over 10,000 square kilometers of mangrove forests have been converted into brackish water ponds (called *tambaks*) for the cultivation of prawns and fish. Valuation of intact tropical mangroves estimated at US$ 1000 per ha drops to US$ 200 per ha due to clearance by shrimp farming [26]. Although some successful restoration efforts have taken place, these are not keeping pace with mangrove destruction.

3.4.6 ESTUARIES

Estuaries are partially enclosed bodies of water along coastlines where fresh water and salt water meet and mix. Most scientists accept the definition given by D.W. Pritchard in 1967 as: "*An estuary is a semi-enclosed coastal body of water which has a free connection with the open sea and within which seawater is measurably diluted with fresh water derived from land drainage*" [20]. Estuaries act as a transition zone between oceans and continents. Fresh water input from land sources (usually rivers) dilutes the estuary's salt content. An estuary is typically the tidal mouth of a river and is made up of brackish water. Freshwater from the river is prevented from flowing directly into the open sea by one or more land formations, such as peninsulas and barrier islands. Estuaries are often characterized by sedimentation or silt carried in from terrestrial runoff and, frequently, from offshore [21]. The pH, salinity, and water levels of estuaries vary; depending on the river that feeds the estuary and the ocean from which it derives its salinity. An estuary retains many nutrients derived from both land and sea, and it protects water quality. It thus forms an ecosystem that is filled with a rich variety of living organisms.

Estuaries are vital habitats for thousands of marine species, often called the *nurseries of the sea* because the protected environment and abundant food provide an ideal location for fish and shellfish to reproduce. Most commercially important fish species spend some part of their life cycle in estuaries. Besides fish, many species of birds depend on estuaries for food and nesting areas. Marine mammals also use estuaries as feeding grounds and nurseries. All these marine organisms feed in estuaries because a healthy estuary produces between 4 and 10 times as much organic matter as a cornfield of the same size. Estuaries provide a wide range of habitats leading to a great diversity of marine life [20].

According to USAID [39], complicated interconnections exist between the quality, quantity and timing of fresh water inflows and the health of estuaries. All of the goods and services that estuaries provide are threatened when freshwater inflows are changed. Even a small change in the flow of freshwater may affect the fundamental functioning of an estuary, which in turn will have ramifications on the animals and plants, as well as on human populations dependent upon the estuary. In many cases, upstream altera-tions to the volume, timing and quality of freshwater inflows have resulted in catastrophic destruction of downstream habitats, loss of species and deg-radation of ecosystems adapted to a certain range of freshwater inflows.

3.4.7 MACROPHYTES

A macrophyte represents a group of aquatic plants that grow in or near water and is emergent, submergent, or floating. In lakes macrophytes pro-vide cover for fish and substrate for aquaticinvertebrates, produce oxygen, and act as food for some fish and wildlife. A decline in a macrophyte population may indicate water quality problems. Such problems may be the result of excessive turbidity, herbicides, or salinization. Conversely, overly high nutrient levels may create an overabundance of macrophytes, which may in turn interfere with lake processes [47].

3.4.8 REMAINING SHELF

The Continental shelf is the extended perimeter of each continent and associated coastal plain, and was part of the continent during the glacial periods, but is undersea during interglacial periods such as the current epoch by relatively shallow seas (known as shelf seas) and gulfs. The continental rise is below the slope, but landward of the abyssal plains

(Figure 3.6). Its gradient is intermediate between the slope and the shelf, of the order of 0.13–2.5. Extending as far as 500 km from the slope, it consists of thick sediments deposited by turbidity currents from the shelf and slope. Sediment cascades down the slope and accumulates as a pile of sediment at the base of the slope, called the continental rise [48].

An 'Ice shelf' is a thick, floating platform of ice that forms where a glacier or ice sheet flows down to a coastline and onto the ocean surface. Ice shelves are found in Antarctica, Greenland and Canada only. The boundary between the floating ice shelf and the grounded (resting on bedrock) ice that feeds it is called the grounding line. When the grounding line retreats inland, water is added to the ocean and sea level rises. In contrast, for Ross Ice Shelf sea ice is formed on water, is much thinner, and forms throughout the Arctic Ocean. It also is found in the Southern Ocean around the continent of Antarctica [49].

3.5   EROSION AND POLLUTION VERSUS POPULATION GROWTH

3.5.1   POPULATION GROWTH AS THE DRIVER

The earth is now home to some 6.5 billion people and is projected to have 9 billion by 2050. World population is increasing with time at an

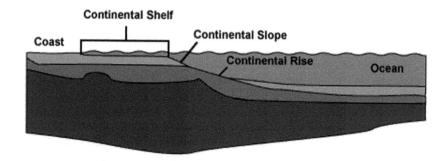

FIGURE 3.6   Continental shelf in relation to ocean [48].

accelerated pace and the population will grow even faster along various coastlines and in already densely populated developing countries. The number of people living within 100 km of coastlines will increase by about 35% in 2050 as compared to that in 1995. This type of migration will expose 2.75 billion people to coastal threats from global warming such as sea level rise and stronger hurricanes in addition to other natural disasters like tsunamis [15]. The expected change of the population (or population density) from 2000 to 2025 region wise shows increase in almost each coastal area [9]. The estimates (population within 100 km of the coastline) show increase by 25% in Asia (except Middle East), 52% in Middle East and North Africa, 81% in Sub-Saharan Africa, 20% in North America, 31% in Central America and Caribbean Islands, and 32% in each South America and Oceania, while there may be decrease by 2.5% in Europe. In India, according to the Department of Ocean Development, there are 40 heavily populated areas along the Indian coast [8].

Apart from climate change, population growth is possibly the single most factor impacting on damage to properties in the coastal ecosystem. Around the world, maximum persons die by drowning by storm surge.

3.5.2 SEDIMENTATION AND EROSION

The dynamics of alluvial landscapes and natural sedimentation patterns that determine the nutrient and energy flows in coastal areas are increasingly being modified by human activities, in particular those that affect water flows (dams, increased water extraction, deviation of rivers) and erosion, especially due to deforestation. This prevents or slows down vertical accretion, thus aggravating salt water intrusion and impairing drainage conditions in riverine, delta or estuarine areas. It reduces or blocks sediment supply to the coast itself, which may give rise to the retreat of the coastline through wave erosion. Beach erosion is a growing problem and affects tourism revenue, especially in island nations. In the Caribbean, as much as 70% of beaches studied over a ten-year period were eroded. Yet, the long-term success of tourism in the region is dependent on excellent beaches, a pristine marine environment, and warm weather.

3.5.3 EUTROPHICATION, HYPOXIA, DEAD ZONES AND NUTRIENT CYCLE

The urban developments are increasingly expanding to fertile agricultural lands and leading to pollution of rivers, estuaries and seas by sewage as well as industrial and agricultural effluents. In turn, this is posing a threat to coastal ecosystems, their biological diversity, environmental regulatory functions and role in generating employment and food. Overuse of fertilizer can result in eutrophication, and in extreme cases, the creation of 'dead zones.' Dead zones occur when excess nutrients—usually nitrogen and phosphorus—from agriculture or the burning of fossil fuels seep into the water system and fertilize blooms of algae along the coast. As the microscopic plants die and sink to the ocean floor, they feed on bacteria, which consume dissolved oxygen from surrounding waters. This limits oxygen availability for bottom-dwelling organisms and the fish that eat them. In dead zones, huge growths of algae reduce oxygen in the water to levels so low that nothing can live. There are now more than 400 known dead zones in coastal waters worldwide, compared to 305 in the 1990s, according to a study undertaken by the Virginia Institute of Marine Science. Those numbers were up from 162 in the 1980s, 87 in the 1970s, and 49 in the 1960s. In the 1910s, four only dead zones had been identified [19].

The occurrence of hypoxia in shallow coastal and estuarine areas has been increasing worldwide, most likely accelerated by anthropogenic activities. Hypoxia in the Northern Gulf of Mexico, commonly named the 'Gulf Dead Zone,' has doubled in size since researchers first mapped it in 1985, leading to very large depletions of marine life in the affected regions [25]. He studied changes in microbial communities as a result of oxygen depletion, the potential contribution of increasing hypoxia to marine production and emission of N_2O and CH_4, and the effect of hypoxic development on methyl mercury formation in bottom sediments at the Gulf of Mexico's Texas-Louisiana Shelf during the summer months.

The World Resources Institute (WRI) reported that driven by a massive increase in the use of fertilizer, the burning of fossil fuels, and a surge in land clearing and deforestation, the amount of nitrogen available for uptake at any given time has more than doubled since the 1940s. In other words, human activities now contribute more to the global supply of fixed nitrogen each year than natural processes do, with human-generated nitrogen totaling about 210 million metric tons per year, while natural processes contribute about 140 million metric tons (Table 3.1).

This influx of extra nitrogen has caused serious distortions of the natural nutrient cycle, especially where intensive agriculture and high fossil fuel use coincide. In some parts of northern Europe, for example, forests are receiving 10 times the natural levels of nitrogen from airborne deposition, while coastal rivers in the Northeastern United States and Northern Europe are receiving as much as 20 times the natural amount from both agricultural and airborne sources. Recently, a new class of chemical substances with toxic and persistent properties was detected in the environment – the polyfluorinated compounds (PFCs). At the Institute for Coastal Research, scientific studies were performed on the PFC-contamination of coastal waters, marine mammals and the atmosphere with emphasis on the mechanisms of global transport and distribution of PFCs [5].

3.6 CLIMATE CHANGE

3.6.1 THE GENERIC ISSUE

Climate change is called for any long-term significant change in the expected patterns of average weather of a specific region over an appropriately significant period of time. A number of factors have been identified

TABLE 3.1 Global Sources of Biologically Available (fixed) Nitrogen [51]

Anthropogenic sources	Annual release of fixed nitrogen (teragram)
Biomass burning	40
Fertilizer	80
Fossil fuels	20
Land clearing	20
Legumes and other plants	40
Wetland draining	10
Total from human sources	**210**
Total from natural sources: Soil bacteria, algae, lightning, etc.	**140**

which collectively or individually impact on the build-up of greenhouse gases (GHGs, like carbon dioxide, methane and nitrous oxide) that threaten to set the earth inexorably on the path to an unpredictably different climate. The Intergovernmental Panel on Climate Change (IPCC) [17] indicates from observations since 1961 that the ocean has been absorbing more than 80% of the heat added to the climate system, and that ocean temperatures have increased to depths of at least 3000 m. It has also been predicted that sea surface temperature would increase in the range of about 1–3°C to result in more frequent coral bleaching events and widespread mortality unless there is thermal adaptation or acclimatization by corals. According to them, many parts of the planet will be warmer, as a result of which droughts, floods and other forms of extreme weather will become more frequent, threatening food supplies. Plants and animals which cannot adjust will die out.

Sea levels would rise and will continue to do so, forcing hundreds of thousands of people in coastal zones to migrate. One of the main GHGs which human populations are adding to the atmosphere, carbon dioxide (CO_2), is increasing rapidly. Around 1750, i.e., at the start of the Industrial Revolution in Europe, there were 280 parts per million (ppm) of CO_2 in the atmosphere. Today the overall amount of GHGs has topped 390 ppm CO_2e (parts per million of carbon dioxide equivalent – all GHGs expressed as a common metric in relation to their warming potential) and the figure is rising by 1.5–2 ppm annually.

Reputable scientists believe the earth's average temperature should not rise by more than 2°C over pre-industrial levels. Among others, the European Union (EU) indicated that this is essential to minimize the risk of what the UN Framework Convention on Climate Change (UNFCCC) calls as dangerous climate change and keep the costs of adapting to a warmer world bearable. Scientists say there is a 50% chance of keeping to 2°C if the total GHG concentration remains below 450 ppm [37].

3.6.1   EFFECTS ON SEA LEVEL RISE AND INUNDATION OF LAND

Coastal areas are prone to threats from natural causes such as tidal surges and sea level rise (Figure 3.7). Each year an estimated 46 million

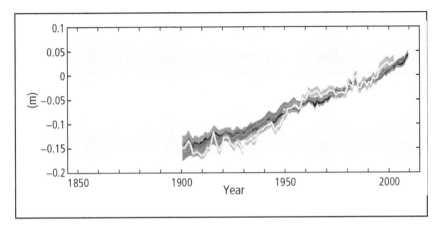

FIGURE 3.7   Global average sea level change (Annually and globally averaged sea level change relative to the average over the period 1986 to 2005; sheds indicate different data sets used, [17: used with permission].

people risk flooding from storm surges. Coasts in many countries currently face severe problems of sea level rise as a consequence of climate change, leading to potential impacts on ecosystems and human coastal infrastructure. The worst scenario projects a sea level rise of 95 cm by the year 2100, with large local differences (resulting from tides, wind and atmospheric pressure patterns, changes in ocean circulation, vertical movements of continents, etc.) in the relative sea level rises [24]. The impacts on sea level rise are therefore expected to be more local than global. The relative change of sea and land is the main factor. Many cities, for instance, even suffer land subsidence as a result of ground water withdrawal. This may be compounded with sea level rise, especially since rates of subsidence may exceed the rate of sea level rise between now and 2100.

Under the worst scenario, the majority of the people who would be affected in different countries are China (72 million), Bangladesh (13 million people and loss of 16% of national rice production), and Egypt (6 million people and 12 to 15% loss of agricultural land), while between 0.3% (Venezuela) and 100% (Kiribati and the Marshall Islands) of the population are likely to be affected. In India, potential impacts on 1 m sea level rise might lead to inundation of 5,763 km² of land [23].

According to IPCC, regions especially at risk are low lying areas of North America, Latin America, Africa, populous coastal cities of Europe, crowded delta regions of Asia, like *Ganges-Brahmaputra delta* facing flood risks from both large rivers and ocean storms, and many small islands, whose very existence is threatened by rising seas. In North America, current preparedness for rising seas, more frequent severe weather, and higher storm surges is low.

The Greenland and West Antarctic ice sheets face substantial melting if the global average temperature rises more than ~2 to ~7°F (1 to 4°C) relative to the period 1990–2000 eventually contributing to an additional sea level rise of ~13 to ~20 ft (4 to 6 m) or more. This would result in the inundation of low lying coastal areas, including parts of many major cities. Even more significant than the direct loss of land caused by the sea rising are the associated indirect factors, including erosion patterns and damage to coastal infrastructure, salinization of wells, sub-optimal functioning of the sewage system of coastal cities (with resulting health impacts), loss of littoral ecosystems and loss of biotic resources. In coastal areas, and particularly deltas, factors such as modified ocean circulation patterns (and their impact on building and erosion of the coast), climate change in the catchment basin and change in coastal climate, not to mention changes in the frequency of extreme events, should be taken into account.

Destruction of habitats in coastal ecosystem is also caused by natural disasters, such as cyclones, hurricanes, typhoons, volcanism, earthquakes and tsunamis causing colossal losses worldwide [50]. The frequency of natural disasters is increasing with time (Figure 3.8), predictably due to climate change, as sea level rise also follows almost the similar trend (Figure 3.7). Trenberth [36] argues that higher sea surface temperatures in the Atlantic Ocean and increased water vapor in the lower atmosphere—caused by global warming—are to blame for the past decade's intense storms. These factors are causing significant physical damage to reefs or move large amounts of bottom material, thus altering habitat, biological diversity, and ecosystem function. There is however no denying of the fact that human induced activities are to a significant extent responsible for climate change accelerating the pace of natural disasters with time resulting in such damages to coastal lives and properties.

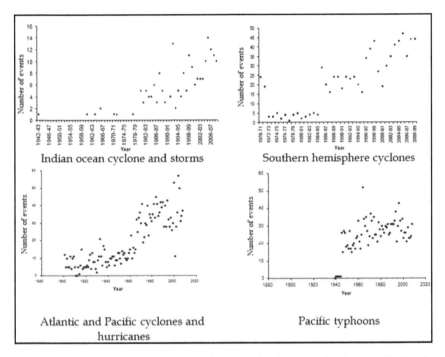

FIGURE 3.8 Trend of change of cyclonic storms, hurricanes and typhoons with time in different oceans in the world [13].

3.6.2 EFFECTS ON AGRICULTURE

Climate change directly affects sensitive sectors like agriculture, forestry and fishery and thereby the livelihoods of millions of coastal communities [33, 35]. Wide array of impacts due to climate change on factors affecting food production has been predicted. One can stipulate the following changes in forcing variables as likely to materialize sometime during the next century [4]:

- A gradual, continuing rise in atmospheric CO_2 concentration entail in increased photosynthetic rates and water-use efficiencies of vegetation and crops hence increase in organic matter supplies to soils.
- Minor increases in evapotranspiration in the tropics to major increases in high latitudes caused both by temperature increase and by extension of the growing period.

- Increases in amount and in variability of rainfall in the tropics; possible decrease in rainfall in a band in the subtropics pole ward of the present deserts; and minor increases in amount and variability in temperate and cold regions could take place. Peak rainfall intensities could increase in several regions.
- A gradual sea level rise could cause deeper and longer inundation in river and estuary basins and on levee back-slopes, and brackish water inundation leading to encroachment of vegetation that accumulates pyrite in soils near the coast.

3.6.2.1   Low Lying Coastal Soils

The probable effects on soil characteristics of a gradual eustatic rise in sea level will vary from place to place depending on a number of local and external factors, and interactions between them [1]. In principle, a rising sea level would tend to erode and move back existing coastlines. However, the extent to which this actually happens will depend on the elevation, the resistance of local coastal materials, the degree to which they are defended by sediments provided by river flow or long shore drift, the strength of long shore currents and storm waves, and on human interventions which might prevent or accelerate erosion.

3.6.2.2   Sediment Supply and Deltaic Aggradations

In major deltas, such as those of the Ganges-Brahmaputra and the major Chinese rivers, sediment supplies delivered to the estuary will generally be sufficient to offset the effects of a rising sea level. Such deltaic aggradations can decrease, however, under three circumstances:

- Where human interventions inland, such as large dams or successful soil conservation programs, drastically reduce sediment supply to the delta, e.g., the construction of the Aswan high dam in 1964 has led to coastal erosion and increased flooding of lagoon margins in the Nile delta [34];
- Where construction of embankments within the delta interrupts sediment supply to adjoining back swamps, exposing them to sub-

mergence by a rise in sea level, e.g., embankments along the lower Mississippi river have cut off sediment supplies to adjoining wetlands which formerly offset land subsidence occurring due to compaction of underlying sediments [6]; and

- Where land subsidence occurs due to abstraction of water, natural gas or oil, e.g., as is presently happening in Bangkok and in the northern parts of the Netherlands.

3.6.2.3 Tidal Flooding

In coastal lowlands which are insufficiently defended by sediment supply or embankments, tidal flooding by saline water will tend to penetrate further inland than at present, extending the area of perennially or seasonally saline soils. Where *Rhizophora* mangrove or *Phragmites* vegetation invades the area, would over several decades lead to the formation of potential acid sulfate soils. Impedance of drainage from the land by a higher sea level and by the correspondingly higher levels of adjoining estuarine rivers and their levees will also extend the area of perennially or seasonally reduced soils and increase normal inundation depths and durations in river and estuary basins and on levee back slopes. In sites which become perennially wet, soil organic matter contents will tend to increase, resulting eventually in peat formation. On the other hand, where coastal erosion removes an existing barrier of mineral soils or mangrove forest, higher storm surges associated with a rising sea level could allow seawater to destroy existing coastal eustatic peat swamps, which eventually may be replaced by fresh or saltwater lagoons.

3.6.2.4 Subsidence of Land

The probable response of low lying coastal areas to a rise in sea level can be estimated on the basis of the geological and historical evidence of changes that occurred during past periods when sea level was rising eustatically or in response to tectonic or isostatic movements, e.g., around the Southern North Sea [18]; in the Nile delta [34]; on the coastal plain of the Guyanas [3] and in the Musi delta of Sumatera [2]. Contemporary evidence is available in areas where land levels have subsided as a result of recent abstraction of water, natural gas or oil from sediments underlying coastal lowlands. Further studies of such contemporary and palaeo-environments are needed together with location specific studies in order to better understand the change processes, identify appropriate responses and assess their technical, ecological and socio-economic implications [41].

3.6.2.5   Trends in Vegetation Growth

Some major and widespread soil changes expected as a result of any global change are positive, especially the gradual increases in soil fertility and physical qualities consequent on increased atmospheric CO_2. The increased productivity and water-use efficiency of crops and vegetation, and the generally similar or somewhat higher rainfall indicated by several global circulation models, not fully counteracted by higher evapotranspiration, would be expected to lead to widespread increases in ground cover, and consequently better protection against runoff and erosion [4].

3.6.2.6   Changes in Soil Properties

According to Brinkman and Sombroek [4], major but less widespread soil changes, including greater biological activity and increased extent of periodic reduction in soils, would be expected where permafrost would disappear. In unprotected low lying coastal areas, gradual encroachment of *Rhizophora* mangroves or *Phragmites* following more extensive brackish water inundation may give rise to the formation of potential acid sulfate soil layers after several decades. Deeper and longer duration flooding of basins and levee back slopes in adjacent river and estuary plains could lead to more extensive reducing conditions and increased organic matter contents, and locally to peat formation.

Other changes due to climate change (temperature and precipitation) are expected to be relatively well buffered by the mineral composition, the organic matter content or the structural stability of many soils. However, decreases in cover by vegetation or annual or perennial crops, caused by any locally major declines in rainfall not compensated by CO_2 effects, could lead to soil structure degradation and decreased porosity, as well as increased runoff and erosion on sloping sites and by the concomitant more extensive and rapid sedimentation. Changes in options available to land users because of climate change may have similar effects [4].

In most cases, changes in soils by direct human action, on-site or off-site (whether intentional or unintended), are far greater than the direct climate induced effects. Soil management measures designed to optimize the soil's sustained productive capacity would therefore, be generally adequate to counteract any degradation of agricultural land by climate change. Soils of these areas, or other land with a low intensity of management such as semi-natural forests used for extraction of wood and other products, are less readily protected against the effects of climate change but such soils too are threatened less by climate change than by human actions – off-site, such as pollution by acid deposition, or on-site, such as excessive nutrient extraction under very low-input agriculture.

To armor the world's soils against any negative effect of climate change, or against other extremes in external circumstances, such as nutrient depletion or excess (pollution), or drought or high intensity rains, the best that land users could do [4], would be:

- To manage their soils to give them maximum physical resilience through a stable, heterogeneous pore systems by maintaining a closed ground cover as much as possible.
- To use an integrated plant nutrient management system to balance the input and off take of nutrients over a cropping cycle or over the years, while maintaining soil nutrient levels low enough to minimize losses and high enough to buffer occasional high demands.

3.6.2.7 Erosion

Coasts are also exposed to increasing risks due to erosion as a result of climate change and sea level rise [17]. It is an important area influenced by the climate change with rising temperatures leading to rise in the sea's water mass. In India, the mainland consists of 43% sandy beaches, 11% rocky coast with cliffs, and 46% mud flats and marshy coast [27]. The damages caused by sea erosion in different coastal states in India alone show a staggering annual loss of Rs. 3,683.87 million (1 US$ = Rs. 65). Following preventive and mitigation measures have been suggested and being adopted:

- Structural measures such as construction of sea wall/revetment, groins and off-shore breakwater;
- Non-structural/soft measures such as artificial nourishment of beaches, vegetation cover and sand bypassing at tidal inlets.

Government of India (GOI) provides assistance for protection of vulnerable coastal states from sea erosion mainly through two schemes, namely (i) Centrally sponsored scheme for protection of critical stretches (through state sector), and (ii) National Coastal Protection Project (NCPP) for protection of the maritime states/ UTs with a view to explore possibilities of funding through external resources or other domestic resources.

3.7   DETERIORATING WATER FLOWS IN RIVER GANGA: A CASE STUDY

Ghorai and Sen [12], Sen [28], Sen et al. [30] and Sen and Ghorai [31] have raised serious concerns over deteriorating water flow in the river Ganga through a long stretch of run within India owing mainly to unplanned anthropological factors arising out of scores of hydro-electric and irrigation projects already commissioned and many others in the pipeline on the river in addition to adverse climate change impacts, a typical example risking the coastal ecosystem in the lower delta across India and Bangladesh. The lower Ganga delta of both India (south of Farakka Barrage) and Bangladesh (south-west) share the same ecology and face threats due to dwindling water diversion via Farakka Barrage and deteriorating water quality of the river in the upstream at different places in India. This being a matter of common concern to both countries, there is need for a holistic and focused attention for which the following suggestions were made with immediate effect to seek for a lasting solution [31]:

- There appears to be a need for revisiting the design of the Farakka Barrage, as well as the discharge and distribution norms of water in the interest of the two countries, keeping in view the predicted flow of upstream Ganga water in long- term perspectives, and if necessary, fresh norms to be decided.

- Predicted flow of water through Ganga-Brahmaputra river system on account of retreat of glaciers and other parametric uncertainties due to climate change needs to be refined with appropriate climate models in deciding the future norms for distribution of water via Farakka Barrage with higher precision in different time scales.
- Need for fresh installation of hydro-electric power and irrigation projects in India must be given extremely careful consideration with stringent norms for discharge of river water in the upstream so that ecology of the area is not disturbed.
- Past hydro-electric power and irrigation projects in the upstream already commissioned need to be reviewed in terms of the norms for discharge of water decided, and if necessary, to be revised, scientifically so that ecology of the area is not disturbed.
- Provisions should be mandatory to make impact analysis of the discharge of water from different projects, be it hydro-electric power and irrigation projects or any others, on the ecology of the area for all past and future installations in India.
- Strict administrative vigilance to be maintained to stop acts of unscrupulous diversion of water forthwith by private agencies in India.
- Location specific integrated water development and management schemes at strategic points over the entire flow length in different time scales to be prepared and their methods of implementation be worked out, with adequate participation and vigilance from the local inhabitants, to ensure maintaining prescribed water quality throughout the year in India.
- In India, in particular, impacts of the water flow at different strategic points into lower delta in respect of salinity in soil and water, flow rate, tidal amplitude and fluctuations, sedimentation/hydrological parameters, navigation through rivers, ground water table depths and qualities, all important components of biodiversity, and any other related parameters should be taken up and monitored with a holistic plan, over minimum five year phases, through a central task force comprising of scientists, NGOs, government officials, local inhabitants, and the same placed in public domain. Similar programs should be simultaneously planned and taken up by Bangladesh. A core team consisting of key members drawn

from both countries should interact and monitor the progress once in each year and suggest for improvement with respect to targets fixed.

Sen and Ghorai [31] have cautioned that the concerned lower Ganga delta of the two countries is largely coastal and therefore fragile in nature subject to the increasing vagaries due to climatic disasters beyond possibly anybody's control to prevent. Additional factors originating from the deteriorating E-flows of the Ganga river network water contribute significantly further to the woes of the inhabitants of the area. In conclusion, they remarked that there might be no shortcuts to improve the ecology for sustained livelihood of the inhabitants in this area across the two countries other than ensuring E-flows via Farakka Barrage, for which careful considerations must be given to the suggestions made above. It is fortunate to observe that Government of India has of late taken cognizance of the fact in the interest of the country to limit future hydro-electric and irrigation projects in hand although a holistic approach as envisaged is still warranted in the interest of the entire ecosystem.

3.8   SUMMARY

Coastal areas with high population density along with potentially high esthetic benefit supporting human lives and livelihoods, are confronted with high risk factors on a multiple of issues viz., anthropogenic factors, topography and related soil properties, protection measures required to be undertaken, and natural disasters caused by the sea and through its interaction with climate and under-sea tectonic movement of the earth, among the major ones – influencing agricultural production and vulnerability of the ecosystem *per se* in a related country. The resultant effect is that the livelihood in the ecosystem remains uncertain and the masses remain poverty-stricken in majority of the countries having reasonably large coastal boundaries due to the lack of concerted efforts to address the risk factors and their interactions in a holistic manner. Soil in the coastal ecosystem *per se* does not have separate significance as far as its productivity is concerned unless it is considered in association with other relevant ecological factors describing the ecosystem. The chapter presents an overview of the distribution of different coastal components in the world along with analysis of relevant risk factors.

The impact of climate change, being a major risk factor, has been discussed as a generic phenomenon and with special emphasis on predictability of sea level rise, oceanic disasters including cyclones, storms, hurricanes, typhoons and inundation of coastal land in different parts of the world. Its role on different aspects in agriculture particularly in low-lying coastal lands and on sediment supply and deltaic aggradations, tidal flooding, subsidence of land, and changes in soil properties is also explained. The influence of climate change on erosion of soil in the coastal areas has been discussed along with important structural and non-structural measures suggested. Finally, a case study on deteriorating water flow in the river Ganga risking the ecosystem in lower delta across India and Bangladesh has been discussed and future lines of action suggested.

KEYWORDS

- bio-shield
- carbon dioxide
- climate change
- coastal ecosystem
- continental shelf
- coral reefs
- delta
- ecology
- erosion
- inundation
- lagoons
- livelihood
- mangrove ecosystem
- natural disasters
- nutrient imbalance

- low-lying soils
- risk factors
- salinization
- salt marshes
- sea level change
- sedimentation
- sediment supply
- subsidence
- sustainable development
- The Ganga
- tidal flooding
- wetlands

REFERENCES

1. Brammer, H., & Brinkman, R. (1990). Changes in soil resources in response to a gradually rising sea-level. In: *Developments in Soil Science* (Scharpenseel et al., Eds.), pp. 145–156.
2. Brinkman, R. (1987). Sediments and soils in the Karang Agung area. In: *Some Aspects of Tidal Swamp Development with Special Reference to the Karang Agung Area, South Sumatra Province* (R. Best, R. Brinkman & J. J. van Roon, Eds.), pp. 12–22. Mimeo, World Bank, Jakarta.
3. Brinkman, R., & Pons, L. J. (1968). *A Pedo-geomorphological Classification and Map of the Holocene Sediments in the Coastal Plain of the Three Guyanas.* Soil Survey Paper No. 4, Soil Survey Institute (Staring Centre), Wageningen. 40 pages.
4. Brinkman. R., & Sombroek, W. G. (2007). *The Effects of Global Change on Soil Conditions in Relation to Plant Growth and Food Production* (http://www.fao.org/docrep/W5183E/w5183e05.htm).
5. Coastal Wiki (2008). *Polyfluorinated Compounds – A New Class of Global Pollutants in the Coastal Environment* (http://www.Polyfluorinated compounds PFC – pollutants in coastal water.htm).
6. Day, J. W., & Templet, P. H. (1989). Consequences of sea level rise: implications from the Mississippi delta. *Coastal Management, 17*, 241–257.
7. Dirk, B., Burke L., McManus, J., & Spalding, M. (1998). In: *Reefs at Risk: A Map-Based Indicator of Threats to the World's Coral Reefs*, World Resources Institute, USA.

8. Dubey, M. (1993). *Population Explosion Hits Coastal Ecosystems.* (http://www.indiaenvironmentportal.org.in/node/5370)
9. Duedall, I. W., & Maul, G. A. (2005). Demography of coastal populations, In: *Encyclopedia of Coastal Science* (M. Schwartz, Ed.), 1211 p. Encyclopedia of Earth Sciences Series, Springer, The Netherlands. pp. 368–374.
10. Encyclopedia Britannica (2009). *Mangrove Forest: Global Distribution* (http://www.britannica.com/science/world-map/images-videos/Major-wetland-areas-and-worldwide-distribution-of-salt-marshes-and/40).
11. Gedan, B. K., Silliman, B. R., & Bertness, M. D. (2009). Centuries of human-driven change in salt marsh ecosystems. *Annual Review of Marine Science*, 1, 117–141.
12. Ghorai, D., & Sen, H. S. (2014). *Living out of Ganga: A traditional yet imperiled livelihood on bamboo post harvest processing and emerging problems of Ganga.* In: *Our National River Ganga: Lifeline of Millions*, Rashmi Sanghi (Ed.), Springer International Publishing Switzerland, pp. 323–338.
13. Ghorai, D., & Sen, H. S. (2015). Role of climate change in increasing occurrences of oceanic hazards as a potential threat to coastal ecology. *Nat. Hazards*, 75, 1223–1245.
14. Giri, C., Ochieng, E., Tieszen, L., Zhu, Z., Singh, A., Loveland, T., Masek, J., & Duke, N. (2011). *Status and distribution of mangrove forests of the world using earth observation satellite data. Global Ecol. Biogeogr.*, 20, 154–159.
15. Goudarzi, S. (2006). *Flocking to the Coast: World's Population Migrating Into Danger.* (http://www.livescience.com/environment/060718_map_settle.html).
16. IFPRI (2007). 2020 Conference, held at Beijing, China, 17–19 October (http://www.ifpri.org/2020ChinaConference/index.htm).
17. IPCC (2014). Climate Change 2014. *Synthesis Report—Contribution of Working Groups I, II and III to the Fifth Assessment Report of the Intergovernmental Panel on Climate Change* [Core Writing Team, R. K. Pachauri and L. A. Meyer (Eds.)]. IPCC, Geneva, Switzerland, 151 p.
18. Jelgersma, S. (1988). A future sea-level rise: its impacts on coastal lowlands. In *Geology and Urban Development*, Atlas of Urban Geology, Vol. 1, UN-ESCAP, 61–81.
19. Minard. A. (2008). *Dead Zones Multiplying Fast, Coastal Water Study Says.* (http://www.DeadZonesMultiplyingFast.htm)
20. Narragansett Bay Commission.(2009) (http://omp.gso.uri.edu/ompweb/doee/science/descript/asia.htm)
21. New World Encyclopedia (2008). *Estuary* (http://www.newworldencyclopedia.org/entry/Estuary)
22. OzCoasts. (2009). *Salt Marsh and Salt Flat Areas.* (http://www.ozcoasts.org.au/indicators/changes_saltmarsh_area.jsp)
23. Pacahuri, R. K. (2008). *Climate change—Implications for India*, presented New Delhi, 25 Apr 2008 (http://164.100.24.209/newls/bureau/lectureseries/pachauri.pdf)
24. Pachauri, R. K. (2008). Climate change: what is next? Managing the interconnected challenges of climate change, energy security, ecosystems and water. Presented in *International Conference, held at ETH University, Zurich, 6 November,* (http://www.cluboframe.org/eng/meetings/winterthur_2008).
25. Portier, R. J. (2003). Trends in Soil Science, Technology and Legislation in the USA. *Journal of Soils and Sediments*, 3, 257.

26. Poyya, M. G., & Balachandran, N. (2008). Strategies for conserving ecosystem services to restore coastal habitats. Paper presented in UNDP-PTEI Conference on *"Restoration of Coastal Habitats,"* held at Mahabalipuram, Tamil Nadu, 20–21 August.

27. SAARC Disaster Management Centre, New Delhi (2009). *Coastal and Sea Erosion.* (http://www.saarc.sdmc.nic.in/coast.asp).

28. Sen, H. S. (2010). Drying up of River Ganga: an issue of common concern to both India and Bangladesh, *Curr. Sci.,* 99, 725–727.

29. Sen, H. S., Bandyopadhyay, B. K., Maji B., Bal, A. R., & Yadav, J. S. P. (2000). Management of coastal agro-ecosystem. In: *Natural Resource Management for Agricultural Production in India* (J. S. P. Yadav & G. B. Singh, Eds.), Indian Society of Soil Science, New Delhi, pp. 925–1022.

30. Sen, H. S., Burman, D., & Mandal, S. (2012). *Improving the Rural Livelihoods in the Ganges Delta through Integrated, Diversified Cropping and Aquaculture, and through Better Use of Flood or Salt Affected Areas,* Lap Lambert Academic Publishing, GmbH & Co. KG, Germany, 61 p.

31. Sen, H. S., & Ghorai, D. (2014). Ensuring environmental water flows in the river Ganga for sustainable ecology in the lower delta across India and Bangladesh. *Yojana,* 14–19.

32. Sen, H. S., Sahoo, N., Sinhababu, D. P., Saha, S., & Behera, K. S. (2011). Improving agricultural productivity through diversified farming and enhancing livelihood security in coastal ecosystem with special reference to India. *Oryza,* 48, 1–21.

33. Sinha, S. K., & Swaminathan, M. S. (1991). Deforestation, climate change, and sustainable nutrition security: a case study for India. *Climate Change,* 19, 201–209.

34. Stanley, D. J. (1988). Subsidence in Northeastern Nile delta: rapid rates, possible causes, and consequences. *Science,* 240, 497–500.

35. Swaminathan, M. S. (1996). *Sustainable Agriculture: Towards Food Security.* Konark Publishers Pvt. Ltd, Delhi, India.

36. Trenberth, K. (2005). Uncertainty in hurricanes and global warming. *Science,* 308, 1753–1754.

37. United Nations Environment Program (2008). What is climate change? (http://www.unep.org/themes/climatechange/whatis/index.asp).

38. US Fish & Wildlife Service (2009). *Coastal program.* (http://www.fws.gov/coastal)

39. USAID (2007). *Estuarine Ecosystems.* (http://www.usaid.gov/our_work/environment/water/estuarine.html).

40. Velayutham, M., Sarkar, D., Reddy, R. S., Natarajan, A., Shiva Prasad, C. R., Challa, O., Harindranath, C. S., Shyampura, R. L., Sharma, J. P., & Bhattacharya, T. (1998). Soil resources and their potentials in coastal areas of India. Paper presented in *"Frontiers of Research and its Application in Coastal Agriculture,"* Fifth National Seminar of Indian Society of Coastal Agricultural Research, held at Gujarat Agricultural University, Navsari, Gujarat, 16–20 September.

41. Warrick, R., & Farmer, G. (1990). The greenhouse effect, climatic change and rising sea level: implications for development. *Transaction of Institute Br. Geography,* 15, 5–20.

42. Wikipedia (2015). *Coastal Management.* (http://en.wikipedia.org/wiki/Coastal_management).

43. Wikipedia (2015). *List of Countries by Length of Coastline.* (http://en.wikipedia.org/wiki/List_of_countries_by_length_of_coastline).
44. Wikipedia (2015). *Coral Reef.* (http://en.wikipedia.org/wiki/Coral_reef).
45. Wikipedia (2015). *Lagoon.* (http://en.wikipedia.org/wiki/Lagoon).
46. Wikipedia (2015). *Mangrove.* (http://en.wikipedia.org/wiki/Mangrove).
47. Wikipedia (2015). *Macrophyte.* (http://en.wikipedia.org/wiki/Macrophyte).
48. Wikipedia (2015). *Continental Shelf.* (http://en.wikipedia.org/wiki/Continental_shelf).
49. Wikipedia (2015). *Ice Shelf.* (http://en.wikipedia.org/wiki/Ice_shelf).
50. Wikipedia (2015). *List of Natural Disasters by Death Toll.* (http://en.wikipedia.org/wiki/List_of_natural_disasters_by_death_toll).
51. World Resources Institute (2006). *Environment Information Portal.* (http://www.NutrientOverloadUnbalancingtheGlobalNitrogenCycle.htm).
52. Zeng, T. Q., Peter Cowell, J., & Deanne, H. (2009). *Predicting Climate Change Impacts on Salt Marsh and Mangrove Distribution: GIS Fuzzy Set Methods.* (http://www.coastgis.org/cgis06/Papers/Zeng_Cowell_Hickey.pdf).

CHAPTER 4

GROUND WATER CONTAMINATION: RECENT ADVANCES IN IDENTIFYING SOURCES

DEEPESH MACHIWAL and MADAN KUMAR JHA

CONTENTS

4.1 INTRODUCTION

On large areas over the globe, water shortages are caused due to its quality rather than the quantity, quality being the key factor in maintaining health of both humans and animals. Water quality besides having a great impact on human health, is also important because of its intrinsic value [78]. As per reports of the UNICEF and the WHO [113], about 2.5 billion people in the world lack adequate sanitation, while 884 million people are without access to safe water. Half of the population of the developing world is exposed to contaminated sources of water. During the decade (1991–2000), more than 665,000 people lost their life in 2,557 natural disasters, of which 90% were water-related disasters and a large number of victims (97%) were from developing countries [44].

The comprehensive and adequate knowledge of contamination levels and correct assessment of trends and patterns in water quality is necessitated in order to protect this vital resource from pollution as well as to suggest efficient and cost-effective measures to control present and expected future threats of contamination at local or regional scales [7]. Apparently, availability of reliable and long-term water quality data is the basic requirement for such assessments and to design appropriate control measures.

Water quality refers to the chemical, physical, biological, and radiological characteristics of including the alterations caused by human activities [24]. It is measured by several factors, such as the concentration of dissolved oxygen, bacteria levels, the amount of salt (or salinity), the amount of material suspended in the water (turbidity), the concentration of microscopic algae and quantities of pesticides, herbicides, heavy metals, and/or other contaminants of interest. It is a measure of the condition of water relative to the requirements of one or more biotic species and or to any human need including agricultural, industrial and other purposes. In fact the water quality is socially defined depending on the desired use of water. Water quality standards and criteria are used to evaluate the suitability of water for various uses being different for different sectors or purposes. These criteria provide guidelines for desirable characteristics and acceptable levels of different constituents of water for various intended uses [24, 25, 78, 112]. Therefore, to establish criteria for water quality, the physical, chemical, and biological or other constituents must be evaluated using standard methods for comparing results of water quality analyzes [26, 78, 80, 81, 112]. The quality of water (Surface as well as ground water) is typically affected by many interacting processes [67]. Quality of surface water in a region is largely controlled by natural processes, e.g., weathering and soil erosion, as well as by anthropogenic factors, e.g., municipal and industrial wastewater discharge. The anthropogenic factors may have a continuous effect on contamination as wastewater discharges constitute a constant polluting source. On the contrary, natural factor may be a seasonal process as the surface runoff only flows during the rainy season in the catchment depending upon the climate [23, 104, 117]. On the other hand, ground water quality in an area is mainly determined by the natural processes, e.g., geology, ground water movement, recharge water quality, and soil/rock interactions with water. Besides, anthropogenic activities including agricultural production, industrial growth, urbanization with increasing exploitation of water resources, etc. along with atmospheric input also play a major role in defining the ground water quality [12, 39]. The pollution of water therefore, results in changes of water quality conditions that negatively impact the integrity of the water for beneficial purposes [48].

Several conventional tools exist for water quality assessment, varying from graphical to statistical in nature, which are described in standard textbooks on ground water hydrology [25, 53, 95]. But, mere assessment of the water quality by knowing its composition alone is not enough to identify the natural and anthropogenic sources of the ground water contamination. Therefore, there is a need of employing modern tools and techniques for distinguishing between the natural and the anthropogenic factors that affect the ground water quality. The application of modern tools and techniques such as multivariate statistical techniques and geographical information system (GIS) for efficient ground water quality interpretation has been highlighted in some of the recent studies [47, 58, 69, 94, 107]. Recently, Machiwal and Jha [74] proposed a standard methodology by integrating multivariate statistical technique, i.e., principal component analysis with GIS-based geo-statistical modeling technique for identification of natural and anthropogenic contamination sources in hard-rock aquifer systems of Udaipur, India.

4.2 CLASSIFICATION OF TOOLS/TECHNIQUES FOR WATER QUALITY INTERPRETATION

Interpretation of the water quality is an important step in the whole task of water quality analyzes. The interpretation involves presentation of water quality data or measures of concentration of different chemical constituents present in the water in either pictorial/graphical form or numerical values. Over the years, plentiful tools and techniques varying from conventional to modern have been evolved for water quality interpretation. An excellent review of graphical, statistical, GIS, geostatistical and modeling techniques for water quality interpretation is given in Machiwal and Jha [69]. However, for the benefit of the readers, the water quality interpretation methods are divided into two broad categories viz., 'conventional techniques' and 'modern techniques,' and available methods/techniques under each category are summarized in Tables 4.1 and 4.2, respectively.

4.3   RECENT APPROACHES OF INTEGRATING PCA, GIS AND GEOSTATISTICAL MODELING TECHNIQUES

The modern tools of the multivariate statistical techniques, i.e., principal component analysis, hierarchical cluster analysis, etc. are valuable tools for evaluating spatial and temporal variability and interpreting large datasets of water quality, identifying sources of natural and anthropogenic contamination and designing a cost-effective monitoring network for quick solution for contamination problems [15, 19, 34, 66, 74, 100, 116]. However, these tools are mostly used independently without integrating with geo-statistical and GIS modeling techniques. Recently, integration of the PCA with GIS-based geo-statistical modeling technique is accomplished to identify sources of the ground water contamination although the applications of this recent advanced approach are very limited for ground water quality studies [74]. A brief description of the PCA and geo-statistical modeling techniques along with their theory and procedure is provided in subsequent sub-sections.

4.3.1   PRINCIPAL COMPONENT ANALYSIS

The principal component analysis (PCA) employs the correlation matrix approach that rearranges the ground water quality data in such a manner so as to better explore the structure or process of the underlying system controlling/affecting the water composition. The overall aim of applying the PCA is to develop a new set of factors/components, called as principal components (PCs) from the initial/original water quality parameters. The generated PCs are linear combination of the initial parameters. At

TABLE 4.1   Conventional Methods for Water Quality Interpretation

Technique	Brief Description	Ref.
Analysis of Variance	Analysis of variance (ANOVA) technique is used to separate differences due to random errors and statistical differences in the observations. The ANOVA can be used to test hypothesis whether there is significant difference in values of a water quality parameter for different sampling sites.	[69]
Box and Whisker Plot	An excellent representation of statistical summary of water quality parameters can be made by Box and Whisker plot. It provides information about five statistical properties, i.e., minimum, maximum, 25th and 75th percentile and median along with the identification of 'outliers.'	[70, 114]
Collins Diagram	In this diagram, concentrations of major cations (Ca, Mg, Na+K) and anions (SO_4, Cl, HCO_3+CO_3) in equivalents per million (epm) are shown by patterns or colors on adjoining vertical bars. Concentration of ions is represented on some scale on the bar.	[16]
Correlation Matrix	It is a bivariate method that determines how well one parameter predicts the other. The relationship between various parameters is studied using Spearman's correlation coefficient. The correlations between water quality parameters indicate chemical processes and possible sources of ions in the water.	[61]
Cumulative Probability Curve	These plots are very useful to identify whether chemistry of a water quality parameter is affected by a single/multiple physical process (population in statistical terms), i.e., anthropogenic pollution or salinization. If the water is affected by a single process, its distribution will be unimodal normal or lognormal and linear distribution will be obtained on probability diagram. On the contrary, the parameter can be considered to be affected by more than one process, if the parameter deviates from linear distribution on probability curve. In such a case, each process or population can be differentiated by the intersection points of two neighboring linear populations.	[65, 102, 103, 111]

TABLE 4.1   (Continued)

Technique	Brief Description	Ref.
Durov Diagram	This diagram, introduced into the Soviet literature by S.A. Durov, plots percentage concentrations of cations and anions in separate triangles. It is somewhat similar to Piper diagram but removes some of its shortcomings. The intersection point of lines extended from points in two triangles to the central rectangle represents the major ion composition on a percent basis.	[123]
Gibbs Diagram	It distinguishes among the influences of rock-water interaction, evaporation and precipitation on water chemistry. It evaluates the relative importance of each of three mechanisms by plotting the weight ratio $Na^+/(Na^++Ca^{2+})$ versus salinity.	[27]
Giggenbach Triangle	It is used to visually determine the water-rock equilibrium based on K-Mg-Na triangle comprising of three zones: immature waters at the base, partially equilibrated waters in the middle, and fully equilibrated waters along the upper curve. The extent of rock-water equilibrium can be estimated depending on where the composition of a given sample lies within the triangle.	[28]
Histogram/ Frequency Plot	Histogram is basically group of bars showing number of classes or frequency of data points in different classes of data values. In this method, distribution of data is assessed visually by looking at location, shape, spread and distribution of data.	[70, 114]
Langelier-Ludwig Diagram	This diagram represents patterns and correlations between major cations and anions for multiple samples on rectangular coordinates. Suitable groupings of cations and anions are plotted as percentages but without any cation or anion triangles.	[62]
Pie Chart/Circular Diagram	It represents concentration of individual ions in epm by circles with their areas proportional to the total concentrations of the ions. Percentage compositions of individual ions or groups of ions are represented by segments of the circle.	[69]

TABLE 4.1 (Continued)

Technique	Brief Description	Ref.
Piper Diagram	Piper diagram shows relative percentage concentrations of major cations, i.e., Ca, Mg, Na and K and anions, i.e., CO_3, HCO_3, Cl and SO_4. It has potential to determine chemical relationships and understanding geochemical evolution of ground water. It is advantageous in representing a large number of samples without confusion and is also convenient for showing the effects of mixing two waters from different sources. Basically, the diagram consists of two triangular fields and a diamond-shaped field sub-divided into ten areas. The overall characteristic of a water sample is represented in the diamond-shaped field by projecting the position of the plots in the triangular fields.	[87]
Radial or Vector Diagram	This diagram utilizes six coordinates to show the concentrations of the six major cations and anions in epm proportional to the length of each coordinate. If the vertex of adjacent ions is connected, the plot is known as a vector-pattern diagram.	[69]
Scholler Diagram	It is a semi-logarithmic plot where total concentrations of the cation and anion compositions of many samples are represented on a single graph. It allows visual detection of minor groupings or trends in the data.	[97, 98]
Stiff Diagram	Stiff diagram represents epm values of cations and anions at some scale along horizontal axes at regular intervals. On joining all the plotted points together, a pattern is produced with shapes characteristic of the composition of the waters presented.	[108]
Student's t-test	The Student's t-test is used to compare the mean values of different groups of parameter observations.	[69]
Time Series Analysis	Time series analysis involves application of statistical tests to find out time series characteristics as listed below.	[70]
	Normality can be tested by graphical (histograms, stem-and-leaf plots, box and whisker plots, normal probability plots, etc.) as well as statistical techniques (Chi-square test, Kolmogorov-Smirnov test, Lilliefors test, Anderson-Darling test, Cramer-von Mises test, Shapiro-Wilk test, Jarqua-Bera test, and D'Agostino-Pearson omnibus test).	

TABLE 4.1   (Continued)

Technique	Brief Description	Ref.
Time Series Plots	In time series plots, value of a water quality parameter is plotted over time scale. Few of the examples include scatter, line, areal or bar diagrams. These plots are used for understanding temporal changes in water quality parameters.	[69]
USSL Diagram	The US Salinity Laboratory Staff proposed this diagram based on salinity and sodium hazards for evaluating the suitability of water for irrigation use. The diagram has four classes of each salinity and alkalinity: low, medium, high and very high. Thus, a water sample may fall in any one of the total sixteen classes based on different combination of salinity and sodicity (alkalinity). Salinity hazard of irrigation water is determined by the total dissolved solids, measured in terms of electrical conductivity. Likewise, if the percentage of Na to Ca+Mg+Na is significantly larger than 50 in the water, then soils containing exchangeable Ca and Mg take up Na in exchange for Ca and Mg causing deflocculation and impairment of the tilth and permeability of soils. The sodium hazard is evaluated by determining sodium adsorption ratio (SAR) given as [53]: $$ SAR = \frac{Na}{\sqrt{(Ca + Mg)/2}} $$ where, the concentrations are expressed in meq/L (milliequivalents per liter).	[115]
Wilcox Diagram	This diagram plots the sodium percentage of water with salinity. It is used to assess the ground water quality for irrigation use. It has four categories: excellent to good, permissible to doubtful, doubtful to unsuitable and unsuitable depending on the different combinations of sodium percentage and salinity. The sodium percent (Na%) is calculated as: $$ Na\% = \frac{Na^+}{Na^+ + Mg^{2+} + Ca^{2+} + K^+} \times 100 $$ where, the quantities of Na, Mg, Ca and K are in meq/L.	[120]

TABLE 4.2   Modern and Advanced Methods for Interpreting Water Quality

Technique	Brief Description	Ref.
Multivariate Statistical Techniques	Multivariate statistical analyzes are useful to identify the major processes that control geochemical evolution of the ground water. Among the several multivariate statistical techniques, principal component analysis (PCA), hierarchical cluster analysis (HCA), discriminant analysis (DA), canonical correlation analysis (CCA), correspondence analysis (CA) are widely-used in the ground water studies.	[74]
	The PCA is used to reduce multidimensional datasets to lower dimensions for analysis. It makes orthogonal linear transformation of the data to a new coordinate system such that the greatest variance by any projection of the data comes to lie on the first coordinate or first principal component (PC), the second greatest variance on the second coordinate, and so on. Each PC represents some specific source of contamination or process influencing water quality.	[20, 22]
	Cluster Analysis, an unsupervised pattern recognition technique, uncovers intrinsic structure of a dataset without making *a-priori* assumption about the data in order to classify the objects of the system into clusters based on their similarities. In HCA, clusters are formed sequentially, starting with the most similar pair of objects and forming higher clusters step by step. The clustering process is repeated until a single cluster including all the water samples is formed. The HCA results are displayed by a dendrogram providing visual summary of the clustering process.	[35, 85]
	The DA is used to determine the parameters, which discriminate between two or more naturally occurring groups. In this technique, raw data are used and a discriminant function $f(G_i)$ for i^{th} group is constructed as follows:	

$$f(G_i) = k_i + \sum_{j=1}^{n} w_{ij}\, p_{ij}$$

where G = total number of groups, k_i is the constant inherent to each group, n is the number of parameters used to classify a set of data into a given group, and w_j is the weight coefficient assigned by DA to the selected parameter (p_j).

TABLE 4.2   (Continued)

Technique	Brief Description	Ref.
	The DA can be either forward stepwise mode where one-by-one parameters are included starting with more significant or backward stepwise mode DA where parameters are removed step-by-step beginning with the less significant. Spatial and temporal variations of the water quality parameters can be evaluated by applying the DA to the raw dataset using standard and stepwise modes to construct discriminant functions.	[49, 122]
	The CCA belongs to the family of correlation techniques, e.g., product moment correlation, multiple regression analysis, etc. However, the CCA is quite different from the PCA despite certain similarities related to concept and terminology. The CCA investigates inter-correlation between two sets of parameters, whereas the PCA identifies the pattern of relationship within one set of data. The CCA may be used to identify any patterns that tend to occur simultaneously in two different datasets and to find out the correlation that occurs between associated patterns.	[43]
	In the CA technique, water quality data are plotted from the rows and columns of a two-way contingency table as points in low-dimensional vector spaces. The geometry of the parameters, i.e., the rows, is related to the geometry of the attributes, i.e., columns, and hence there exists a correspondence between the rows and columns. In performing the CA, inter-variable relationships are determined between attributes and other attributes, parameters and other parameters, and between parameters and attributes.	[10, 33, 84]
Partial Least Squares	Partial least squares (PLS) technique finds relationship between a dependent or response parameter (y) and an independent or descriptor parameter (x), and looks to be similar to CCA technique. The PLS maximizes the covariance between two matrices x and y, which are decomposed into corresponding score matrices and loading matrices. A special type of the PLS modeling is discriminant PLS that is used to find parameters and directions in the multivariate space that discriminate known classes in a calibration set. This technique explores the spatial and temporal variations in water quality composition and also assesses the influence of anthropogenic activities on water quality.	[52]

TABLE 4.2   (Continued)

Technique	Brief Description	Ref.
Remote Sensing and GIS	Remote sensing (RS) technique overcomes the constraints of time and cost involved with *in-situ* collection of water samples and their measurements by allowing identification of spatial and temporal trends of water quality over large areas. The RS is also advantageous to retrieve water quality data of ungauged surface water bodies in inaccessible areas. Few of the optically-active surface water quality parameters include chlorophyll-a, suspended matter, and dissolved organic matters, which have been monitored by Landsat-MSS/Thematic Mapper imagery in inland and estuarine water quality monitoring.	[8, 14, 18, 54, 63, 92, 118]
	Geographical information system (GIS) is a powerful tool to store, organize, quantify, analyze, interpret and display large volume of spatial data with high accuracy and minimum error, and to use these data for decision-making. GIS has been effectively used for analyzing spatial and temporal variability/trends of water quality that is very helpful in water quality monitoring and modeling, pollution hazard assessment, and environmental change detection.	[9, 30, 36, 106]
Geo-statistics	Geo-statistics involves a set of statistical estimation techniques that uses the quantities which vary in space, i.e., spatial parameters. The basic aim of the geo-statistical techniques is to interpolate spatially-correlated data to describe spatial distribution over the surface. It takes advantage of a general observation that, on average, values closer together in space will be more similar than those farther from each other. It involves developing experimental semi-variogram model, fitting of theoretical models, selecting the best-fit model and finally, to estimate parameter value at unknown locations using the best-fit model.	[11, 32, 50, 83, 96]
Water Quality Index	It is used to evaluate the quality of both ground water and surface water, and is particularly important tool for developing nations whose economy is poor and the beneficiaries are poorly educated. There is no standard methodology for constructing a water quality index (WQI). However, two common steps are: (i) selection of a set of parameters that measure the important physical, chemical, and microbiological water characteristics, and (ii) formulate a rule to summarize all the information in a unique number, i.e., WQI. In literature, there are large numbers of studies where WQI has been employed to interpret water quality in different parts of the globe.	[1, 2, 6, 42, 75, 79, 89, 91, 99, 105, 109]

TABLE 4.2   (Continued)

Technique	Brief Description	Ref.
Geochemical Modeling	This modeling tool is very useful to understand effect of geochemical reactions on the water quality. The geochemical modeling is an advance technique as it serves multiple purposes, e.g., determining prevailing geochemical reactions, quantifying the extent to which these reactions occur, predicting the fate of inorganic contaminants, and estimating direction and rates of ground water flow. It is used by two approaches: (a) inverse modeling where observed ground water quality data are used to realize the geochemical reactions, and (b) forward modeling that uses hypothesized geochemical reactions to predict ground water quality data. The main intent of applying inverse modeling is to determine the net chemical reactions that quantitatively measure the chemical and isotopic data of ground water samples.	[5, 88]
Artificial Neural Network	Artificial neural network (ANN) is a cost-effective and attractive technique and has been widely applied for simulation and prediction of quality of both surface water and ground water. An ANN resembles biological neural networks of the human brain and is a massive and parallel-distributed information processing system. Processing units of the ANN are known as neurons joined by weighted connections. The nodes receive input from other nodes or an external stimulus, and then the input signal is processed using an activation/transfer function. Finally, a transformed output signal is generated, which may be the final output from the network or the input to another node. There exist many ANN architectures, but multilayer networks are commonly used for forecasting.	[3, 13, 29, 38, 41, 60, 90, 93, 101, 119]

the beginning of the PCA, Eigen values and eigenvectors are extracted from the correlation matrix, and then the least significant eigenvectors are discarded from the group of total extracted eigenvectors [17]. Once the important eigenvectors are selected, these are suitably transformed into the PCs of the water quality dataset. In general, the structure of the extracted significant PCs is like that the first PC explains the highest part of the variance of the system, the second PC smaller than the first one, and the last PC contains the least variance of the system. Extent of relationship, i.e., strong, moderate or week, between the water quality parameter and its PC is characterized by the value of PC loading. The relationship can be either positive or negative depending upon the 'plus' or 'minus' sign associated with the PC loadings. All the extracted PCs cumulatively describe entire variance of the system; however, as the number of PC increases, their individual contribution to system's variance decreases. Therefore, in order to select the number of PCs to be retained, the Kaiser Normalization Criterion [51] is used.

The Kaiser criterion suggests retaining only those PCs, who best describe the variance of the analyzed data and can reasonably be interpreted. For the PCs best-explaining variance of the system, the eigenvalue remains more than one [37]. The adequacy of the explanation of variance of a particular water quality parameter by a given set of components is quantified by the term 'communality' [46]. The number of parameters represented by the PCs or communalities is obtained by squaring the parameter values of the PC matrix and summing the total within each parameter. For an ideal and successful application of the PCA, there will be small number of significant PCs with high communalities (close to 1) and the PCs will be easily and quickly interpretable in terms of particular source or process of the contamination [22]. Two steps of the PCA involve standardization of the initial water quality dataset and extraction of the significant PCs. The water quality data are standardized by using z-scale transformation as shown below:

$$z_{ji} = \frac{x_{ji} - \overline{x}_j}{s_j} \tag{1}$$

where, x_{ji} = value of the jth water quality parameter measured at i^{th} site, $\overline{x}_j$ = mean (spatial) value of the jth parameter, and s_j = standard deviation of the jth parameter.

4.3.2 GEO-STATISTICAL MODELING

A number of techniques are available to interpolate point values over the space, i.e., kriging, inverse distance weighting, deterministic splines, etc. Among these techniques, geo-statistical modeling or kriging has been the widely used in various fields including mining, geology [50] and hydrology [11, 32, 83, 96] as it is considered as the best linear unbiased estimation

[56]. The geo-statistical modeling is also advantageous over other inter-polation methods due to its more flexibility than that of the other methods [56]. Furthermore, accuracy of prediction is relatively high as the weights are not simply decided by some arbitrary rule that may be true sometimes and false at other times, rather the weights are selected depending upon the variability of the function in space. The appropriate weights are deter-mined by utilizing the prior experience and analyzing the data following a systematic and objective-based approach in order to derive a variogram. All the weights may be equal or with high variations depending upon the scale of variability. The geo-statistical modeling is further advantageous as it enables evaluating the magnitude of the estimation error [56] such as root mean square error to determine the reliability of the estimate depending upon the variogram and the location of the measurements. A brief descrip-tion of the theory is presented in subsequent sub-section while details can be accessed in many textbooks [31, 45, 56].

4.3.2.1   Theoretical Description

In geo-statistical modeling of PC loadings, a random function $Z(x)$ is considered to represent PC loadings known at n points/locations in space $Z(x_i)$, i = 1, 2, ... n and the PC loadings of the function Z is estimated at the point x_0, where value of the PC loadings is not known, the value of function at unknown point is defined as [50, 56]:

$$Z^*(x_0) = \sum_{i=1}^{n} \lambda_i Z(x_i)$$ (2)

where, $Z^*(x_0)$ = estimated value of function $Z(x)$ at point x_0 and λ_i = weighting factors that minimize the variance of the estimation error (ordi-nary kriging weights).

The value of the function Z estimated by Eq. (2) should strictly fulfill two important conditions of unbiased and optimality. The unbiased condi-tion is fulfilled when the expected value of estimation error defined by the mean difference between the estimated $z^*(x_0)$ and the true (unknown) $z(x_0)$ value of the PC loadings becomes zero. On the other side, the condition of optimality is meant with the minimum variance of the estimation error. The spatial structure can be defined by a theoretical variogram, which is expressed by a kriging system of linear equations combining neighboring information. It can be defined as:

$$\sum_{j=1}^{n} \lambda_j \, C\!\left(x_i, x_j\right) - \mu = C\!\left(x_i, x_0\right) \quad i = 1, 2, \ldots, n \tag{3}$$

The Eq. (3) is subjected to the constraint on weights:

$$\sum_{j=1}^{n} \lambda_j = 1 \tag{4}$$

where, μ is the Lagrangian multiplier and $C(x_i, x_j)$ is the value of covariance between two points x_i and x_j.

In case of an intrinsic case i.e., working with variogram, the kriging Eqs. (3) and (4) may be rearranged and modified as follows [4, 76]:

$$C\!\left(x_i, x_j\right) = C(0) - \gamma\!\left(x_i, x_j\right) \tag{5}$$

$$C\!\left(x_i, x_0\right) = C(0) - \gamma\!\left(x_i, x_0\right) \tag{6}$$

The Eqs. (5) and (6) hold good only when both the covariance and the variogram exist, i.e., variables are stationary.

4.3.2.2   Step-by-Step Procedure

A flowchart showing step-by-step procedure of integrated application of PCA, GIS and geo-statistical modeling techniques to know spatial distribution of principal component loadings is illustrated in Figure 4.1. Four important steps of the procedure are explained in the following sections.

Step I: Computing Experimental Semi-variogram Model
The first step in geo-statistical modeling is computing experimental semi-variogram model using the observed dataset. The experimental geo-statistical or semi-variogram model is defined as the function of separation vector between two points i and j. The values of separation vectors, e.g., h_1, h_2, etc. are decided as:

$$h = \left| x_i - x_j \right| \tag{7}$$

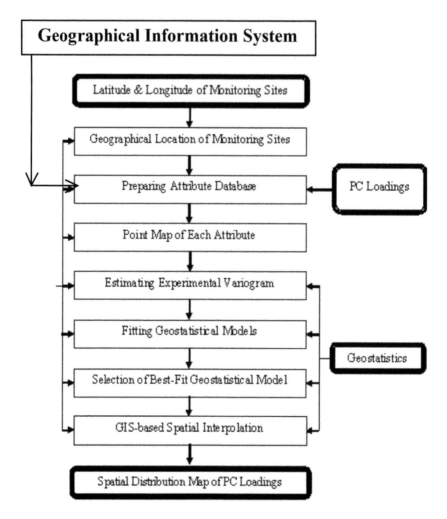

FIGURE 4.1 Flowchart showing step-by-step procedure for integrated application of principal component analysis and GIS-based geo-statistical modeling.

The observed data are grouped into different pairs depending upon the value of h, and some function is averaged to obtain a semi-variogram (γ_{ij}) as defined below [31]:

$$\gamma(h) = \frac{1}{2N_h} \sum_{i=1}^{N_h} \{z(x_i) - z(x_i + h)\}^2 \tag{8}$$

where, N_h = number of pairs for a given lag distance h.

The theoretical geo-statistical model is defined by two parameters, i.e., sill and range (Figure 4.2). The sill parameter is the constant value on the y-axis of the experimental semi-variogram model around which a semi-variogram stabilizes after a large distance. Whereas, the range parameter is the value at x-axis of the semi-variogram model at which the semi-variogram becomes constant or nearly constant. The sill parameter remains very close to the variance of the PC loadings [4, 77].

There is another term, called as nugget that is explained by the sudden and apparent jump near the origin of semi-variogram model that occurs in some cases. The shape of the semi-variogram model between origin and the point of stabilization varies for different variables, and entirely depends on the nature of variability [77]. For understanding the spatial structure of the PC loadings, experimental PC loadings are classified into different groups of lag distances with approximately the similar number of data, and then semi-variogram values are calculated for each class (denoted by individual points shown in Figure 4.2) using geo-statistical modeling technique.

Step II: Fitting of Theoretical Semi-Variogram Models

The experimental semi-variogram model usually in the form of an erratic curve [56, 57] is calculated with the help of Eq. (8) using the observed PC loadings of the PCA technique. This experimental semi-variogram curve is inconsistent, and thus, it is not used for the estimation purpose. Therefore, the second step of the geo-statistical modeling is executed where the curve of the experimental variogram is approximated by another theoretical curve having some specific properties and mathematical expression. The fitted smooth and well-structured curve is called as theoretical semi-variogram model. In the field of hydrology and hydrogeology, the widely-used theoretical semi-variogram models are spherical, circular, Gaussian, and exponential models [45, 56]. These four semi-variogram models are expressed below:

(i) Spherical Model

$$\gamma(h) = C_0 + C\left(\frac{3h}{2a} - \frac{h^3}{2a^3}\right), \text{ for } 0 < h \leq a \tag{9}$$

$$\gamma(h) = C_0 + C, \text{ for } h > a \tag{10}$$

(ii) Circular Model

$$\gamma(h) = C_0 + C \left[1 - \frac{2}{\pi} arc\,cos(h/a) + \frac{2h}{\pi a} \sqrt{1 - (h/a)^2} \right], \text{ for } 0 < h \le a \tag{11}$$

$$\gamma(h) = C_0 + C, \text{ for } h > a \tag{12}$$

(iii) Gaussian Model

$$\gamma(h) = C_0 + C \left[1 - e^{-(h/a)^2} \right] \tag{13}$$

(iv) Exponential Model

$$\gamma(h) = C_0 + C \left(1 - e^{-h/a} \right) \tag{14}$$

where, $C_0 + C$ is the sill, a is the range, and h is the separation vector or lag distance. An example of fitting of theoretical semi-variogram model (line curve) to the experimental semi-variogram model (point data) is illustrated in Figure 4.2.

Although this kind of fitting/modeling can be accomplished through multiple ways but the visual matching is mostly adopted. In this process several trials are made by modifying the model parameters, i.e., sill, range and nugget, and minimizing the error of matching between the two semi-variogram models. Few automatic fitting and modeling procedures are also available but their efficacy is not excellent and success rate is low.

Step III: Selection of the Best-Fit Geostatistical Model
The next step is selection of the best-fit geo-statistical model by using few goodness-of-fit criteria, e.g., mean error (ME), root mean squared error (RMSE), correlation coefficient (R), mean standard error (MSE), mean reduced error (MRE), reduced variance ($S_{R_e}^2$), and coefficient of determination (R^2).

The goodness-of-fit criteria based on correlation measures such as R and R^2 are considered to be over-sensitive to extreme values and insensitive to additive and proportional differences between observations

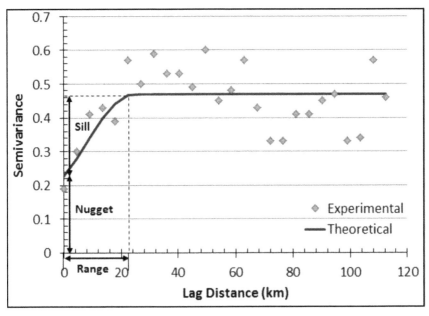

FIGURE 4.2   Theoretical semi-variogram model (shown by line curve) fitted to experimental semi-variogram model (shown by point data).

and regression-based predictions [82]. Therefore, two additional criteria namely, modified Nash-Sutcliffe efficiency, MNSE [64], and modified index of agreement, MIA [121] may be employed to ensure better evaluation as reported by Machiwal and Jha [71, 72].

Step IV: Spatial Interpolation Using the Best Fit Model
The last step is to generate spatially-distributed surface map of the variable by interpolating the values at unknown points using the best-fit geo-statistical model selected in the previous step. In general, entire geo-statistical modeling technique is employed in GIS platform on raster-based maps where all computations are performed on pixel-by-pixel basis.

4.4   EXAMPLE FOR THE APPROACH: A CASE STUDY

This section demonstrates the techniques of principal component analysis and GIS-based geo-statistical modeling through a case study for identifying sources of the ground water contamination in a hard-rock aquifer system of western India.

4.4.1 STUDY AREA

The study area (Udaipur district), extending over 12698 km², is situated in hard-rock terrain of Aravalli hills in western India (Figure 4.3). Geographical coordinates of the area range from 23°45′ to 25°10′ N latitude and from 73°0′ to 74°35′ E longitude. The study area receives an average annual rainfall of 675 mm and has tropical and semi-arid climate. Surface topography has large undulations with ground elevation ranging from 160–1350 m from mean sea level. All rivers have seasonal flows and main rivers being Ahar, Berach, Gomati, Jakham, Mahi, Sabarmati, Sei, Som and Wakal. The Ahar and Berach rivers drain the most urbanized and densely-populated northern and northeast portions of the area. Geology of the hard-rock aquifer system consists of phyllite-schist, gneiss, schist, granite and quartzite. The area has major land use/land cover in the form of cultivable land, forest, pasture, waste land, water bodies and built-up

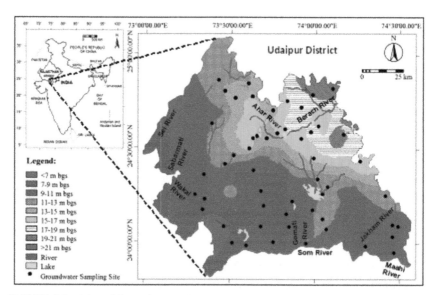

FIGURE 4.3   Map of the study area showing location of ground water sampling sites and depth to ground water levels (modified from Machiwal and Jha [74]).

land. The south and southeast portions of the area have a good network of canals, which provide surface irrigation to command areas.

4.4.2   DATA AND METHODOLOGY

Ground water quality data for a period of 11 years (1992–2002) of 53 sampling sites were collected from the Central Ground Water Board, Rajasthan, India. A total of 15 water quality parameters, i.e., calcium, magnesium, sodium, potassium, iron, sulfate, chloride, carbonate, bicarbonate, nitrate, silica, pH, EC, total dissolved solids, and hardness were considered in this study. The data were collected during the pre-monsoon season during March-May months.

In the study, mean annual values of the ground water quality parameters were computed and used for the analysis, and hence the results will be representative of mean over the 11-year period (1992–2002). Furthermore, the values of ground water quality parameters were standardized before employing the PCA. The PCA was applied using STATISTICA software. Number of significant PCs was found out through the scree plot using the Kaiser [51] criterion that the eigenvalue should be more than one. In order to understand the dominant factors influencing the ground water quality of the aquifer system in the area, unit circle plots of different pairs of the significant PCs were plotted. Coordinates or loadings of the significant principal components, explaining the highest variation in the system, for 53 sites were subjected to the geo-statistical modeling. Experimental semi-variogram of the observed PC loadings was prepared and fitted to four geo-statistical models (i.e. spherical, circular, Gaussian, and exponential). The best-fit model was selected using five goodness-of-fit criteria discussed in earlier section. Finally, the selected best-fit geo-statistical model was utilized to determine spatial distribution of the PC loadings in the study area.

4.4.3   RESULTS AND DISCUSSION

4.4.3.1   Factors Influencing Ground water Quality

The scree plot with eigenvalues of different PCs shown in Figure 4.4 reveals that the eigenvalue of first three PCs is more than 1, which explains

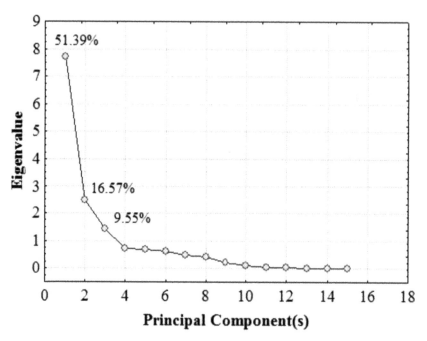

FIGURE 4.4   Scree plot showing eigenvalues for different principal components.

51.39, 16.57 and 9.55% of total variance, respectively of the system. Thus, these three principal components, cumulatively describing more than 75% of the variance, are considered to adequately understand the major factors/ components influencing the ground water quality in the area.

The values of PC loading of these three PCs are presented in Table 4.3. The PC loadings were classified into three groups of strong, moderate and weak with corresponding to absolute loadings of more than 0.75, 0.75–0.50 and 0.50–0.30, respectively following the criteria proposed by Liu et al. [68]. Following the criteria, it is seen that the first PC contains significant negative PC loadings of calcium, magnesium, sodium, sulfate, chloride, total dissolved solids (TDS), hardness and electrical conductivity (EC), along with moderate loadings of nitrate and potassium, and weak loadings of bicarbonate and silica (Table 4.3). Likewise, the second PC has the strong loadings of bicarbonate, fluoride and pH along with weak loadings of calcium, sodium and iron (Table 4.3). In the similar manner, the third PC is not having any parameter with strong loadings; however,

TABLE 4.3   Principal Component Loadings of the Ground water Quality Parameters

Water quality parameter	Principal component loadings of the parameters		
	PC 1	**PC 2**	**PC 3**
Bicarbonate	–0.334#	**–0.755**	0.236
Calcium	**–0.822**	0.430#	0.134
Chloride	**–0.886**	0.072	–0.361#
Fluoride	–0.296	**–0.769**	–0.045
Iron	0.075	0.488#	–0.561*
Magnesium	**–0.878**	–0.098	0.107
Nitrate	–0.716*	0.273	0.338#
Potassium	–0.615*	–0.141	0.288
Silica	–0.326#	0.287	0.641*
Sodium	**–0.786**	–0.307#	–0.424#
Sulphate	**–0.942**	0.012	–0.008
Total dissolved solids	**–0.953**	–0.076	–0.221
Electrical Conductivity	**–0.961**	–0.080	–0.202
Hardness	**–0.931**	0.172	0.146
pH	0.267	**–0.757**	0.037

Note: Bold face values indicate PC loadings; * indicates moderate PC loadings; # indicates weak PC loadings.

iron and silica has moderate loadings, and sodium, chloride and nitrate have weak loadings on the third PC.

Furthermore, unit circle plots of PC loadings were drawn for different pairs of the three significant PCs in order to explore distinct patterns/processes dominating the ground water quality, as shown in Figures 4.5(a–c). Herein, three groups of the water quality parameters are delineated depending upon their closeness on the unit circle plots and their similar type of association (strong, moderate or weak) with two given principal components. The first group of the parameters include major ions, i.e., calcium, magnesium, sodium, potassium, chloride, sulfate and nitrate along with EC, TDS and hardness. The strong loadings of sodium, magnesium and potassium indicate ion exchange processes and natural weathering of rock minerals [21, 59, 110]. In addition, strong loadings of EC, chloride, and sulfate may be attributed to contribution from precipitation and deposition from dust material [59].

    The dominance of nitrate in ground water is apparently the effect of anthropogenic contamination, e.g., use of fertilizers and disposal of sewage wastes. In the second group, three parameters, i.e., pH, bicarbonate and fluoride are included. The pH and bicarbonate in the ground water are contributed from the reaction of soil carbon dioxide along with dissolution of silicate minerals [86, 110]. The presence of fluoride in ground water may be due to leaching of fluoride containing minerals, ion exchange of fluoride and hydroxyl ion in the clay minerals, high evapotranspiration, longer residence of water in the aquifer, and heavy use of fertilizers. The factor governing the ground water quality through elements of the PC 1

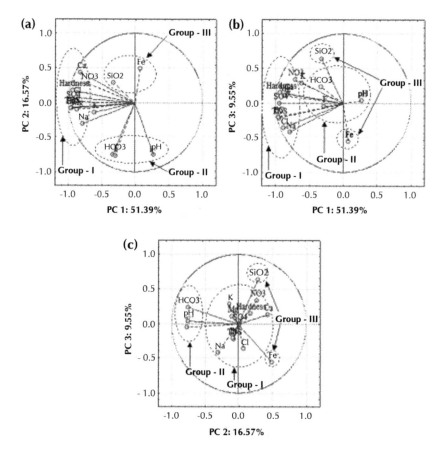

FIGURE 4.5 Unit circle plots drawn between different pairs of three principal components: (a) PC1 and PC2, (b) PC1 and PC3, and (c) PC2 and PC3.

may be characterized as 'major ion contamination, ' and the factor responsible for ground water quality due to PC 2 may be termed as 'soil leaching contamination.' The third group contains only two parameters, i.e., iron and silica.

Moreover, it is seen that 8 of total 10 parameters of group I have strong association (PC loadings >0.75) with the first PC (Figure 4.5a, b), and therefore, this group or factor has strong influence on the ground water quality or geochemical processes of the aquifer system. Likewise, three elements of the second group or factor have strong dominance on the second PC (Figure 4.5a, c). However, iron and silica have moderate loadings or association with third PC (Figure 4.5b, c).

4.4.3.2   Spatial Distribution of Principal Component Loadings

Prior to employing geo-statistical modeling to develop spatial distribution maps of loadings of three significant PCs, spatial trends of the PC loadings were checked by performing regression analysis between PC loadings and geographical coordinates of the 53 sampling sites. The results of goodness of fit criteria revealed absence of any significant spatial trends, and therefore, the PC loadings have stationarity. The model parameters, i.e., nugget, sill and range of four geo-statistical models were visually adjusted by trial and error method. The final values of the model parameters for four geo-statistical models are given in Table 4.4. It is seen that nugget and sill values for the third PC is relatively less than that for the first and second PC. However, the range value of the second and third PC is higher than that for the first PC.

Experimental semi-variogram of the observed values of PC loadings for three significant PCs were drawn and fitted with four models by adjusting final values of the model parameters (Figure 4.6a–c) to select the best-fit model using goodness-of-fit criteria (Table 4.5). It is obvious from Table 4.5 that the values of goodness-of-fit criteria favor the selection of three geo-statistical models, i.e., spherical, circular and exponential, for spatial interpolation of PC loadings of three significant PCs. However, the exponential model was selected for PCs 1 and 2, and the spherical model was selected for PC 3 based on lowest RMSE value and highest values of R, R^2, MNSE and MIA.

TABLE 4.4   Value of Parameters of Four Geo-Statistical Models for Three Principal Components

Principal component	Model parameter	Spherical	Circular	Exponential	Gaussian
I	Nugget (m)	2	2.5	2	3.5
	Sill (m)	9	9	10.5	9
	Range (km)	50	50	35	30
II	Nugget (m)	1	1	1	1.2
	Sill (m)	2.7	2.7	3	2.8
	Range (km)	75	65	45	40
III	Nugget (m)	0.4	0.4	0.5	0.4
	Sill (m)	1.4	1.4	1.7	1.4
	Range (km)	70	60	50	30

TABLE 4.5   Value of Goodness-of-Fit Criteria for Four Geo-Statistical Models Fitted to the Loadings of Three Significant Principal Components

Principal Component	Criteria	Spherical	Circular	Exponential	Gaussian
1	RMSE (m)	1.03	1.25	**1.00**	1.77
	R	0.95	0.92	**0.96**	0.79
	R^2	0.9	0.85	**0.91**	0.62
	MNSE	0.66	0.59	**0.67**	0.41
	MIA	0.81	0.76	**0.82**	0.65
2	RMSE (m)	0.95	0.96	**0.90**	1.26
	R	0.86	0.86	**0.89**	0.63
	R^2	0.74	0.74	**0.8**	0.39
	MNSE	0.38	0.38	**0.42**	0.18
	MIA	0.61	0.60	**0.63**	0.46
3	RMSE (m)	**0.56**	**0.56**	0.59	0.70
	R	**0.91**	0.90	0.90	0.82
	R^2	**0.82**	**0.82**	0.8	0.67
	MNSE	**0.51**	0.50	0.48	0.35
	MIA	**0.72**	0.71	0.70	0.63

RMSE is the root mean square error; R = correlation coefficient; R^2 = coefficient of determination; MNSE = modified Nash-Sutcliffe efficiency; MIA = modified index of agreement. Figures in bold face indicate the best-fit model.

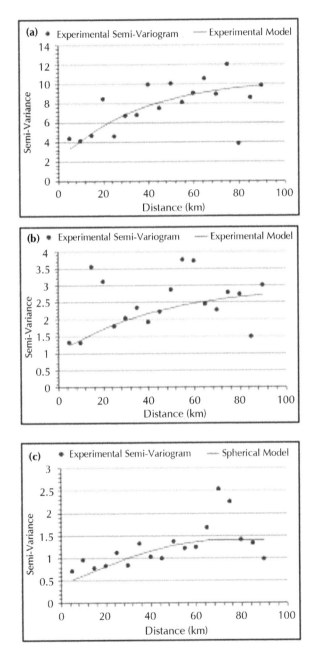

FIGURE 4.6 Fitting of the best-fit geo-statistical models to experimental semi-variogram models of loadings for (a) first, (b) second and (c) third principal components.

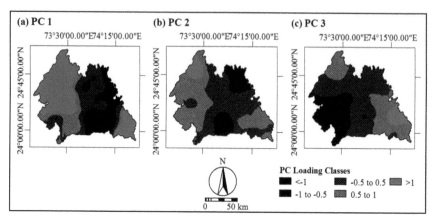

FIGURE 4.7   Spatial distribution of loadings of PCs 1–3 generated using the Kriging technique.

Spatial distribution of the PC loadings was computed using the selected best-fit geo-statistical models. The spatially-distributed PC loadings for three significant PCs are shown in Figure 4.7(a–c). It is seen that the PC loading values are highly negative (< −1) for PC 1 in the northeast and southern portions of the area. The northeast portion is densely-populated with existence of large urban areas. On the other side, the southern portion contains large irrigation command areas. Thus, the type of land use in both the portions clearly evidenced the chances of anthropogenic contamination due to unsewered contamination from the urban built-up lands and excessive fertilizer application in the irrigation commands. Similar to PC1, the loadings of PC2 depicted negative values in the northeast and southern portions of the area. Therefore, it appears that the PCs 1 and 2 are indicative of the anthropogenic sources of the ground water contamination, which is mainly governed by the dominance of the major ions along with EC, TDS, hardness, nitrate, fluoride, pH and bicarbonate.

The spatial distribution of the PC loading values is somewhat different for PC3 compared to that for PCs 1 and 2 (Figure 4.7c). The PC loading values are negative for PC3 in the southwest portion of the area. The PC3 or third group of the ground water quality parameters contains silica, which is reported to be significantly decreasing at most sites in the area [74]. This reduction in silica content can be well-understood by relating it with the declining ground water level trends [73]. Actually, the silica content in the ground water is directly proportional to residence time of water below the ground and relatively low amount of silica indicates less interaction of the water with rocks [40, 55]. Thus, it seems that large amount of the ground water in hard-rock aquifer system of the area has been replaced with recent freshwater recharge over a large period of time in response to rainfall, which has relatively low silica content [55]. It is observed from Figure 4.7(a–c) that the area having positive PC loading values (ranging from 0.5 to >1) represents naturally-occurring geochemical processes. In these areas, no clear evidences of anthropogenic contamination are seen as the built-up lands and agricultural lands, having potential to influence ground water quality by human activity, do not exist. In this portion, natural contamination occurs due to geogenic factors active in the aquifer system.

Finally, the results of the spatial distribution of PC loadings are found in good agreement with results of the ground water quality index (GWQI) reported by Machiwal et al. [75]. A comparison of PC loading maps of three significant PCs with that of the GWQI revealed that areas having negative PC loadings (ranging from –0.5 to <–1) matched with the areas having low ground water quality (GWQI ranging from 0–10%). Similarly, the portions with positive values of the PC loadings matched with high ground water quality (GWQI ranging from 80–100%). These interpretations and comparisons confirmed that recently employed advance approach of integrating the results of multivariate statistical techniques with GIS and geo-statistical modeling has the great potential to identify anthropogenic and geogenic factors causing ground water contamination.

4.5 SUMMARY

Ground water quality in an area is greatly controlled by natural processes (e.g., geology, ground water movement, recharge water quality, and soil-rock interactions with water), anthropogenic factors (e.g., agricultural activities, increasing industrialization and urbanization, mining activities, and exploitation of water resources) and atmospheric inputs. Several conventional tools and techniques are available to interpret water quality of an aquifer system. However, the natural and anthropogenic factors may not be easily distinguished from chemical composition of ground water alone. Hence, the need for integrated application of modern approaches and tools such as multivariate statistical techniques and GIS-based geo-statistical modeling techniques has been emphasized for the efficient management of ground water quality. Multivariate statistical techniques offer a valuable tool for the evaluation of spatio-temporal variations and interpretation of complex water quality datasets, apportionment of pollution sources/factors (natural or anthropogenic) and design of a monitoring network for the effective and reliable management of water resources as well as for rapid solution to pollution problems.

This chapter aims at providing an overview of conventional and modern techniques used for the assessment of ground water quality and identification of natural and anthropogenic sources of ground water contamination. Furthermore, the chapter includes a case study in a semi-arid region of India where the proposed approach is demonstrated by applying a methodology for combining factor scores of the PCA with GIS-based geo-statistical modeling precisely. The spatial distribution of the PC loadings revealed that negative PC loadings indicate sources of anthropogenic contamination. On the contrary, positive PC loadings suggested natural or geogenic geochemical processes influencing the ground water quality of the hard-rock aquifer systems. The whole methodology is illustrated in a comprehensive manner to identify natural and anthropogenic factors governing ground water quality.

KEYWORDS

- communality
- eigen values
- eigen vector
- geo-statistical modeling
- GIS
- goodness-of-fit criteria
- ground water contamination

- hard rock aquifers
- kaiser normalization criterion
- kriging
- major ion contamination
- multivariate statistical techniques
- natural and anthropogenic contamination
- nugget
- optimality condition
- principal component analysis
- principal component loading
- range
- scree plot
- semi-variogram
- sill
- soil leaching contamination
- spatial interpolation
- spatial trends
- spatially-distributed
- standardization
- stationarity
- unbiased condition
- unit circle plots
- variance
- water quality

REFERENCES

1. Abassi, S. A. (1999). Water quality indices: State-of-the art. *Journal of the Institutions of Public Health Engineers*, 1, 13–24.
2. Adak, M. D. G., Purohit, K. M., & Datta, J. (2001). Assessment of drinking water quality of river Brahmani. *Indian Journal of Environmental Protection*, 8, 285–291.

3. Aguilera, P. A., Frenich, A. G., & Torres, J. A. (2001). Application of the Kohonen neural network in coastal water management: Methodological development for the assessment and prediction of water quality. *Water Research*, 35, 4053–4062.

4. Ahmed, S. (2006). Application of Geostatistics in Hydrosciences. In: *M. Thangarajan (editor), Ground water Resource Evaluation, Augmentation, Contamination, Restoration, Modeling and Management.* Capital Publishing Company, New Delhi, India, 78–111.

5. Appelo, C. A., & Postma, D. (1996). *Geochemistry, Ground water and Pollution.* Rotterdam, AA Balkema, 536 p.

6. Babiker, I. S., Mohamed, M. M. A., & Hiyama, T. (2007). Assessing ground water quality using GIS. *Water Resources Management*, 21, 699–715.

7. Bartram, J., & Ballance, R. (1996). *Water Quality Monitoring: A Practical Guide to the Design and Implementation of Freshwater Quality Studies and Monitoring Programs.* The United Nations Environment Program (UNEP) and the World Health Organization (WHO), 348 p.

8. Brivio, P. A., Giardiano, C., & Zilioli, E. (2001). Determination of concentration changes in Lake Garda using image-based radioactive transfer code for Landsat TM images. *International Journal of Remote Sensing*, 22, 487–502.

9. Burrough, P. A., & McDonnell, R. A. (1998). *Principles of Geographical Information Systems.* Oxford University Press, Oxford, U.K., 333 p.

10. Carr, J. R. (1990). CORSPOND: a portable FORTRAN-77 program for correspondence analysis. *Computers and Geosciences*, 16, 289–307.

11. Castrignanò, A., Giugliarini, L., Risaliti, R., & Martinelli, N. (2000). Study of spatial relationships among some soil physicochemical properties of a field in central Italy using multivariate geostatistics. *Geoderma*, 97, 39–60.

12. Chan, H. J. (2001). Effect of land use and urbanization on hydrochemistry and contamination of ground water from Taejon area, Korea. *Journal of Hydrology*, 253, 194–210.

13. Chaves, P., & Kojiri, T. (2007). Conceptual fuzzy neural network model for water quality simulation. *Hydrological Processes*, 21, 634–646.

14. Choubey, V. K. (1994). Monitoring water quality in reservoirs with IRS-1A-LISS-I. *Water Resources Management*, 8, 121–136.

15. Cloutier, V., Lefebvre, R., Therrien, R., & Savard, M. M. (2008). Multivariate statistical analysis of geochemical data as indicative of the hydrogeochemical evolution of ground water in a sedimentary rock aquifer system. *Journal of Hydrology*, 353, 294–313.

16. Collins, W. D. (1923). Graphic Representation of Water Analyses. *Industrial and Engineering Chemistry*, 15, 394.

17. Davis, J. C. (1986). *Statistics and Data Analysis in Geology.* Wiley, Toronto, 646 pp.

18. Dekker, A. G., Zamurovic-Nenad, Z., Hoogenboom, H. J., & Peters, S. W. M. (1996). Remote sensing, ecological water quality modeling and in-situ measurements: A case study in shallow lakes. *Hydrological Sciences Journal*, 41, 531–547.

19. Demirel, Z., & Güler, C. (2006). Hydrogeochemical evolution of ground water in a Mediterranean coastal aquifer, Mersin-Erdemli basin (Turkey). *Environmental Geology*, 49, 477–487.

20. Dillon, R., & Goldstein, M. (1984). *Multivariate Analyses: Methods and Applications*. Wiley, New York.

21. Drever, I. J. (1997). *The Geochemistry of Natural Waters*. 3rd edition, Prentice Hall, Inc., Englewood Cliffs, New Jersey.

22. Dunteman, G. H. (1989). *Principal Component Analysis*. Sage, Thousand Oaks, CA.

23. Edmunds, W. M., Shand, P., Hart, P., & Ward, R. S. (2003). The natural (baseline) quality of ground water: a UK pilot study. *Science of the Total Environment*, 310(1–3), 25–35.

24. Fetter, C. W. (1994). *Applied Hydrogeology*. 4th edition, Prentice Hall, NJ, 592p.

25. Freeze, R. A., & Cherry, J. A. (1979). *Ground water*. Prentice-Hall, Inc., Englewood Cliffs, New Jersey, pp. 237–302, 383–462.

26. GESAMP (1988). *Report of the Eighteenth Session*, Paris 11–15 April 1988. GESAMP Reports and Studies No. 33, United Nations Educational, Scientific and Cultural Organization, Paris.

27. Gibbs, R. J. (1970). Mechanisms controlling world water chemistry. *Science*, 170, 1088–1090.

28. Giggenbach, W. F. (1988). Geothermal solute equilibria. Derivation of Na-K-Mg-Ca geoindicators. *Geochimica et Cosmochimica Acta*, 52, 2749–2765.

29. Gobindraju, R. S., & Ramachandra Rao, A. (2000). *Artificial Neural Network in Hydrology*. Kluwer, Dordrecht, 329 p.

30. Goodchild, M. F., Parks, B. O., & Steyaert, L. T. (editors) (1993). *Environmental Modeling with GIS*. Oxford University Press, New York.

31. Goovaerts, P. (1997). *Geostatistics for Natural Resources Evaluation*. Oxford University Press, New York.

32. Goovaerts, P. (1999). Geostatistics in soil science: State-of-the-art and perspectives. *Geoderma*, 89, 1–45.

33. Greenacre, M. J. (1984). *Theory and Application of Correspondence Analysis*. Academic Press, London, 364 p.

34. Güler, C., & Thyne, G. D. (2004). Hydrologic and geologic factors controlling surface and ground water chemistry in Indian Wells-Owens Valley area, southeastern California, USA. *Journal of Hydrology*, 285, 177–198.

35. Güler, C., Thyne, G. D., McCray, J. E., & Turner, A. K. (2002). Evaluation of graphical and multivariate statistical methods for classification of water chemistry data. *Hydrogeology Journal*, 10, 455–474.

36. Gurnell, A. M., & Montgomery, D. R. (eds.) (2000). *Hydrological Applications of GIS*. John Wiley & Sons Ltd., Chichester, U.K., 176 p.

37. Harman, H. H. (1960). *Modern Factor Analysis*. University of Chicago Press, Chicago, USA.

38. Haykin, S. (1999). *Neural Networks: A Comprehensive Foundation*. 2nd Edition, Prentice Hall Inc., Upper Saddle River, New Jersey.

39. Helena, B., Pardo, R., Vega, M., Barrado, E., Fernandez, J. M., & Fernandez, L. (2000). Temporal evolution of ground water composition in an alluvial aquifer (Pisuerga River, Spain) by principal component analysis. *Water Research*, 34, 807–816.

40. Hem, J. D. (1986). *Study and Interpretation of the Chemical Characteristics of Natural Waters, 3rd edition*. U.S. Geological Survey Water-Supply Paper 2254, 263 p.

41. Hong, Y. S., & Rosen, M. R. (2001). Intelligent characterization and diagnosis of the ground water quality in an urban fractured-rock aquifer using an artificial neural network. *Urban Water*, 3, 193–204.

42. Horton, R. K. (1965). An index number system for rating water quality. *Journal of Water Pollution Control Federation*, 37, 300–305.

43. Hotelling, H. (1936). Relations between two sets of variates. *Biometrika*, 28, 312–377.

44. IFRC (2001). *World Disasters Report 2001*. International Federation of Red Cross and Red Crescent Societies, Geneva, Switzerland.

45. Isaaks, E., & Srivastava, R. M. (1989). *An Introduction to Applied Geostatistics.* Oxford University Press, New York.

46. Jackson, J. E. (1991). *A User's Guide to Principal Components.* John Wiley & Sons, New York.

47. Jha, M. K., Chowdhury, A., Chowdary, V. M., & Peiffer, S. (2007). Ground water management and development by integrated remote sensing and geographic information systems: Prospects and constraints. *Water Resources Management*, 21, 427–467.

48. Johnson, L. E. (2009). *Geographic Information Systems in Water Resources Engineering.* CRC Press, Boca Raton, 286 p.

49. Johnson, R. A., & Wichern, D. W. (1992). *Applied Multivariate Statistical Analysis*, 3rd edition. Prentice-Hall International, Englewood Cliffs, New Jersey, USA, 642 p.

50. Journel, A. G., & Huijbregts, C. J. (1978). *Mining Geostatistics.* Academic Press, New York.

51. Kaiser, H. F. (1958). The varimax criterion for analytic rotation in factor analysis. *Psychometrika*, 23, 187–200.

52. Kallio, M. P., Mujunen, S. P., Hatzimihalis, G., Koutoufides, P., Minkkinen, P., Wilki, P. J., & Connor, M. A. (1999). Multivariate data analysis of key pollutants in sewage samples: A case study. *Analytica Chimica Acta*, 393, 181–191.

53. Karanth, K. R. (1987). *Ground Water Assessment: Development and Management.* Tata McGraw-Hill Publishing Company Limited, New Delhi, 720p.

54. Keiner, L. E., & Yan, X. H. (1998). A Neural network model for estimating sea surface chlorophyll and sediments from Thematic Mapper imagery. *Remote Sensing and Environment*, 66, 153–165.

55. Khan, M. M. A., & Umar, R. (2010). Significance of silica analysis in ground water in parts of Central Ganga Plain, Uttar Pradesh, India. *Current Science*, 98(9), 1237–1240.

56. Kitanidis, P. K. (1997). *Introduction to Geostatistics: Applications in Hydrogeology.* Cambridge University Press, New York, 249 p.

57. Kitanidis, P. K. (1999). Geostatistics: Interpolation and inverse problems. In: *J. W. Delleur (Editor-in-Chief), The Handbook of Ground water Engineering*, CRC Press, Florida, pages 12–1 to 12–20.

58. Kolsi, S. H., Bouri, S., Hachicha, W., & Dhia, H. B. (2013). Implementation and evaluation of multivariate analysis for ground water hydrochemistry assessment in arid environments: a case study of HajebElyoun–Jelma, Central Tunisia. *Environmental Earth Sciences*, 70, 2215–2224.

59. Kumar, M., Ramanathan, A. L., Rao, M. S., & Kumar, B. (2006). Identification and evaluation of hydrogeochemical processes of Delhi, India. *Environmental Geology*, 50, 1025–1039.

60. Kuo, V., Liu, C., & Lin, K. (2004). Evaluation of the ability of an artificial neural network model to assess the variation of ground water quality in an area of blackfoot disease in Taiwan. *Water Research*, 38, 148–158.

61. Kurumbein, W. C., & Graybill, F. A. (1965). *An Introduction to Statistical Models in Geology*. McGraw-Hill, New York.

62. Langelier, W., & Ludwig, H. (1942). Graphical methods for indicating the mineral character of natural waters. *Journal of American Water Association*, 34, 335–352.

63. Lathrop, R. (1992). Landsat Thematic Mapper monitoring of turbid inland water quality. *Photogrammetric Engineering and Remote Sensing*, 58, 465–470.

64. Legates, D. R., & McCabe Jr., G. J. (1999). Evaluating the use of "goodness-of-fit" measures in hydrologic and hydroclimatic model validation. *Water Resources Research*, 35, 233–241.

65. Lepeltier, C. (1969). A simplified statistical treatment of geochemical data by graphical representation. *Economic Geology*, 64, 538–550.

66. Lin, C. Y., Abdullah, M. H., Praveena, S. M., Yahaya, A. H. B., & Musta, B. (2012). Delineation of temporal variability and governing factors influencing the spatial variability of shallow ground water chemistry in a tropical sedimentary island. *Journal of Hydrology*, 432–433, 26–42.

67. Lischeid, G., & Bittersohl, J. (2008). Tracing biogeochemical processes in stream water and ground water using nonlinear statistics. *Journal of Hydrology*, 357, 11–28.

68. Liu, C. W., Lin, K. H., & Kuo, Y. M. (2003). Application of factor analysis in the assessment of ground water quality in a blackfoot disease area in Taiwan. *Science of the Total Environment*, 313, 77–89.

69. Machiwal, D., & Jha, M. K. (2010). Tools and techniques for water quality interpretation. In: *Advances in Water Quality Control*. Krantzberg, G., Tanik, A., Antunes do Carmo, J. S., Indarto, A., & Ekdal, A. (Editors-in-Chief), Scientific Research Publishing, Inc., California, USA, 211–252.

70. Machiwal, D., & Jha, M. K. (2012). *Hydrologic Time Series Analysis: Theory and Practice*. Springer, Germany and Capital Publishing Company, New Delhi, India, 303 p.

71. Machiwal, D., & Jha, M. K. (2014a). GIS-based water balance modeling for estimating regional specific yield and distributed recharge in data-scarce hard-rock regions. *Journal of Hydro-Environment Research*, DOI: 10.1016/j.jher.2014.07.004.

72. Machiwal, D., & Jha, M. K. (2014b). Role of geographical information system for water quality evaluation. In: *Nielson, D. (Editor), Geographic Information Systems (GIS): Techniques, Applications and Technologies, Nova Science Publishers, USA, 217–278.*

73. Machiwal, D., & Jha, M. K. (2014c). Characterizing rainfall-ground water dynamics in a hard-rock aquifer system using time series, geographic information system and geostatistical modeling. *Hydrological Processes*, 28, 2824–2843.

74. Machiwal, D., & Jha, M. K. (2015). Identifying sources of ground water contamination in a hard-rock aquifer system using multivariate statistical analyzes and GIS-

based geostatistical modeling techniques. *Journal of Hydrology: Regional Studies*, doi: 10.1016/j.ejrh.2014.11.005.

75. Machiwal, D., Jha, M. K., & Mal, B. C. (2011). GIS-based assessment and characterization of ground water quality in a hard-rock hilly terrain of western India. *Environmental Monitoring and Assessment*, 174, 645–663.

76. Marsily, G. de (1986). *Quantitative Hydrogeology: Ground water Hydrology for Engineers*. Academic Press, CA, 440 p.

77. Matheron, G. (1965). *Les Variables Regionalisèes et leur Estimation*. Masson, Paris, France.

78. McCutcheon, S. C., Martin, J. L., & Barnwell, T. O. (1993). Water quality. In: *Handbook of Hydrology*. D. R. Maidment (editor-in-chief), McGraw-Hill, Inc., New York, pages 11.1–11.73.

79. Melloul, A. J., & Collin, M. (1998). A proposed index for aquifer water quality assessment: The case of Israel's Sharon region. *Journal of Environmental Management*, 54, 131–142.

80. Meybeck, M., & Helmer, R. (1992). An Introduction to Water Quality. In: *Water Quality Assessments: A Guide to Use of Biota, Sediments and Water in Environmental Monitoring*, 2nd Edition, D. Chapman (editor), the United Nations Educational, Scientific and Cultural Organization (UNESCO), the World Health Organization (WHO), and the United Nations Environment Program (UNEP), E&FN Spon, U. K., 609 p.

81. Meybeck, M., Kimstach, V., & Helmer, R. (1992). Strategies for Water Quality Assessment. In: *D. Chapman (editor), Water Quality Assessments: A Guide to Use of Biota, Sediments and Water in Environmental Monitoring*, 2nd Edition, the United Nations Educational, Scientific and Cultural Organization (UNESCO), the World Health Organization (WHO), and the United Nations Environment Program (UNEP), E&FN Spon, U.K., 609 p.

82. Moore, D. S. (1991). *Statistics: Concepts and Controversies*. 3rd edition, W. H. Freeman, New York, 439 p.

83. Mouser, P. J., Hession, W. C., Rizzo, D. M., & Gotelli, N. J. (2005). Hydrology and Geostatistics of a Vermont, USA Kettlehole Peatland. *Journal of Hydrology*, 301, 250–266.

84. Oleson, S. G., & Carr, J. R. (1990). Correspondence analysis of water quality data: Implications for fauna deaths at Stillwater Lakes, Nevada. *Mathematical Geology*, 22, 665–698.

85. Otto, M. (1998). Multivariate methods. In: *Analytical Chemistry*, R. Kellner, J. M. Mermet, M. Otto, and Widmer H. M. (Eds.), Wiley-VCH, Weinheim, Germany, 916 p.

86. Ozler, H. M. (2003). Hydrochemistry and salt-water intrusion in the Van aquifer, east Turkey. *Environmental Geology*, 43, 759–775.

87. Piper, A. M. (1944). A graphical procedure in the geochemical interpretation of water analysis. *American Geophysical Union Transactions*, 25, 914–928.

88. Plummer, L. N. (1984). *Geochemical Modeling: A Comparison of Forward and Inverse Methods*. National Water Well Association, Banff, Alberta, Canada.

89. Pradhan, S. K., Patnaik, D., & Rout, S. P. (2001). Water quality index for the ground water in and around a phosphatic fertilizer plant. *Indian Journal of Environmental Protection*, 21, 355–358.

90. Raman, H., & Chandramouli, V. (1996). Deriving a general operating policy for reservoirs using neural networks. *Journal of Water Resources Planning and Management*, ASCE, 122, 342–347.

91. Ramesh, S., Sukumaran, N., Murugesan, A. G., & Rajan, M. P. (2010). An innovative approach of Drinking Water Quality Index – A case study from Southern Tamil Nadu, India. *Ecological Indicators*, 10, 857–868.

92. Ritchie, J. C., Charles, M. C., & Schiebe, F. R. (1990). The relationship of MSS and TM digital data with suspended sediments, chlorophyll, and temperature in Moon Lake, Mississippi. *Remote Sensing of the Environment*, 33, 137–148.

93. Rogers, L. L., & Dowla, F. U. (1994). Optimization of ground water remediation using artificial neural networks with parallel solute transport modeling. *Water Resources Research*, 30, 457–481.

94. Sánchez-Martos, F., Jiménez-Espinosa, R., & Pulido-Bosch, A. (2001). Mapping ground water quality variables using PCA and geostatistics: A case study of Bajo Andarax, southeastern Spain. *Hydrological Sciences Journal*, 46, 227–242.

95. Sara, M. N., & Gibbons, R. (1991). Organization and Analysis of Water Quality Data. In: *Practical Handbook of Ground-Water Monitoring*, D. M. Nielsen (Ed.), Lewis Publishers, Michigan, USA, 541–588.

96. Schaefer, J. A., & Mayor, S. J. (2007). Geostatistics reveal the scale of habitat selection. *Ecological Modeling*, 209, 401–406.

97. Schoeller, H. (1955). Geochimie des eaux souterraines. *Rev. de l'Inst. Francais du Petrole*, Paris, 10, 181–213.

98. Schoeller, H. (1962). *Les Eaux souterraines*. Mason et Cie, Paris.

99. Schultz, M. T. (2001). A Critique of EPA's Index of Watershed Indicators. *Journal of Environmental Management*, 62, 429–442.

100. Selle, B., Schwientek, M., & Lischeid, G. (2013).Understanding processes governing water quality in catchments using principal component scores. *Journal of Hydrology*, 486, 31–38.

101. Sharma, V., Negi, S. C., Rudra, R. P., & Yang, S. (2003). Neural networks for predicting nitrate-nitrogen in drainage water. *Agricultural Water Management*, 63, 169–183.

102. Sinclair, A. J. (1974). Selection of thresholds in geochemical data using probability graphs. *Journal of Geochemical Exploration*, 3, 129–149.

103. Sinclair, A. J. (1976). *Application of Probability Graphs in Mineral Exploration*. Association of Exploration Geochemists, Rexdale, Ontario, 95 p.

104. Singh, K. P., Malik, A., Mohan, D., & Sinha, S. (2004). Multivariate statistical techniques for the evaluation of spatial and temporal variations in water quality of Gomtiriver (India): A case study. *Water Research*, 38, 3980–3992.

105. Soltan, M. E. (1999). Evaluation of ground water quality in Dakhla Oasis (Egyptian Western Desert). *Environmental Monitoring and Assessment*, 57, 157–168.

106. Stafford, D. B. (Ed.) (1991). *Civil Engineering Applications of Remote Sensing and Geographic Information Systems*. ASCE, New York.

107. Steube, C., Richter, S., & Griebler, C. (2009). First attempts towards an integrative concept for the ecological assessment of ground water ecosystems. *Hydrogeology Journal*, 17, 23–35.

108. Stiff, H. A. (Jr.). (1951). The interpretation of chemical water analysis by means of patterns. *Journal of Petroleum Technology*, 3, 15–17.

109. Stigter, T. Y., Ribeiro, L., & Carvalho, Dill, A. M. M. (2006). Application of a ground water quality index as an assessment and communication tool in agro-environmental policies: Two Portuguese case studies. *Journal of Hydrology*, 327, 578–591.

110. Subba Rao, N., John Devadas, D., & Srinivasa Rao, K. V. (2006). Interpretation of ground water quality using principal component analysis from Anantapur district, Andhra Pradesh, India. *Environmental Geosciences*, 13, 239–259.

111. Tennant, C. B., & White, M. L. (1959). Study of the distribution of some geochemical data. *Economic Geology*, 54, 1281–1290.

112. Todd, D. K. (1980). *Ground water Hydrology*. 2nd edition, John Wiley & Sons, NY, pp. 111–163.

113. UNICEF and WHO (2008). *Progress on Drinking Water and Sanitation: Special Focus on Sanitation*. UNICEF, New York and WHO, Geneva, 54 p.

114. USEPA (1998). *Guidance for Data Quality Assessment: Practical Methods for Data Analysis*. Quality Assurance Division, EPA QA/G-9, version QA97, United States Environmental Protection Agency (USEPA), Washington, 3–5.

115. USSL (1954). *Diagnosis and improvement of saline and alkaline soils*. United States Salinity Laboratory (USSL), Handbook 60, United States Department of Agriculture (USDA), Washington.

116. Valdes, D., Dupont, J.-P., Laignel, B., Ogier, S., Leboulanger, T., & Mahler, B. J. (2007). A spatial analysis of structural controls on Karst ground water geochemistry at a regional scale. *Journal of Hydrology*, 340, 244–255.

117. Vega, M., Pardo, R., Barrado, E., Deban, L., 1996. Assessment of seasonal and polluting effects on the quality of river water by exploratory data analysis. *Water Research*, 32, 3581–3592.

118. Wang, X. J., & Ma, T. (2001). Application of remote sensing techniques in monitoring and assessing the water quality of Taihu Lake. *Bulleting of Environmental Contamination and Toxicology*, 67, 863–870.

119. Wen, C. W., & Lee, C. S. (1998). A neural network approach to multiobjective optimization for water quality management in a river basin. *Water Resources Research*, 34, 427–436.

120. Wilcox, L. V. (1955). *Classification and Use of Irrigation Water*. Circular 696, United States Department of Agriculture (USDA), Washington, DC.

121. Willmott, C. J., Ackleson, S. G., Davis, R. E., Feddema, J. J., Klink, K. M., Legates, D. R., O'Donnell, J., & Rowe, C. M. (1985). Statistics for the evaluation and comparison of models. *Journal of Geographical Research*, 90, 8995–9005.

122. Wunderlin, D. A., Diaz, M. P., Ame, M. V., Pesce, S. F., Hued, A. C., & Bistoni, M. A. (2001). Pattern recognition techniques for the evaluation of spatial and temporal variations in water quality. A case study: Suquia river basin (Cordoba-Argentina). *Water Research*, 35, 2881–2894.

123. Zaporozec, A. (1972). Graphical interpretation of water quality data. *Ground Water*, 10, 32–43.

PART II

TOLERANCE TO SOIL SALINITY

CHAPTER 5

REVIEW OF PHYSIOLOGY OF SALT TOLERANCE OF *SALVADORA PERSICA* AND HALOPHYTIC GRASSES IN SALINE VERTISOLS

G. GURURAJA RAO, J. C. DAGAR, SANJAY ARORA, and ANIL R. CHINCHMALATPURE

CONTENTS

5.1 INTRODUCTION

Massive food requirement for the teeming world population can only be met through expansion of irrigation facilities and intensive agricultural.

On the contrary, both these practices practiced non-scientifically have resulted in degrading the lands and bringing to the fore several second-generation problems around the globe. It has been assessed that large chunks of the arable lands around the world have been damaged as a result of water logging, soil salinity and other forms of physical and chemical degradation. As per FAO-AGL [12] over 6% or 831 million-ha (M ha) of world's land and nearly 1/10 of world's irrigated lands have been adversely impacted by water logging and/or soil salinization. In India, soil salinity has been observed to have charted a parallel path with irrigation development. The situation has assumed a serious dimension in arid and semiarid regions of the world where evapotranspiration exceeds precipitation. Preliminary area statistics of degraded and wastelands of India puts the data of waterlogged lands at 1.66 million ha (M ha) due to surface stagnation and 4.75 M ha due to subsurface water logging [28]. Similar estimate for saline/sodic soils reveal that 6.73 M ha area is saline/sodic in nature, 2.96 M ha being saline and 3.77 M ha sodic [34]. Extreme events, climatic aberrations and anthropogenic interventions are likely to further aggravate the aerial extent of these soils. CSSRI [9] projects that the extent of salt affected soils in India may treble to 20.0 M ha by 2050. Besides, the encroachment of the best quality lands by domestic and industrial sectors is pushing agricultural to marginal areas.

Salinity-related land degradation is a serious challenge for food and nutritional security in the world especially in the developing countries. Since the development of water logging and soil salinization is intricately related to hydrologic cycle, soil type, climate, vegetation, parent rock materials and anthropogenic interference, Vertisols are quite prone to the development of these problems. The Vertisols and associated soils cover nearly 257 M ha of the earth's surface out of which about 76.4 M ha occurs in India [36] which indicates that nearly 22% of the total geographical area of the country is occupied by Vertisols. Vertisols, also called as black soils in India because of their color, are base rich soils capable of continuous cropping and have a considerable potential for agricultural production yet special management practices (tillage and water management) are needed to secure sustained production. For example, Vertisols, because of their very hard consistency when dry and very plastic and sticky when wet, have limited workability to very short periods of optimal water status. In India, saline and sodic Vertisols have formed under irrigation and are rare under natural conditions [8]. The management of Vertisols in general and saline Vertisols in particular is of great concern that needs a holistic approach for their management.

Another problem associated with saline environment is the ground water quality. Usually, arid and semiarid regions along with coastal regions are endowed with saline ground water, which poses problems in the cultivation of glycophytes. Therefore, growing salt tolerant plants or economic halophytes is one of the most viable options for management of the degraded habitats and associated problem of saline ground water.

This chapter discusses the response of cultivating Meswak (*Salvadora persica*) and halophytic grasses in highly saline Vertisols of India or by using highly saline ground water; and explores the physiological aspects of their salt tolerance under abiotic stress of salinity.

5.2 MESWAK (*SALVADORA PERSICA*)

Salt affected black soils constitute a major portion of saline soils in the Gujarat State in India and pose serious threat to the state economy. Besides, the highly saline soils present a desolate look and salt laden winds damage the air and the overall environment. While surveying the large tracts of this nature, it was noticed that *Salvadora persica*, a facultative halophyte and a potential source for seed oil, is a predominant species in highly saline habitats of coastal and inland saline black soils. It is also called *khari* (saline) *Jal* against *Salvadora oleoides* [*meethi* (sweet) *Jal*] found in the dry and arid regions of India. The genus *Salvadora* belongs to the family *Salvadoraceae,* comprising of three genera (*Azima, Dobera* and *Salvadora*) and 10 species distributed mainly in the tropical and subtropical region of Africa and Asia [31]. It is also a medicinal plant, its bark containing resins and an alkaloid called Salvadoricine. The leaves, root bark, fruits and seeds are used for the treatment of cough, fever, asthma and as purgative. Roots are also used for chest diseases while, sap used for treating sores [33, 41]. Around the world *S. persica* L., is more popular by the brand name of Meswak and the tree is referred as the toothbrush tree and extensively used as tooth brush [2–4]. The seeds are good source of nonedible oil rich in C-12 and C-14 fatty acids having immense applications in soap and detergent industry [16, 19, 20].

The species grows well on saline black soils having salinity up to 65 dS m^{-1} and found to yield well. The natural habitats are near mangroves, saline lands, swamps, thorn shrubs, desert flood plains, seasonally wet sites and along drainage lines in arid zones. Apparently, this observation helped us in deciding to pursue its cultivation and tolerance characteristics in saline inhospitable environments.

5.2.1 CULTIVATION TECHNOLOGY

For the cost effective cultivation of the *Salvadora persica*, the following technical interventions were found promising:

- Saplings are grown in 1 kg polyethylene bags placed in 20 cm deep trenches. Gururaja Rao [16] and Gururaja Rao et al. [18] reported that in the absence of good quality water saplings can be grown by using saline water of EC 15 dS m^{-1}. Irrigate at about fortnight intervals until 90 days old at which saplings are ready for transplanting (Figure 5.1).
- The saplings should be transplanted in 30×30×30 cm^3 pits at a spacing of 4 m × 4 m, which is ideal planting geometry for saline black soils [18]. Farmyard manure at the rate of 1 kg pit^{-1} and di-ammonium phosphate (DAP) at 50 g pit^{-1} need be added before transplanting.
- The sapling can be transplanted for cultivation on highly saline soils having salinity up to 55 dS m^{-1}.
- Use of saline ground water at flowering stage proved beneficial in increasing seed and oil yield (Figure 5.2).

The studies in Gujarat (India) revealed that plantations start bearing by 2nd year and by fifth year, the plants yield about 1800 kg ha^{-1}. The mean oil content varied in a narrow range of 31–32% although it decreased with increasing salinity from more than 32% at 25–35 dS m^{-1} to less than 30% at 55–65 dS m^{-1}. Reddy et al. [40] also reported that *S. persica* can be cultivated in saline alkali soils and the oil content of its seeds may range from 30–40%.

FIGURE 5.1 *Salvadora persica* on highly saline black soils: Seedlings (a, left) and Mature (b, right) plants.

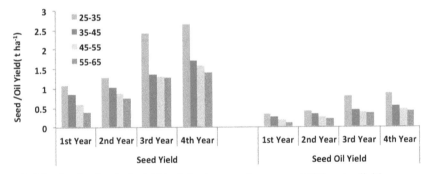

FIGURE 5.2   Seed and oil yield in *Salvadora persica* grown at different salinities.

The data in Figure 5.3 reveal that although there is reduction in height and canopy spread with increasing salinity, yet the growth is quite promising and sustainable even at high salinity. This specie provides good economic returns. *Salvadora persica* being a large, well-branched evergreen shrub or small tree having soft whitish yellow wood with numerous branches, drooping, glabrous and shining, it has the potential of re-greening the highly saline soils that cannot be put under arable farming. The monitoring of soil properties revealed that soil salinity is considerably reduced by the 4th year and as such one can take up intercropping with less

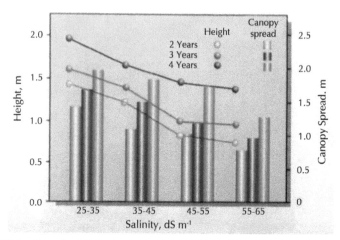

FIGURE 5.3   Growth of *Salvadora persica* on highly saline black soil.

tolerant crops/forages (See section on grasses). Greening with plants helps to win back several bird species as the trees provide a dwelling place for them [18]. Realizing the overall impact of this technology, the National Bank for Agriculture and Rural Development (NABARD), Mumbai, India in association with the Research Station has developed a bankable model scheme in refinancing mode for cultivation of *Salvadora persica* on salt affected black soils.

5.3   PHYSIOLOGICAL BASIS OF SALT TOLERANCE

5.3.1   GROWTH

Although salinity in the root medium manifests its effects on growth and physiology of plants, halophytes like *Salvadora* generally do not get substantially affected by salinity and maintain their growth. Growth measured in terms of plant height and canopy coverage showed that the plants attained a height of 1.95 m by 4th year at 25–35 dS m^{-1} and 1.48 m at 55–65 dS m^{-1} salinity levels (Figure 5.3). The canopy coverage of 1.72 m and 1.12 m at 25–35 dS m^{-1} and 55–65 dS m^{-1} salinity, respectively indicate that this species is capable of withstanding high salinity of 65 dS.m^{-1} (Figure 5.3).

The higher canopy coverage facilitates better light interception. Higher growth of the plant is also ascribed to higher synthesis of osmotic substances such as proline, amino acids, sugars which facilitate osmotic adjustment [20], results being in conformity with the results reported by Maggio et al. [32]. Elfeel and Al-namo [11] reported that drought conditions enhanced growth adaptive traits in the three species namely *Acacia tortilis, Leptadenia pyrotechnica* and *S. persica*, although survival was better in the former two species while growth was better in *S. persica*.

5.3.2 SALT COMPARTMENTATION

Halophytes have no special metabolic adaptation to high salinity but they have the ability to compartmentalize Na^+ to very high concentrations in vacuoles. Distribution of sodium and chloride ions (Figure 5.4) studied in different plant parts of *S. persica* growing at different *in-situ* salinities indicated that maximum amount of Na^+ and Cl^- ions were retained in bark, roots and senescing leaves, which acted as the potential sinks for these toxic ions thereby sparing other plant parts like immature leaves, partially mature and physiologically mature leaves to perform normal physiological activities to help in normal growth and development (Table 5.1). It enables the plants to remain lush green even at high salinity. Further, senescing leaves act as potential sinks for toxic ions thereby reducing the load on other photosynthesizing tissues, which remain by and large salt free [18]. Concentration of sodium and chloride in plant parts increased

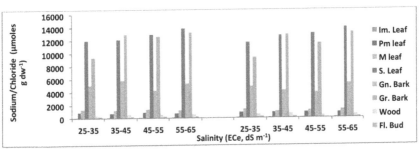

FIGURE 5.4 Compartmentation of sodium and chloride in *Salvadora persica* grown at different salinities on a highly saline black soil.

with increase in salinity of the soil. The capacity of the sink also increased with age of the plant as well as increase in salinity which indicates that *S. persica* has very well developed salt compartmentation mechanism [20].

5.3.3 ION TRANSPORT (FLUX) AND UPTAKE

The rate and ion transport (flux) from root to shoot and to whole plant was calculated using the formula:

$$Js = (M_2 - M_1) \ln [(W_2/W_1/(t_2 - t_1)(W_2 - W_1)] \tag{1}$$

where, Js is the rate of transport (flux), $\mu g\ g^{-1}$ root dry weight d^{-1}, M_1 and M_2 are the amounts of ion in the shoot/whole plant given by content per unit dry weight X dry weight, and W_1 and W_2 are the fresh weights of the roots at the harvest times t_2 and t_1 [37].

Though Na^+ concentration increased with increase in salinity, the total Na uptake showed a decreasing trend which may be obviously due to decrease in the biomass yield with increase in salinity. Similarly, chloride uptake in root is much higher than that of the shoot. The rate of flux of Na^+ and Cl^- ions to the whole plant while increase with increase in salinity showed a decreasing trend with age (Table 5.2). The flux of these ions from root to the shoot was a fraction of that to the whole plant indicating that roots accumulate more ions than shoots. In this case, roots act as both Na^+ and Cl^- accumulator (Table 5.2).

The K^+/Na^+ ratio was observed to be more in immature as compared to senescing leaves, indicating the translocation of N^+ from immature leaves, a mechanism to avoid salt stress. The high K^+/Na^+ ratio in expanding leaves helps in metabolic functions [7, 35].

5.3.4 SEASONAL VARIATION IN IONIC CONCENTRATION

Data presented in Figure 5.5 indicate that immature, partially mature and mature leaves possess 5–8% of total ion uptake during dry and wet moths, while wood had very low amounts of 0.5–1.5%. Contrary to this, senescing leaves accounted 40–48% during dry months and 25–26% during

TABLE 5.1 Concentration of Na^+ and Cl^- (%) Ions in Different Plant Parts of *Salvadora persica* Grown on Saline Black Soils

Plant Part	Salinity Range (dS m⁻¹)											
	25 – 35			35 – 45			45 – 55			55 – 65		
	2nd year	3rd year	4th year	2nd year	3rd year	4th year	2nd year	3rd year	4th year	2nd year	3rd year	4th year
Na^+												
Root	1.13	1.62	1.86	1.61	2.11	2.30	1.91	2.37	2.57	1.93	2.39	2.60
Wood	0.02	0.03	0.03	0.03	0.04	0.04	0.03	0.04	0.05	0.04	0.06	0.06
Bark	1.59	1.72	1.95	2.12	2.29	2.60	2.33	2.52	2.86	2.73	2.95	3.34
Im. leaf	0.02	0.02	0.02	0.02	0.02	0.03	0.02	0.02	0.03	0.03	0.03	0.03
M. leaf	0.18	0.18	0.21	0.19	0.20	0.23	0.21	0.22	0.25	0.22	0.24	0.27
S. leaf	1.66	1.81	2.06	2.11	2.30	2.61	2.30	2.38	2.71	2.39	2.51	2.83
L S D p≤0.05	0.36	0.13	0.18	0.46	0.21	0.22	0.89	0.16	0.30	0.69	0.46	0.71
Cl^-												
Root	2.13	2.65	2.70	2.65	3.36	3.60	2.88	3.00	3.90	2.94	4.00	4.01
Wood	0.04	0.05	0.06	0.05	0.06	0.06	0.05	0.07	0.07	0.07	1.00	0.10
Bark	2.49	2.69	3.05	3.43	3.70	4.01	3.74	4.04	4.41	4.26	4.62	5.14
Im. leaf	0.03	0.04	0.04	0.04	0.04	0.05	0.04	0.04	0.04	0.5	0.05	0.06
M. leaf	0.28	0.29	0.32	0.31	0.31	0.38	0.33	0.34	0.39	0.35	0.40	0.42
S. leaf	2.63	2.82	3.19	3.17	3.67	2.71	3.58	3.76	4.23	3.58	4.04	4.59
LSD p≤0.05	0.40	0.16	0.39	1.05	0.28	0.30	0.58	0.22	0.40	0.95	0.98	0.83

Im = immature leaf; M= physiologically mature leaf, S = senescing leaf.

TABLE 5.2 Uptake and Flux of Na+ and Cl- Ions in S. *persica* on Saline Black Soils

Salinity Class (dS m⁻¹)	Uptake (g)				Flux ($\mu g\ g^{-1}\ d^{-1}$)			
	Shoot		Root		Shoot		Root	
	Na⁺	Cl⁻	Na⁺	Cl⁻	Na⁺	Cl⁻	Na⁺	Cl⁻
2ⁿᵈ Year								
25 – 35	6.44	10.18	8.40	15.86				
35 – 45	5.12	8.53	9.31	15.29				
45 – 55	4.10	6.57	6.58	9.91				
55 – 65	3.68	5.64	4.97	7.56				
CD 0.05	1.21	1.88	1.93	2.12				
3ʳᵈ Year					**Between 3ʳᵈ and 2ⁿᵈ Year**			
25 – 35	16.01	25.90	27.36	44.93	29.9	46.1	9.8	16.2
35 – 45	14.21	22.95	27.69	44.08	39.0	61.3	12.9	20.4
45 – 55	10.13	16.21	18.56	29.43	50.2	81.3	16.8	26.9
55 – 65	9.82	15.59	13.62	22.84	78.8	131.4	19.5	52.6
CD 0.05	2.11	2.88	3.58	5.35	10.5	13.8	4.3	5.8
4ᵗʰ Year					**Between 4ᵗʰ and 3ʳᵈ Year**			
25 – 35	22.31	34.71	38.33	56.66	10.8	12.9	3.9	5.5
35 – 45	18.42	28.69	37.64	58.73	12.3	17.8	3.7	5.0
45 – 55	14.43	22.30	37.23	37.23	17.8	23.5	7.3	9.7
55 – 65	13.51	20.84	29.35	29.35	29.7	410.2	11.9	16.9
LSD p≤0.05	3.95	4.23	0.53	1.88	1.88	3.50	1.20	1.70

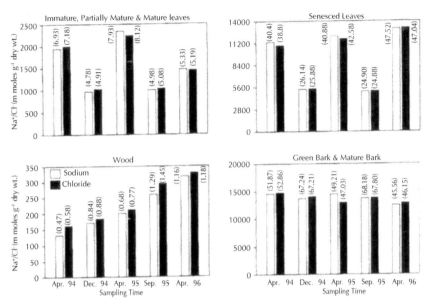

FIGURE 5.5 Seasonal variations in ion uptake in S. persica grown on highly saline black soils.

wet season. However, bark accounted 46–53% in dry and 67–68% in wet season. These results indicated that bark tissues retain more sodium and chloride during wet season, thus lowering the salt injury to the leaves, and hence low degree of senescence when the soils had adequate moisture coupled with low salinity. Leaves of *S. persica* also showed some degree of succulence which may also facilitate the dilution of salts within the tissues [18, 24, 43]. Khatak et al. [29] also documented the seasonal changes in amino acids and mineral ions in *S. persica*.

5.3.5 TISSUE TOLERANCE: RELATION BETWEEN SODIUM AND CHLOROPHYLL

Data on tissue tolerance as studied by combined chlorophyll and sodium estimates indicate that about 2250 μmoles of sodium is needed to reduce chlorophyll content of leaves by 50% compared to leaves of plants growing in non-saline environment. The scatter diagram (Figure 5.6) indicates that the high degree of scatter could be due to both variations in

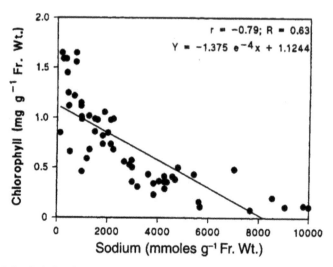

FIGURE 5.6   Relation between leaf sodium and chlorophyll in *S. persica* (pooled data for all leaves).

individual plants and from leaf-to-leaf. Linear regression fit accounts for 58% of the variability although the subjective appearance of the fit is poor at higher sodium concentration. The scatter diagrams showed that as Na concentration increased, chlorophyll content decreased both in immature and mature leaves. The LC_{50} (lethal concentration) value for Na analogous to LD_{50} (lethal dose) of toxicology is defined here as the mean Na concentration in the individual leaves having more than 50% chlorophyll of non-salinity healthy leaves. The salinity induced decrease in chlorophyll can be attributed to a weakening of pigment-protein-lipid complex [42].

The LC_{50} values for the immature leaves are much less than that of the senescing leaves. Such a high value of LC_{50} in senescing leaves indicted that the leaves of *S. persica* possesses a fair degree of tissue tolerance to Na. The linear regression equation for immature and senescing accounts for 82.1 and 99.0% variability, respectively (Figure 5.7), while the earlier studies in *S. persica* indicated 58% variability both in immature and mature leaves when data of all leaves were pooled in the scattered diagram [22, 23].

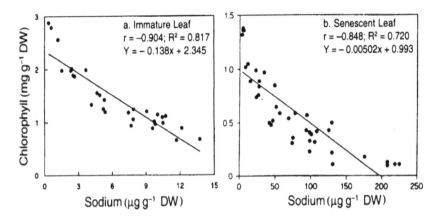

FIGURE 5.7 Relationship between leaf sodium and chlorophyll in *S. persica* (immature and senescing leaves).

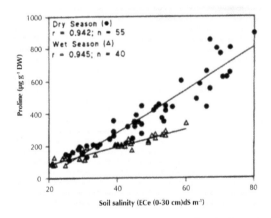

FIGURE 5.8 Relationship between leaf sodium and proline in *S. perisca* under salinity.

5.3.6 FREE PROLINE

Proline was found to increase with salinity, increase being three-folds at 55–65 dS m^{-1} when compared to low salinity of 25–35 dS m^{-1} (Figure 5.8) and accounted for about 26.6% of the total amino-acid content. Very high amount of free proline accumulated in dry season (200–890 µgg^{-1} dry weight) compared to wet season (70–400 µg g^{-1} dry weight) helping in osmo-regulation as reported earlier [13,44]. Similar increase in keto acids with increasing salinity and amino acids such as aspartic acid, asparagines, glutamic acid and glutamine play an important role in amination and transamination reactions and osmotic adjustments respectively. The accumulation of keto acids with concomitant development of leaf succulence under high salinity suggests the possible turgor maintenance that results in better plant growth.

5.3.7 EPICUTICULAR WAXES

Another adaptive feature exhibited by *S. persica* is its ability to synthesize very high amounts of leaf epicuticular waxes (LEW) which increased with salinity being highest at 55–65 dS m^{-1}, showing a four-fold increase over the lowest salinity of 25–35 dS m^{-1}. Compared to plants growing in non-saline environment, the increase was 33, 71, 126 and 287% at salinities of 25–35, 35–45, 45–55 and 55–65 dS m^{-1} respectively. LEW showed a distinct seasonal variation being low in wet than dry season (Figure 5.9)

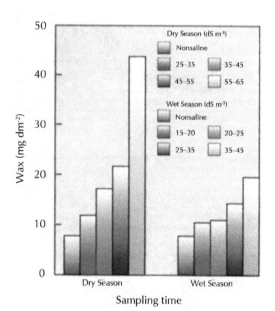

FIGURE 5.9 Leaf epicuticular waxes in *S. persica* during wet and dry seasons.

in which plants tend to develop a thick waxy layer when the soil salinity in 0–30 cm soil profile is 60–70 dS m^{-1}. Composition of LEW indicates the presence of high amounts of β-diketones and OH-β-diketones which impart glaucousness to the leaves resulting in the lowered heat load on leaves. LEW with high amounts of primary and secondary alcohols and aldehydes (Table 5.3) also plays an important role in reducing cuticular transpiration, a water conservation strategy to cope up stress induced by salinity [25].

5.3.8 SOIL SALINITY UNDER PLANTATIONS

The soils grouped as deep, clay loam, hyperthermic, montmorillonitic family of Vertic Haplustepts showed high degree of spatial and temporal variation in soil salinity initially ranging from 65 to 70 dS m^{-1} in the top layer. Salinity of the soil decreased with depth, i.e., from surface to 90 cm depth (Figure 5.10). Cultivation of *S. persica* up to 5 years resulted in slight decline in soil salinity as compared to the pre-planting salinity. Changes in surface salinity are partly attributed to the ability of plants to extract the salt and partly due to root activity which improves the physical properties of the soil. However, the magnitude of fluctuation in salinity was not much at lower layers. The ground water table might be contributing to salinity causing only minor changes at lower depths. The spatial variability of surface salinity under 5-year old plantation (Figure 5.11) showed significant difference from the initial salinity prior to planting.

5.4 CULTIVATION OF FORAGE GRASSES ON SALINE BLACK SOILS

Agriculture and animal husbandry in India are interwoven as mixed farming and livestock rearing forms an integral part of rural living. India supports nearly 20% of the world livestock. Most often and especially in arid and resource poor regions, livestock is the only source of cash income for subsistence farms. It insures some livelihood in the event of crop failure. At present, India faces a net deficit of 61.1% green fodder, 21.9% dry crop residues and 64% feeds (http://www.icar.org.in/files/forage-and-grasses.

TABLE 5.3 Composition of LEW (mg dm^{-2}) of *S. persica* Grown on Highly Saline Black Soil

Salinity range (dS m^{-1})	Fatty acids	OH-β-diketones	Primary alcohols	Secondary alcohols	β-diketones	Aldehydes	Hydrocarbons, esters and ketones
25–35	0.84	2.54	2.61	2.47	1.90	0.82	0.82
35–45	2.16	3.14	3.24	2.16	2.16	1.06	1.08
45–55	2.75	3.17	3.20	1.76	1.76	0.80	0.90
55–65	3.12	7.70	7.72	7.68	5.16	4.68	1.90
LSD p≤0.05	0.62	0.53	0.61	0.58	0.48	0.50	0.23

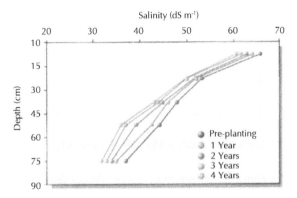

FIGURE 5.10   Soil salinity variations over the years under *S. persica* grown on highly saline black soil (55–65 dS m⁻¹ range).

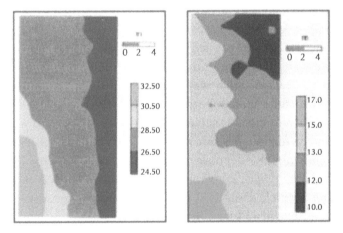

FIGURE 5.11   Spatial variability in soil salinity prior and after planting of *S. persica*.

pdf, [26]). Due to ever-increasing population pressure, arable land will mainly be used for food and cash crops, leaving little chance of having good quality arable lands for fodder production. Therefore, poor quality saline/sodic lands provide a good avenue to increase availability of fodder if some grasses can be identified to suit the saline/sodic environment.

In this scenario, cultivation of forage grasses, *Dichanthium annulatum* and *Leptochloa fusca* was attempted. *Dichanthium*, also known as Delhi grass in India, adapts to drought and thrives well on black cotton soils. It tolerates saline and alkali conditions but not acidity [27]. *Leptochloa fusca* has earlier been reported to be highly tolerant to alkali conditions as well as to water logging. According to Kumar and Abrol [30], this grass can be grown in barren alkali soils (pH 10) without amendment and can tolerate water logging for a period of 8 days showing yield increase from 207 g pot^{-1} for a water stagnation of 2 days to 312 g pot^{-1} for a water stagnation of 8 days in the first year of cultivation. *Leptochloa fusca* was found to be relatively most tolerant, *A. lagopoides* intermediate and *C. dactylon* most sensitive to water logging [6]. Water logging treatments caused a marked but comparable reduction in chlorophyll content in all the three species. Shoot iron and manganese contents increased in all the species with increasing water logging, but least in *L. fusca* which had the ability to accumulate relatively more iron and manganese in or on its roots [6]. Qureshi and Barrett-Lennard [38] observed that this grass, also known as *Kallar* grass in Pakistan can be cultivated on soils having EC_e in the range of 20–30 dS m^{-1}.

Both the grasses were planted through root slips in a ridge-furrow planting system with 50 cm high ridges 1 m apart. *Dichanthium* was planted on ridges while *Leptochloa* was cultivated in furrows forming an ideal proposition to fully utilize the land. Results revealed that *Dichanthium annulatum* is quite suited to saline black soils, as it possessed well-defined salt compartmentation, wherein roots act as potential sinks for toxic ions like sodium and chloride, making the shoot portions relatively salt free [17, 21]. *Dichanthium annulatum* and *Leptochloa fusca* on moderate saline soils can yield about 1.9 t ha^{-1} and 3.2 t ha^{-1}, respectively. Nitrogen applied at the rate of 45 kg ha^{-1} (in the form of urea) at the time of planting increased the forage yield by about 70% in *Dichanthium annulatum* (Table 5.4). It also improved forage quality traits. The two grasses in combination were found ideal in saline black soils having salinity up to 8–10 dS m^{-1} (Table 5.5).

5.5   CULTIVATION OF FORAGE GRASSES WITH SALINE WATER

Aeluropus lagopoides, a perennial grass exists in highly saline habitats uninhabitable to many other plant species. The plant is able to expel the salt it gains from the highly saline soil through glands on the leaves and

TABLE 5.4   Effect of Nitrogen on Growth and Forage Yield of Forage Grosses [21]

Grass species	Height (m)		Tillers plant⁻¹		Green forage yield (t ha⁻¹)	
	+ N	– N	+ N	– N	– N	– N
Leptochloa fusca	1.39	0.99	12.54	4.46	3.21	2.13
Dichanthium annulatum	1.01	0.87	10.24	7.38	2.24	1.32
$CD_{0.05}$	Height		Tillers		Yield	
Planting method	0.13		3.11		0.88	
Grass species	0.22		2.32		0.55	
Planting method x Grass species	NS		NS		NS	

TABLE 5.5   Growth and Yield of Forage Grasses Under Ridge and Furrow Planting System on a Saline Soil Having EC_e 15.4 dS m⁻¹ in 0–30 cm Layer [21]

Grass species	Height (m)		Tillers plant⁻¹		Green forage yield (t ha⁻¹)	
	Ridge	Furrow	Ridge	Furrow	Ridge	Furrow
Leptochloa fusca	1.18	1.02	10.62	9351	3.17	3.73
Dichanthium annulatum	0.91	0.74	6.41	5.32	1.85	1.76
$CD_{0.05}$	Height		Tillers		Yield	
Planting method	0.12		0.91		NS	
Grass species	0.16		1.53		0.82	
Planting method x Grass species	NS		2.24		NS	

as such in itself it has very low salt content [14]. The small waxy leaves and strong root network also help this species to survive in stressful salty environments [1], especially throughout the summer months when there is a three-fold increase in soil salinity. *Eragrostis* is also cultivated and used both as a fresh and dry forage grass but is commonly categorized as medium sensitive to salts with a low salinity threshold although presence of broad intra-specific genetic variation in its accessions and varieties for salt tolerance have been reported being more in the former. Gulzar et al. [15] indicated that *Aeluropus* species can be used to increase forage production in salt affected wastelands because of their high protein content and high salinity tolerance. This grass is used locally as a fodder for livestock and is also useful in coastal sand dune stabilization. The effect of saline water irrigation (EC 10, 20, 30 and 40 dS m^{-1}) given till the grasses showed flower initiation (about 5–6 weeks) is shown in Figure 5.12. It showed that shoot mass decreased while root mass increased with increasing salinity in both the grasses.

Data on plant height and tiller production at the time of harvesting (Table 5.6) indicated that *Aeluropus lagopoides* produced higher number of tillers and showed better growth at all salinity treatments when compared to *Eragrostis* indicating its higher salt tolerance. While plant height decreased with increase in salinity in *Eragrostis, Aeluropus* showed increased growth with salinity. The total tiller production was very high in *Aeluropus.* Though there is an increase in tillers with salinity in both the species, their further growth got slowed down with increase in salinity, particularly in *Eragrostis.* A decrease was recorded in growth with an increase in the level of saline water over a period of 36 days (flower initiation stage) particularly in *Eragrostis.*

However, the difference between 10 and 20 dS m^{-1} saline water treatments over a period of time was almost similar in *Aeluropus lagopoides* while *Eragrostis* showed only little variation (Table 5.6). At highest salinity, it was seen that though the buds producing new tillers emerged, the high

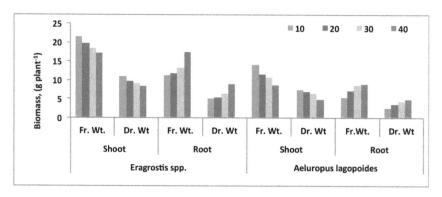

FIGURE 5.12   Biomass of halophytic forage grasses *Aeluropus lagopoides* (right) and *Eragrostis* (left) under saline water irrigation.

TABLE 5.6   Growth of Halophytic Grasses as Influenced by Saline Water Irrigation

Salinity (dS m^{-1})	Height (m)	Tillers plant^{-1}
Eragrostis species		
10	0.49	8
20	0.40	9
30	0.34	15
40	0.28	15
Aeluropus lagopoides		
10	0.45	40
20	0.51	46
30	0.69	47
40	0.74	55
CD$_{(0.05)}$	0.087	3.1

salt content of the shoot system affected their further growth. However, these grasses responded well to the saline irrigation even at 30 dS m^{-1} of irrigation water. Tissue ion content of the leaf and stem tissue samples indicated that higher amount of sodium and chloride accumulated in the tissues at highest salinity indicating their ability to absorb the ions [17]. Irrigation with saline water increased the soil salinity, increasing with increasing number of irrigations and being highest of around 30 dS m^{-1} under both the species when irrigated with 40 dS m^{-1} saline water [17]. The salinity reduced with progress of monsoon.

5.5.1 BIOCHEMICAL CONSTITUENTS: ION CONTENT

The leaf and stem ions Na$^+$ and Cl$^-$ increased with increasing salinity. The shoot (leaf and stem) sodium content after two irrigations increased from 2000 µmoles to 5900 µmoles in leaf of *Eragrostis* specie and 3500 µmoles to 5100 µmoles in *Aeluropus lagopoides*. In stem, the Na$^+$ and Cl$^-$ content were higher when compared to the leaves indicating stem as a potential sink. Between the grasses, Na$^+$ and Cl$^-$ contents were more in *Aeluropus lagopoides* than *Eragrostis* specie. Ion partitioning (Na$^+$ and Cl$^-$) in shoots

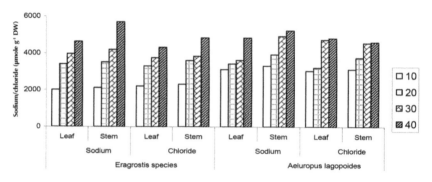

FIGURE 5.13   Tissue sodium and chloride in forage grasses under saline water irrigation.

and roots of two grasses indicated that roots do act as sinks for the toxic ions (Figure 5.13). Uptake and flux of Na^+ and Cl^- and the total Na^+ uptake was higher in *Aeluropus lagopoides* than *Eragrostis* specie though the increase was only marginal (Tables 5.7 and 5.8). The total Na^+ content is less in shoot than in the root in both the grasses irrespective of salinity and age of the plant. Chloride uptake, however, is relatively more in root than in shoot. The rate of flux of Na^+ and Cl^- to the whole plant increased with salinity and age of the plant [17].

5.5.2   CHLOROPHYLL

The Chlorophyll and its components, Chlorophyll *a* and Chlorophyll *b* (Table 5.9) indicated that there was no significant variation in Chlorophyll *a* among 20, 30 and 40 dS m^{-1} irrigations in *Eragrostis* specie. However, the Chlorophyll *b* and total Chlorophyll showed significant variation among different treatments in *Eragrostis* specie. Similarly in *Aeluropus lagopoides* also the Chlorophyll *a* did not show much variation under saline water irrigations of 20, 30 and 40 dS m^{-1}. The Chlorophyll *b* and total Chlorophyll content showed significant variation with saline water treatments. The Chlorophyll *a/b* ratio was also found to increase at higher salinity levels indicating higher stability of Chlorophyll *a* when compared to Chlorophyll *b* [17].

TABLE 5.7   Uptake and Flux of Na⁺ and Cl⁻ Ions in *Aeluropus lagopoides* Under Saline Water Irrigation

Salinity (dS m⁻¹)	Uptake (g)				Flux (µg g⁻¹ d⁻¹)			
	Shoot		Root		To whole plant		To shoot	
	Na⁺	Cl⁻	Na⁺	Cl⁻	Na⁺	Cl⁻	Na⁺	Cl⁻
First week								
10	4.20	4.85	5.82	5.13				
20	3.92	4.60	5.20	4.93				
30	3.61	4.05	4.85	4.14				
40	3.40	3.70	3.85	3.96				
Second week					Between 1ˢᵗ and 2ⁿᵈ week			
10	5.15	4.74	6.40	4.68	8.78	1.88	2.42	3.62
20	4.95	4.44	6.00	4.53	9.92	13.84	3.03	4.81
30	4.45	4.10	5.28	3.96	12.62	14.09	3.84	4.99
40	3.62	3.78	4.62	3.65	13.12	26.32	4.12	8.92
Third week					Between 2nd and 3rd week			
10	5.60	5.66	6.78	5.25	13.42	16.24	3.62	5.14
20	5.05	5.05	6.18	4.94	17.32	20.33	5.14	6.24
30	4.62	4.60	5.85	4.60	18.92	28.75	5.92	8.36
40	4.28	4.20	4.90	4.28	24.32	32.48	7.96	10.62
Fourth week					Between 3rd and 4th week			
10	6.05	6.20	6.90	6.18	19.70	26.78	6.32	8.32
20	5.25	5.00	6.23	5.54	26.30	32.14	8.44	10.64
30	5.05	5.05	5.63	4.84	28.15	39.36	9.63	14.20
40	4.60	4.60	5.00	4.12	30.10	42.74	10.40	16.32

5.5.3   TISSUE TOLERANCE

The combined Na⁺ and Chlorophyll analysis indicated an inverse relationship between tissue sodium and leaf chlorophyll. Data on tissue tolerance indicated that about 8950–9000 µmoles of Na⁺ are needed to reduce the chlorophyll content of leaf when compared to the leaves of the plants. The scatter diagram of Na⁺ and chlorophyll showed a high degree of scatter, which could be due to both variation in induced plant and leaf-to-leaf

TABLE 5.8 Uptake and Flux of Na$^+$ and Cl$^-$ Ions in Eragrostis Under Saline Water Irrigation

Salinity (dS m^{-1})	Uptake (g)				Flux (μg g^{-1} d^{-1})			
	Shoot		Root		To whole plant		To shoot	
	Na$^+$	Cl$^-$	Na$^+$	Cl$^-$	Na$^+$	Cl$^-$	Na$^+$	Cl$^-$
First week								
10	3.91	4.37	5.42	5.14				
20	3.64	4.12	4.86	4.58				
30	3.42	3.90	4.39	4.32				
40	3.08	3.76	4.14	4.24				
Second week					Between 1st and 2nd week			
10	4.78	4.36	5.96	4.35	7.90	9.92	2.18	3.42
20	4.32	4.04	5.78	4.14	8.62	10.42	3.62	3.75
30	4.02	3.92	4.14	3.62	9.36	12.62	3.92	4.62
40	3.64	3.51	4.04	3.44	10.42	18.80	4.14	5.96
Third week					Between 2nd and 3rd week			
10	4.92	4.88	5.84	4.98	10.39	12.82	3.14	4.36
20	4.81	4.64	5.72	4.86	13.86	16.01	3.36	5.62
30	4.32	4.38	4.92	4.64	15.14	22.41	3.92	6.98
40	4.02	4.14	4.44	4.32	19.52	24.62	4.79	9.39
Fourth week					Between 3rd and 4th week			
10	5.64	5.84	6.10	5.14	15.76	19.72	3.79	4.72
20	4.92	4.92	5.72	5.02	16.80	25.6	5.16	6.72
30	4.84	4.63	5.32	4.84	18.44	26.8	6.13	8.42
40	4.12	4.24	4.79	4.36	26.12	28.42	7.14	8.92

TABLE 5.9 Effect of Saline Irrigation on Chlorophyll Content (mg g^{-1} Fresh Weight)

EC$_e$ (dS m^{-1})	*Eragrostis* specie				*Aeluropus lagopoides*			
	Chl a	Chl b	Total Chl	Chl a/b ratio	Chl a	Chl b	Total Chl	Chl a/b ratio
10	0.198	0.296	0.494	0.6689	0.209	0.329	0.538	0.6352
20	0.190	0.274	0.464	0.6934	0.200	0.320	0.520	0.625
30	0.190	0.264	0.454	0.7197	0.200	0.292	0.492	0.6849
40	0.190	0.242	0.432	0.7851	0.196	0.268	0.464	0.7313
C.D $_{0.05}$	0.004	0.008	0.0071		0.004	0.008	0.007	

Chl = Chlorophyll

variations. A significant negative correlation was noticed between tissue Na⁺ and Chlorophyll in both the grasses.

5.6   FORAGE PRODUCTION IN SALINE SOILS

The green forage yield of these grasses under field conditions is given in Figure 5.14. The grasses developed well even on saline soils when irrigated with saline water. The data indicated that *Eragrostis* specie showed higher forage yield under field conditions when compared to *Aeluropus lagopoides* at a salinity of 14.6 dS m^{-1}. Working with *Eragrostis* specie, Asfaw and Danno [5] reported that tef varieties are most affected by salinity than tef accessions. Forage quality was highest when *A. lagopoides* was in the vegetative stage and tended to decrease sharply as the plant matured towards the seed ripening stage [39]. Increase in these forage quality traits with increase in salinity of irrigation water was noticed in both the grass species indicating their higher production at higher salinity levels. Crude fiber, a mixture of cellulose, hemicellulose and lignin gives strength and its higher content indicates higher photosynthate production and its conversion. Higher ash content of *Aeluropus lagopoides* can be ascribed to higher mineral uptake as reported in other grasses as well.

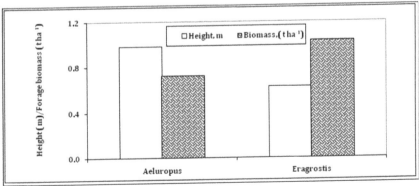

FIGURE 5.14   Growth and forage yield of grasses grown on saline black soil (EC$_e$ 12.8 dS m^{-1}; CD 0.05: Height: 0.26; Biomass: 0.418).

5.6.1 EFFECT OF NITROGENOUS FERTILIZER ON GROWTH AND FORAGE YIELD

Application of nitrogen has been reported to enhance forage biomass and also the uptake of salts from the soil. Moreover, nitrogen being the most limiting nutrient for crop production on saline Vertisols being poor in N and organic matter, nitrogenous fertilizer when given with saline water resulted in significant increase in forage yield of both the grasses. Of the two, *Eragrostis* found to have higher growth, tillers, forage yield with 60 kg ha^{-1} N application when saline water was applied at 15 days interval (Table 5.10). The grasses were found very effective in salt removal from the soil layers, *Aeluropus* removing more salt than *Eragrostis*. Analysis of tissue sodium and chloride indicated their content *per se* decreased when compared to those given no nitrogen. This low tissue sodium and chloride, however, improved the forage quality parameters. Nitrogen given at 60 kg ha^{-1} resulted in lowered tissue ion content, resulting mainly from the increased biomass which resulted in lowered salt distribution per unit weight of the tissue.

5.6.2 ION COMPARTMENTATION

Ion compartmentation at organ level indicated higher amount of sodium in roots followed by stem and old leaves and the least in inflorescence in

TABLE 5.10 Effect of Nitrogen on Biomass (t ha^{-1}) of Halophytic Grasses with Saline Water

	Aeluropus lagopoides			*Eragrostis* species		
	Nitrogen (kg ha^{-1})					
Irrigation	0	30	60	0	30	60
I_1	1.01	1.24	1.29	1.12	1.25	1.34
I_2	1.10	1.28	1.36	1.19	1.28	1.41
I_3	1.15	1.31	1.41	1.22	1.31	1.44
CD$_{0.05}$						
Nitrogen (N)	0.18			0.08		
Irrigation (I)	0.09			0.12		
N x I	0.11			0.11		

TABLE 5.11   Ion Compartmentation in Halophytic Grasses (mmoles g⁻¹ dry weight)

Plant part	*Aeluropus lagopoides*			*Eragrostis species*		
	Na⁺	**K⁺**	**Na/K**	**Na⁺**	**K⁺**	**Na/K**
Inflorescence	2.6	4.4	0.590	4.3	4.9	0.876
Mature foliage	9.2	8.8	1.409	11.6	7.6	1.526
Stem	16.1	10.4	1.548	12.4	7.9	1.570
Old leaves	13.6	7.9	1.722	14.2	7.4	1.972
Root	30.2	8.8	3.432	29.4	9.1	3.231

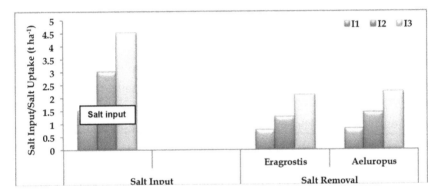

FIGURE 5.15   Salt input and salt uptake by forage grasses under saline water irrigation on saline Vertisols (I1, I2, I3 indicate numbers of irrigations).

both the grasses. Similar trend was observed in potassium in that foliage and roots had higher potassium than inflorescence (Table 5.11). Higher accumulation of sodium in roots, old leaves and stems indicate the physiologically mature foliage had relatively low tissue sodium (Figure 5.5). Of the two forage grasses, Aeluropus had higher potassium in foliage while Eragrostis had higher potassium in roots. Contrary to this, sodium was found to be more in the foliage of Eragrostis while roots of Aeluropus had marginally higher sodium. Once the flowering occurs, higher sodium is found to be more in older leaves in Eragrostis, when compared to Aeluropus while older leaves showed lesser sodium when compared to shoot [17].

5.7   SALT BUDGETING AND SALT REMOVAL

Salt budgeting including contributions of saline water, subsurface salinity and the salt uptake by the halophytic grasses indicated that *Aeluropus* had better salt removal, i.e., 43.9% when compared to *Eragrostis* with 39.7% (Figure 5.15). This feature is highly useful in using these grasses under saline agriculture programs for lowering soil salinity, which over the years will help cultivation of lesser tolerant and more economically potential species.

5.8   UPSCALING POTENTIAL OF HALOPHYTE CULTIVATION

The initial interest in bio-saline agriculture and researches on halophytes began in early 1900's. It was not until the 1950's that Hugo Boyko, one of the modern pioneers of saline agriculture, began to apply these methods in the field and push the limits of salt tolerance in conventional crops. Serious efforts to collect and evaluate wild halophytic germplasm in simulated field experiments were undertaken for the first time, recognizing the commercial viability of halophytes and other salt tolerant crops.

Salicornia cultivation is now catching up as it can be harvested for its selenium-rich animal feed, salad and vegetable salt. On the other hand, halophyte domestication and improving crop salt tolerance have yet to be given a high priority due to the fact that salinity is still perceived as a localized constraint rather than a regional or global stress on agricultural production. As an alternative to improving the salt tolerance of plants that already have desirable commercial traits, the domestication of halophytes should be focused through improving the agronomic characteristics of wild salt tolerant species through selection and breeding. By far, the most critical feature of successful halophyte domestication is the infusion of private and public capital for germplasm collection, breeding programs, and bio-saline applications in the field. Germplasm collection and preservation and its availability for research, breeding, and experimentation must be secured as it will prove invaluable for future domestication when the economic value of halophytes is fully recognized, markets are established, and commercial seed companies begin to take over this function.

The prospect for adopting halophytes in commercial agricultural production depends on a number of economic factors including the cost of saline resources (soil and water) and other inputs, plant yield assessments, harvesting, processing, and marketing requirements, consumer and end-user acceptance as well as the appropriate valuation of associated environmental costs and benefits. For halophytic crops, the cost of more abundant saline resources (i.e. land and water) is often significantly lower than those needed for freshwater cultivation; the design, planting, and management of salt tolerant crops on previously irrigated farms may further reduce costs by taking advantage of existing on and off-farm infrastructure (i.e. irrigation, drainage, and mechanization). Actual yields per hectare and the cost of harvesting, processing, and marketing halophytic produce are also important considerations in evaluating their commercial feasibility. However, until the environmental costs (of chemical monocultures) and benefits (of halophyte cultivation) are properly monetized, traditional cost/benefit analysis will be unable to accurately reflect the economic viability of halophytes.

Phytoremediation, the cultivation of plants for the purpose of reducing soil and water contamination (by organic and inorganic pollutants) that results from the improper disposal of aquaculture, agriculture, and industrial effluents, can be a good and in many cases the only effective and economical method of removing or reducing contaminates. It is particularly so when large areas are to be covered and where physical/chemical treatments and leaching are too expensive or unfeasible. A number of halophytic grasses have proven to be effective in revegetating the brine-contaminated soils that typically result from gas and oil mining. Eid [10] showed the removal of Zn, Cu and Ni by *Sporobolus virginicus* and *Spartina patens, Sporobolus* being more efficient than *Spartina*.

Similarly, halophyte cultivation for carbon sequestration is another area of interest. Since, halophyte biomass yields are comparable to those of glycophytes yet the associated costs of cultivation are often far less particularly in areas where there is an over-abundance of saline resources, halophytic agro-forestry plantations present a cost-effective option for sequestering carbon and reducing their elevated levels in the biosphere. While trying to determine if indeed halophytes can be effectively utilized as carbon sinks, their potential for meeting our more immediate needs for crop alternatives and environmental conservation can also be addressed.

Clearly, with the economics of halophytes especially their medical value being documented besides other uses, there is a great potential of halophytes under various groups to be cultivated in a not too distant future.

5.9 SUMMARY

Salinity of agricultural lands and saline irrigation waters coupled with water scarcity have become the major impediments in agricultural production programs in arid and semi-arid regions resulting in poverty and other related social and economic issues. Saline Vertisols due to their peculiar physical and chemical constraints such as poor hydraulic conductivity, low infiltration rates, high clay content and narrow workable moisture range pose serious threats for crop production especially in the presence of salinity or when irrigated with saline water. Efforts are therefore, needed to find halophytes or halophytic grasses of economic importance to bring such problem soils under production system.

This chapter discusses some of the technological interventions to green the barren saline Vertisols using *Salvadora persica*, a facultative halophyte and a potential source for seed oiland some forage grasses for fodder such as *Dichanthium annulatum* and *Leptochloa fusca* for saline waterlogged soils and *Aeluropus lagopoides and Eragrostis* specie for saline water irrigated lands. Detailed analysis is presented to highlight the physiological basis of salt tolerance in *Salvadora persica* and briefly for the fodder grasses. In the case of *Salvadora persica,* data have been presented on ion transport (flux) and uptake, salt compartmentation, tissue tolerance, relation between sodium and chlorophyll, free proline and epicuticular waxes to conclude that the plant has well developed mechanisms for salt tolerance.

Amongst the forage grasses, it emerged that the two grasses *Dichanthium annulatum* and *Leptochloa fusca* can be an ideal combination for cultivation in saline black soils having salinity up to 8–10 dS m^{-1}. Amongst the other two grasses, salt removal from the soil layers was more in *Aeluropus* than *Eragrostis* although *Eragrostis* specie showed higher forage yield under field conditions when compared to *Aeluropus lagopoides* at a salinity of 14.6 dS m^{-1}. Towards the end of the chapter, an attempt is also made to enlist the potential economic halophytes and salt tolerant plants that can profitably be culivated in saline Vertisols. Salt tolerant plants such as *Salvadora perisca, Salicornia, Calophyllum inophyllum, Pandanus* spp., *Suaeda, Anethum graveolens* etc. can be of great industrial value. Many other species like *Aeluropus, Eragrostis, Leptochloa, Atriplex* have high forage value and thus, can form an integral part of forage production systems.

KEYWORDS

- *Aeluropus lagopoides*
- biomass
- chlorophyll
- compartmentation
- *Dichanthium annulatum*
- economic viability
- epicutical wax
- *Eragrostis*
- forage grasses
- halophytes
- ion transport
- ion uptake
- *Leptochloa fusca*
- lethal dose
- *Meswak*
- nitrogen
- oil yield (*Salvadora persica*)
- osmo-regulation
- proline
- phytoremediation

- **ridge and furrow**
- **saline soils**
- **saline water irrigation**
- **salt budget**
- *Salvadora persica (Meswak)*
- **tissue tolerance**
- **vertisols**
- **water logging**

REFERENCES

1. Ahmed, M. Z., Gilani, S. A., Kikuchi, A., Gulzar, S., Khan, M. A., & Watanabe, K. N. (2011). Population diversity of *Aeluropus lagopoides*: a potential cash crop for saline land. *Pakistan Journal of Botany*, **43**, 595–605.
2. Almas, K. (2002). The effects of Salvadora persica extract (Miswak) and chlorahexidine gluconate on Human Dentin: A SEM study. *J. Contemp. Dent. Prac.*, 3, 27–30.
3. Al-Otaibi, M., M., Al-Harthy, A., Gustafsson, A., Johansson, R. C., & Angmar-Mansson, B. (2004). Subgingival Plaque microbiota in Saudi Arabians after use of miswak chewing stick and tooth brush. *J. Clin. Periodontol.*, 31, 1048–1053.
4. Al-Otaibi, M., M., Al-Harthy, A., Söder U. A., & Angmar-Månsson, B. (2003). Comparative effect of chewing sticks and tooth brushing on plaque removal and gingival health. *Oral Health Prev. Dent.*, 1, 301–307.
5. Asfaw, K. G., & Danno, F.I, (2011). Response of dry matter production of tef (*Eragrostis tef* Zucc) (Trotter) accessions and varieties to NaCl salinity. *Curr. Res. J. Biol Sci.*, 3, 300–307.
6. Ashraf, M., & Yasmin, H. (1991). Differential water logging tolerance in three grasses of contrasting habitats: *Aeluropus lagopoides* (L.) Trin., *Cynodon dactylon* (L.) Pers. and *Leptochloa fusca* (L.) Kunth. *Environmental and Experimental Botany*, 31, 437–445.
7. Bogemans, J., Stassart, J. M., & Neirincex, L. (1990). Effects of NaCl stress on ion translocation in barley. *J. Plant Physiology*, 135, 753–758.
8. Chinchmalatpure A. R., & Gururaja Rao G. (2009). Nutrient management for major cropping systems in salt affected Vertisols. In: *Enhancing Nutrient Use Efficiency in Problem Soils. Gurbachan Singh, Ali Qadar, N. P. S. Yaduvanshi and P. Dey (Eds)*, Central Soil Salinity Research Institute, Karnal, India. 109–131.
9. CSSRI (2014). *CSSRI Vision 2050*. Central Soil Salinity Research Institute, Karnal. 29p.
10. Eid, M. A. (2011). Halophytic plants for phytoremediation of heavy metals contaminated soil. *J. Am. Sci.*, 7, 377–382.

11. Elfeel, A. A., & Al-Namo, M. L. (2011). Effect of imposed drought on seedlings growth, water use efficiency and survival of three arid zone species (*Acacia tortilis, subsp raddiana, Salvadora persica* and *Leptadenia pyrotechnica*). *Agriculture and Biology Journal of North America*, 2, 493–498.

12. FAO-AGL. (2000). *Global Network on Integrated Soil Management for Sustainable Use of Salt-Affected Soils.* FAO-AGL, Land and Plant Nutrition Management Service. (http://www. fao.org /ag/AGL/agll/spush).

13. Flowers, T. J. (1985). Physiology of halophytes. *Plant and Soil*, 89, 40–56.

14. Gulzar, S., & Khan, M. A. (2001). Seed germination of a halophytic grass *Aeluropuslagopoides. Annals of Biology*, 87, 319–324.

15. Gulzar, S., Khan, M. A., & Ungar, I. A. (2003). Effects of Salinity on growth, ionic content and plant water relations of *Aeluropus lagopoides. Comm. Soil Sci. Plant Anal.*, 34, 1657–1668.

16. Gururaja Rao, G. (1995). India discovers a salt species. *Salt Force News*, Victoria (Australia), 45, 4.

17. Gururaja Rao, G., Chinchmalatpure, A. R., Meena, R. L., & Khandelwal, M. K. (2011). Saline agriculture in saline Vertisols with halophytic forage grasses. *J. Soil Salinity and Water Quality*, 3, 41–48.

18. Gururaja Rao, G., Nayak, A. K., Chinchmalatpure, A. R. (2003). *Salvadora persica: A Life Support Species for Salt Affected Black soils.* Tech. Bull. 1/2003, Central Soil Salinity Research Institute, Regional Research Station, Bharuch, Gujarat. 54 p.

19. Gururaja Rao, G., Nayak, A. K., Chinchmalatpure, A. R., Abhay Nath, & Ravindra Babu, V. (2004a). Growth and yield of *Salvadora persica*, a facultative halophyte grown on highly saline black soil. *Arid Land Res. & Management*, 18, 51–61.

20. Gururaja Rao, G., Nayak, A. K., Chinchmalatpure, A. R., Mandal, S., & Tyagi, N. K. (2004b). Salt tolerance mechanism of *Salvadora persica* grown on highly saline black soil. *J. Plant Biol.*, 31, 59–65.

21. Gururaja Rao, G., Nayak, A. K., Chinchmalatpure, A. R., Singh, Ravender, & Tyagi, N. K. (2001). *Resource Characterization and Management options for Salt Affected Black Soils of Agro-ecological Region V of Gujarat State.* Tech. Bull.1/2001, Central Soil Salinity Research Institute, Regional Research Station, Anand, Gujarat, 83 p.

22. Gururaja Rao, G., Polra, V. N.m & Ravindra Babu, V. (1999a). Salt tolerance of *Salvadora persica* – A facultative halophyte. *Indian J. Soil Conser.*, 27, 55–63.

23. Gururaja Rao, G., Polra, V. N., Ravindra Babu, V., & Girdhar, I. K. (1999b). Growth and development of *Salvadorapersica* on highly saline blacks soils: Salt tolerance during immature phase. *Indian J. Plant Physiol.*, 4, 152–156.

24. Gururaja Rao, G., & Rajeswara Rao, G. (1982). Anatomical changes in the leaves and their role in adaptation to salinity in pigeon pea (*Cajanusindicus* Spreng) and gingelley (*Sesamumindicum* L.). *Proc. Indian National Sci. Acad.*, 48, 774–778.

25. Gururaja Rao, G., & Ravindra Babu, V. (1997). Epicuticular wax of *Salvadora persica* grown on saline black soils. *Indian J. Plant Physiol.*, 2, 290–292.

26. http://www.icar.org.in/files/forage-and-grasses.pdf. Opened on 12.07.2015.

27. http://www.fao.org/ag/agp/AGPC/doc/Gbase/data/pf000213.htmOpened on 12.07.2015.

28. ICAR-NAAS, 2010. *Degraded and Wastelands of India: Status and Spatial Distribution.* Indian Council of Agricultural Research and National Academy of Agricultural Research, New Delhi. 158p.

29. Khatak, M., Khatak, S., Siddqui, A. A., Vasudeva, N., Aggarwal, A., & Aggarwal, P. 2010. *Salvadora persica*. *Pharmacogn Rev.*, 4, 209–214.
30. Kumar, A., & Abrol, I. P. 1986. *Grasses in Alkali Soils*. Bulletin No. 11. Central Soil Salinity Research Institute, Karnal. 95 p.
31. Mabberley, D. J. (2008). *Mabberley's Plant-book, A Portable Dictionary of Plants, Their Classifications and Uses*. 3rd Edition, University Press, Cambridge.
32. Maggio, A., Reddy, M. P., & Jolly, R. J. (2000). Leaf gas exchange and solute accumulation in the halophyte, *Salvadora persica* grown under moderate salinity. *Environ. Exp. Bot.*, 44, 31–38.
33. Mahar, A. Q., & Malik, A. R. (2001). A study on the medicinal plants of Kotdiji, District Khairpur Sindh, Pakistan. *Scient. Sindh J. Res.*, 8, 31–38.
34. Mondal, A. K., Obi Reddy, G. P., & Ravisankar, T. (2011). Digital database of salt affected soils in India using Geographic Information System. *Journal of Soil Salinity and Water Quality*, 3, 16–29.
35. Munns, R., Fisher, D., & Tonnet, M. (1986). Na and Cl transport on the phloem from leaves of NaCl-treated barley. *J. Plant Physiology*, 13, 757–766.
36. Murthy, R. S., Bhattacharjee, J. C., Lande, R. J., & Pofali, R. M. (1982). Distribution, characterization and classification of Vertisols. *Transaction 12th International Congress Soil Science*, New Delhi, India, 2, 3–22.
37. Pitman, M. G. (1975). Ion transport in plant cells and tissues. In: *Whole Plant*, Baker, D. A., & Hall, J. L. (Eds.), pp. 267–308, North-Holland, Amsterdam-Oxford.
38. Qureshi, R. H., & Barrett-Lennard, e.g., (1998). *Saline Agriculture for Irrigated Land in Pakistan: A Handbook*. Australian Centre for International Agricultural Research, Canberra, Australia. 142 p.
39. Rad, M. S., Rad, J. S., Teixeira da Silva, L. A., & Mohsenzadeh, S. 2013. Forage quality of two halophytic species, *Aeluropus lagopoides* and *Aeluropus littoralis*, in two phenological stages. *International journal of Agronomy and Plant Production*, 4, 998–1005.
40. Reddy, M. P., Shah, M. T., & Patolia, J. S. (2008). *Salvadora persica*, a potential species for industrial oil production in semiarid saline and alkali soils. *Industrial Crops and Products*, 28, 273–278.
41. Savithramma, N. Ch., Sulochana, & Rao, K. N. (2007). Ethnobotanical survey of plants used to treat asthma in Andhra Pradesh. *India. J. Ethnopharmacol.*, 113, 54–61.
42. Strogonov, B. P. (1964). Structure and function of plant cells in saline habitats. New trends in the study of salt tolerance. *Israel Program for Scientific Translations*, Sci. Transl. Jerusalem.
43. Tiku, B. L. (1976). Effect of salinity on photosynthesis of the halophyte, *Salicornia rubra* and *Distichlis stricta*. *Physiol. Plant*, 37, 25–28.
44. Weretilnyk, E. A., Bodnanek, S., McCue, K. F., & Hanson, A. D. (1982). Comparative biochemical and immunological studies of the glycinebetaine synthesis pathways in diverse families of dicotyledons. *Planta*, 178, 342–352.

CHAPTER 6

EFFECT OF SOIL SALINIZATION ON PLANT GROWTH AND PHYSIOLOGY OF *PLECTRANTHUS* SPECIES

G. V. RAMANA, CH. RAMAKRISHNA, SK. KHASIM BEEBI, and K. V. CHAITANYA

CONTENTS

6.1 INTRODUCTION

Plant system grows well and produces optimally only under a set of optimum conditions of temperature, light, water, soil nutrients and atmospheric conditions etc. On the other hand, plants being sessile, are continuously exposed to various abiotic stress factors such as drought, salinity, temperature (both high and low), radiation, dust, which make huge impact on the biomass production, yield and survival of agriculturally important crops and plants. Such an exposure leads to severe losses in productivity reaching up to 70% in most crop plants [21, 55]. Among the abiotic stresses, soil salinity has been a major threatening factor that is affecting the global agriculture production and destabilizing the balance of need and demand of burgeoning population [59]. Approximately 953 million ha constituting about 8% of global agriculturally important land have been affected by salinity [22] where the concentration of salts in soil water is sufficiently high to impair plant growth. Both anthropogenic and natural causes have caused build-up of salts in the soil [8, 12, 57]. Adversity of the situation is compounded when irrigation water is also saline as is the case with croplands in arid and semiarid climatic zones. Aridity in climate is proportional to intensity of salinization in top layers of soil with diverse distributions of salts in the soil profiles [54].

Plants differ widely in salt tolerance from sensitive glycophytes, whose normal growth is inhibited at low concentration of salts to the most resistant halophytes, which grow profusely on saline habitats. While barley, cotton, rye, and Bermuda grass are classified as salt tolerant (a relative term), wheat, oats, sorghum, and soybeans are classified as moderately salt tolerant.

Maize, alfalfa, pulses and most vegetables are moderately sensitive to sensitive to salt. Voluminous data on salt tolerance is available that lists the soil salinity or salinity of the irrigation water that could be tolerated by a given crop for a pre-decided yield reduction [7, 40, 46]. Recent studies following Tsunami in India revealed that many fruit yielding trees such as mango, rambutan and cocoa are quite sensitive to soil salinization. The yield of coconut and rubber trees also reduced as a result of cation imbalance in the ground water inhibiting potassium ion uptake along with the variations in soil pH [44].

Salinity or sodicity stress impacts the plants in a number of ways affecting their morphology, physiology and biochemical processes (Figure 6.1). These invariably induce wide range of changes in the plants as their regular metabolism is seriously affected. These adversely affect plant growth, yield and even quality [1]. It is observed that crucial physiological processes such as transpiration, stomatal conductance and photosynthesis are reduced in response to increasing exposure to salinity stress [72].

Reduced plant growth under increasing salinity stress is attributed to several factors such as accumulation of Na^+, which in turn restrict K^+

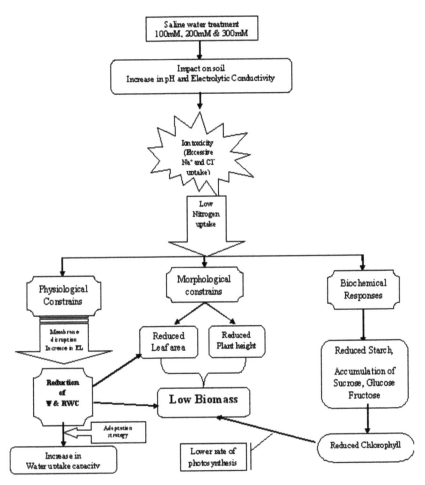

FIGURE 6.1 Impact of soil salinization on the morpho-physiology and growth of *Plectranthus* species.

uptake [67] and alterations in the osmotic potential, ion balance and nutrition availability [39]. Grattan and Grieve [26], in fact mentioned that soil salinity and its impact on nutrient availability, uptake and distribution are highly complex topics in the context of crop plants nutrition. Plants respond to the changes in these processes through several adaptation strategies (Figure 6.1). Mechanisms of salinity tolerance in plants have been extensively studied and in the recent years these studies focus on the function of key enzymes and plant morphological traits. This chapter reports the results of a study on *Plectranthus* species response to salinity stress caused by the saline irrigation water. The *Plectranthus* species chosen for this study are perennial, aromatic with sheath like stems and belongs to *Lamieaceae* family commonly found in tropical-to-subtropical climatic conditions. These can survive on well-drained fertile soil to dry hill slopes with an altitude ranging from 300 m to 1800 m.

Plectranthus species are grown for their ethno botanical use as ornamental, medicinal and economic plants due to variations in leaf texture, diverse secondary metabolites and ease of cultivation [60]. In ethnic medicines these are used as remedy for stomachache, fever, headache, epilepsy and dyspepsia [38]. The water conservation, medicinal properties and antimicrobial properties of the root, stem and leaf extracts of the *Plectranthus* species have been well studied but not much is known about their abiotic stress responses especially to the irrigation water salinity.

In this chapter, the role of salinity stress on the growth and biomass production of six different *Plectranthus* species as well as on their salinity stress adaptations mechanisms have been reported.

6.2 MATERIALS AND METHODS

Six genotypes of *Plectranthus* species, *Plectranthus amboinicus* (Lour.) Spreng., *Plectranthus scutellarioides* (Linn.) R. Br., *Plectranthus amboinicus Variegatus* Lour., *Plectranthus zeylanicus* (Benth.) L.H. Cramer, *Plectranthus barbatus* (Wild.) Andrews and *Plectranthus caninus* Roth, were identified by Dr. S.B. Padal, Department of Botany, Andhra University, Visakhapatnam having specimen numbers 21901, 21902, 21903, 21904, 21905 and 21906 respectively in the Herbarium of GITAM University, Visakhapatnam, India. These six genotypes were used for the present investigation. All the six species were propagated in 30 cm pots; five replicates each under natural photoperiod [Irradiance (400–700 nm) of 1600–1800 μmol m^{-2} s^{-1}] with day/night temperatures of 30°C/23°C with an average air humidity of 60%. Three-month-old plants with uniform growth were used.

The plant pots were arranged randomly 1 m apart. Irrigation to the pots was applied daily. The irrigation water with three different NaCl concentrations 100 mM, 200 mM and 300 mM were used for two weeks. Na+ and Cl- being the chief ions in the water, these induce salt stress in plants. While sodium is non-essential element, chloride is an essential micronutrient for all plants [45], but both the ions are toxic if present in excessive amount and are capable of triggering disorders in plants. Displacement of calcium and magnesium ions can occur as a result of long-term use of high SAR water [66]. Control plants of all genotypes were maintained separately and irrigated daily with normal water. Third or fourth leaf from the top of the plant was collected for physiological and biochemical studies.

6.2.1 SOIL WATER ANALYSIS

Various soil properties were measured as per standard procedures [4]. Electrical conductivity (EC) and pH were measured in 1:2 soil: water ratio. Organic carbon was determined using titration method [80]. Phosphorous was analyzed using Bray and Kurtz [10] method. Potassium was analyzed using Toth and Prince [77] approach using flame photometer. All the micro nutrients were analyzed using methodology of Lindsay and Norvell [36]. Various cations and anions and nitrate Nitrogen (NO_3-N) in the water were determined as per procedure given by Richard [66]. Sodium Adsorption Ratio (SAR) was calculated to assess the sodicity hazard of the irrigation waters.

6.2.2 ANALYSIS OF PLANT SAMPLES

6.2.2.1 Biomass, Leaf Area and Plant Height

Aerial fresh biomass was measured using a sensitive balance. Plant height was measured using a meter rod. Graph paper method of Pandey and Singh [53] was employed to measure plant leaf area.

6.2.2.2   Sodium, Chloride and Nitrogen Analysis in Plant Samples

Total Na^+ content in leaves was estimated according to Jackson [30] by a micro-processor based flame photometer. Extraction of Cl^- was done with 0.1M sodium nitrate in 1:100 (w/v) ratio [24]. Chloride content in leaves was quantified by mercury (II) thiocyanate method as per Adriano and Doner [2] using UV-VIS spectrophotometer. Total nitrogen in plant samples was analyzed using Kjeldal method.

6.2.2.3   Leaf Water Potential, Relative Water Content, Water Uptake Capacity and Electrolytic Leakage

Leaf water potential (Ψ) of control and salinity stressed *Plectranthus* plants was determined using plant moisture system (SKPM 1400 series, Skye instruments, UK). Relative water content and water uptake capacity of leaves were measured by determining their fresh weight (FW), dry weight (DW) and turgid weight (TW). Relative water content and water uptake capacity were determined using method described by Sangakkara et al. [71]. Leaf membrane damage was determined by recording electrolyte leakage (EL) following the procedure described by Valentovic et al. [78].

6.2.2.4   Chlorophyll, Glucose, Fructose, Sucrose and Starch Contents

Chlorophyll content was measured according to Arnon[5]using acetone extracts. The glucose and fructose contents in the 80% ethanolic extract were determined according to Dubois et al. [19] and Ashwell [6] respectively. The contents of starch and sucrose in the leaf tissues were estimated enzymatically according to the method of Ramachandra Reddy et al. [62].

6.3   RESULTS AND DISCUSSIONS

6.3.1   *CHEMICAL PROPERTIES OF SOIL*

Almost all aspects of plant development, including germination, vegetative growth and reproductive development are affected by soil salinity, which imposes osmotic stress and ion toxicity, reduces nutrient availability (N, Ca, K, P, Fe, Zn) or causes imbalance in uptake and imposes oxidative stresses on plants. Soils analysis revealed that chemical properties of the soil varied widely as a result of irrigation with saline water of 0.1 M, 0.2 M and 0.3 M. Except control, soil EC values were more than 9 dSm^{-1} indicating high salinity levels in the soil, the limit of EC being below 2 dS m^{-1} for normal soils (Table 6.1).

The pH of the soil increased gradually from 7.3 to 8.5 by the end of treatment. While the pH of 7.3 indicates a normal soil, pH of 8.5 of 1:2 soil: water solution represents a border line case of alkali soil, which was recorded in the treatment where soil was exposed to 0.3 M NaCl. These values together substantiate that the soil is hostile for plant growth. In the present study, except nitrate concentration all the remaining elements, i.e., bicarbonates, calcium magnesium ratio, chlorides and residual sodium carbonates in water increased due to the increase in SAR (Table 6.2).

Increased SAR levels followed the pattern of NaCl concentration in treated waters 100 mM, 200 mM and 300 mM NaCl respectively. The 300 mM concentration of water was found to be most unsuitable for irrigation as its SAR value was more than 19. Sodium ions activity was versatile and profound, which altered the soil and water chemistry. Impact of increas-

TABLE 6.1 Salinity Stress-Induced Changes in the Soil Profile of Six Different Plectranthus Species

Parameters	Control	100 mM	200 mM	300 mM
pH	7.30±0.12	8.00±0.20	8.10±0.10	8.50±0.14
EC (dS m^{-1})	0.62±0.02	9.70±1.16	19.50±2.30	22.60±1.16
Organic carbon (%)	0.70±0.04	0.65±0.05	0.55± 0.02	0.40±0.03
Copper (ppm)	1.26±0.61	0.51±0.14	0.42±0.12	0.36±0.09
Iron (ppm)	15.18±0.14	9.83±1.92	8.55±1.51	6.20±1.3
K$_2$O (kg ha^{-1})	627.95±10.6	211.00± 7.6	348.11±6.9	306.17±5.2
Manganese (ppm)	2.30± 0.13	2.10±0.17	1.71± 0.14	0.82±0.19
P$_2$O$_5$ (kg ha^{-1})	91.36±5.2	91.36±3.2	91.36± 4.2	8.91±2.8
Zinc (ppm)	2.06±0.47	2.66±0.34	1.91±0.71	1.58±0.54

Each value is the mean ± SE of five independent determinations, ($t_{(4)}$ = 2.2, $P<0.05$).

TABLE 6.2   Salinity Stress-Induced Changes in the Four Different Water Compositions Used for Irrigation of Six Different Plectranthus Species

Parameters	Control	100 mM	200 mM	300 mM
Nitrate Nitrogen (mg L^{-1})	8.15±0.14	7.56±0.51	7.45±0.39	7.4±0.46
Sodium (mg L^{-1})	0.81±0.12	7.76±0.83	17.9±0.64	33.5±0.76
Bicarbonates and carbonates (meq L^{-1})	3.00±0.04	5.80±0.07	9.20±0.11	10.20±0.09
Calcium and Magnesium (meq L^{-1})	2.20±0.13	3.80±0.67	5.60±0.51	5.80±0.42
Chlorides (meq L^{-1})	4.00±1.18	76.00±2.4	108.00±3.1	200.00±2.8
RSC (meq L^{-1})	1.20±0.03	2.00±0.06	3.60±0.09	4.40±0.07
SAR (meq L^{-1})$^{1/2}$	0.77±0.08	5.66±0.23	10.65±0.31	19.74±0.56

Each value is the mean ± SE of five independent determinations, ($t_{(4)}$ = 2.2, $P<0.05$).

ing sodium ions increased pH, EC and decreased essential organic carbon apart from the micro macro elements (Table 6.1). Apart from the above, organic carbon content was also moderate in all the soil samples except in 300 mM NaCl soil which is less than 0.5% illustrating the reduction of organic material in soils with high alkalinity. Phosphorus levels found to be not much affected due to salinization except in severe salinity.

6.3.2   MORPHOLOGICAL CHARACTERS

Plants exposed to salt stress are known to have suppressed growth and biomass [56]. Gradual reduction in plant height was observed in stressed plants as compared to control. With increasing levels of salinity there was a loss of ~3% reduction in plant height in six genotypes of *Plectranthus*. *Plectranthus amboinicus, Amboinicus variegates and caninus* had higher plant height followed by *Plectranthus zeylanicus and barbatus*, which could sustain equally at higher salinity with a plant height of 27.7 cm and 26.6 cm, respectively (Table 6.3). Increasing soil EC, SAR and pH with increasing salinity of irrigation water influenced the growth and biomass of the plants. Similar results for other crops have been reported in the literature. In *Mangifera indica* increase in the uptake of sodium and chloride ions have been reported to cause reduction in the plant height to more than

3% [16]. The increased salinity of irrigation water resulted in more growth reduction in shoots of *Gossypium arboreum L.* [27], *R. fortuniana* and *R. odorata* [49].

Total aerial biomass was observed to have a downward trend with increasing salinity of irrigation water. Biomass in *P. scutelleriodes* reduced by more than 50% whereas in remaining genotypes it was approximately 20 to 30%, which might be attributed to increasing salinity (Table 6.3). Growth and biomass of pepper plants irrigated with saline water of EC 4.69 dSm^{-1} has been reported to decrease significantly [28]. Loss of biomass could also be ascribed to reduced photosynthesis. It could also be correlated to the availability of nutrients whose balance and increasing plant nutrient disruption is affected due to excess sodium ion activity [70]. Impact of Na$^+$ and Cl$^-$ ions on nitrogen availability is clearly evident from the results obtained from foliar elemental analysis of *Plectranthus* species. Increase of salts content in water is likely to negatively affect the nitrate content of water and nitrogen uptake in the genotypes, which further reduced the biomass of the *Plectranthus* genotypes under salinity treatment. Apart from biomass, total yield also found to have reduced in fruit crops. Salinity impact have reduced the aerial biomass and fruit yield in three different cultivars of *Lycopersicon esculentum* [37] and in two *Cucurbit* species when irrigated with three different water salinities [89].

When the plants are exposed to salinity stress, leaf area is observed to reduce, which could be attributed to variations in osmotic pressure. In the present study, reduction in leaf area to the tune of 30 to 40% occurred in all the six *Plectranthus* species when exposed to 0.3 M NaCl stress (Table 6.3). *P. scutelleriodes* experienced severe reduction in leaf area and *P. amboinicus* sustained with a leaf area of 45.28 sq. cm even at severe salinity (Table 6.3). Leaf area reduction in cotton genotypes from 0 mM NaCl to 240 mM NaCl have been reported earlier [88].

Morphological characters and agriculture yield have been reported to have been negatively affected due to soil salinity accompanied by a wide range of hydrospheric, atmospheric and pedospheric constrains. Sodium being the chief cause of salinity, it terminates the secondary clay minerals in the soil by dispersion. It occurs due to the proliferation of Na$^+$ replacing the Ca^{2+} and Mg^{2+} facilitated by certain physical traits like ionic radii, hydration ability, electric charge and variations in pH [50]. Dispersion of

TABLE 6.3 Salinity Stress-Induced Changes in the Leaf Area, Biomass, Plant Height and Chlorophyll Contents of Six Different *Plectranthus Species*

Plants/treatments	Leaf area (cm^2)	Biomass (g)	Plant height (cm)	Chlorophyll (mg g^{-1})
P. ambonicus				
Control	74.38±2.1	793±1.12	42.4±1.9	0.24±0.017
100 mM	63.8±1.5	663±2.8	39.2±1.7	0.21±0.011
200 mM	47.9±1.69	589±2.2	35.6±2.4	0.183±0.018
300 mM	45.28±2.2	552±1.9	31.8±1.6	0.173±0.013
P. amboinicus Variegatus				
Control	34.69± 2.1	721±1.89	43.8±2.3	0.251±0.021
100 mM	32.89±1.8	698±1.73	41.6±1.9	0.205±0.015
200 mM	28.61± 1.9	621±1.87	39.3±2.7	0.175±0.031
300 mM	22.8±1.6	586±2.3	37.6±2.5	0.168±0.042
P. barbatus				
Control	66.14±2.3	591±1.84	36.4±1.56	0.167±0.016
100 mM	57.94±1.9	582±1.92	34.7±1.4	0.157±0.024
200 mM	45.8±2.16	568±1.76	29.2±1.87	0.149±0.013
300 mM	40.18±1.2	522±2.14	26.6±1.3	0.133±0.021
P. caninus				
Control	32.34±1.5	599±1.47	41.4±1.76	0.162±0.021
100 mM	31.64±1.3	576±1.31	37.8±1.21	0.154±0.019
200 mM	29.14±1.7	551±2.4	34.2±1.54	0.146±0.020
300 mM	26.32±1.9	539±2.12	31.5±1.4	0.112±0.010
P. zeylanicus				
Control	48.06±1.1	572±1.45	43.5±1.12	0.498±0.017
100 mM	44.35±1.8	563±2.6	39.9±1.5	0.427±0.021
200 mM	42.81±1.5	548±2.1	32.5±1.96	0.417±0.019
300 mM	40.21±1.7	532±2.71	27.7±1.8	0.315±0.016
P. scutelleriodes				
Control	22.56±1.06	672±0.55	30.7±1.06	0.245±0.017
100 mM	21.4±1.11	518±1.91	27.6±1.12	0.104±0.018
200 mM	20.68±1.5	412±1.4	22.8±1.18	0.0952±0.011
300 mM	16.54±1.3	389±1.68	18.8±1.97	0.0427±0.01

Each value is the mean ± SE of five independent determinations, ($t_{(4)}$ = 2.2, $P<0.05$).

clay causes pedospheric water logging due to transport of clay particles through the soil leading to their accumulation at some depth as well as resulting in blocking pores in soil horizons [11]. Prolonged dispersion of clay induces topsoil crusting enhancing surface runoff, soil erosion, nutrient leaching and surface water logging, reducing hydraulic properties like infiltration, water permeability, aeration, etc. at root zones and top soil horizons.

6.3.3 NUTRIENT UPTAKE UNDER SALINITY STRESS

The predominant symptom of salinity stress is the loss of biomass, which is also correlated to the availability of nutrients as excess sodium ion activity results in nutrient imbalance and increased plant nutrient disruption [70]. Nutritional imbalance is apparent from decrease in organic carbon, potassium and micro elements like copper, zinc, iron and manganese with increasing salinity (Table 6.1). In *Plectranthus* species, the Na^+ ions accumulated more along with Cl^- ions which negatively affected the nitrogen intake of plants with increasing salinity (Figure 6.2).

Both Na^+ and Cl^- compete with other essential nutrients like K^+, NO_3^- and H_2PO_4 to transport proteins to binding sites and further for translocation and deposition within the plants [76]. Enhanced uptake of Na^+ and Cl^- and decreased uptake of K^+, in radish, muskmelon and in strawberry resulted in the decline of vegetative growth, fruit yield and finally induced plant mortality after 60 days [51]. Na^+ and Cl^- ions uptake increased to more than 2% in *Citrus aurantium* under salt stress which influenced uptake of other nutrients [73]. Antagonistic effect of sodium and chloride has a severe impact on nitrogen uptake in plants as reported for sugarcane plants [41, 61]. Cl^- inhibited NO_3^- uptake more and in response to salinity NO_3^- concentrations are drastically reduced [25].

Soil salinization influenced the potassium levels to considerable extent, reducing its presence in the soil. It might be due to sodium ions blocking the potassium preventing it from reaching the plant. Studies have suggested that increase in organic carbon encouraged positive correlation with the availability of micronutrients in the soil and affected negatively under alkaline pH [74].

Excessive Na$^+$ and Cl$^-$ are reported to cause deficiencies of calcium, potassium and imbalance in the ratio of other nutrients [45]. The solubility of micronutrients like Cu, Fe, Mn and Zn is particularly low in most of the saline-alkali soils. Indian soils are reported to be deficient in zinc (12%), iron (5%) manganese (3%) and copper (33%) [75]. Micronutrients like Cu, Fe and Zn were available to the plants even after two weeks of treatment but their concentration from normal to highly saline soil was observed to be decreasing approximately two fold.

Iron is an important micronutrient as a component of enzymes that are involved in a variety of biochemical processes like nitrogen fixation, photosynthesis and respiration [34]. Plant growth, therefore is highly depressed in case of lack of Fe nutrition or due to gradual depletion of Fe in the sodic soils [58, 86]. Increase in the soil pH above 5.0 reduces the availability of Fe in the soil because availability of iron to plant from soil is much better at low soil pH [35], which means alkaline soil will not allow various forms of iron in the soil to dissolve to restrict their availability. Earlier experiments have revealed the reduced uptake of iron with increasing alkalinity of the soil attributable to hydration and oxidation of higher hydroxides and oxides [15].

Increase in pH, reduction in organic carbon and increase in EC also reduces the availability of zinc as compared to optimum conditions [84]. Similarly, micronutrients like Cu and Mn also exhibited negative correlations under alkaline conditions [73]. Copper as an essential redox active transition metal element have significant role in many phyto-physiological processes. It is a structural element for a diverse regulatory proteins and active in cell wall metabolism, mitochondrial respiration and photosynthetic electron transport [64]. When copper ion is not available in required amounts then typical deficiency symptoms can be observed affecting leaves and reproductive organs [85].The availability of micro nutrients is affected by sodicity and also by the application of synthetic fertilizers comprising of macro nutrients like phosphate which will hamper the uptake of micro nutrients [33]. Considering the interaction between salinity and micronutrients like copper, Eskandari [20] suggested the need to undertake extensive research especially for horticulture crops. Manganese in soil profile was severely affected at moderately

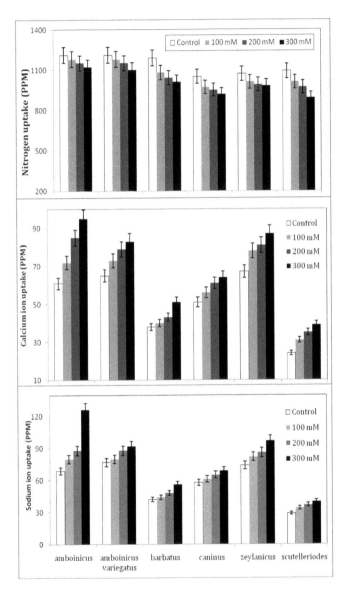

FIGURE 6.2 Na⁺, Cl⁻ and nitrogen uptake in six different *Plectranthus* species. Each point is an average of five independent determinations ± SE, (t(4) = 9.2, P< 0.05).

and high saline soils falling below critical level due to activity of sodium ions (Table 6.1).

6.3.4 PHYSIOLOGY OF PLANTS

Salinity is a common problem of arid and semiarid environments where many plants have developed mechanisms to acclimatize and tolerate salinity apart from water deficit [47]. Osmotic adjustments will vouchsafe in maintaining the cell turgor and water uptake under saline conditions, which is a significant mechanism to adapt to salt stress [13]. Water potential is an important indicator reflecting growth, development, metabolism and stress resistance. During stress conditions plants tend to accumulate solutes to reduce osmotic potential, improve water absorption capacity, maintenance of turgor pressure commonly described as osmotic adjustment. In the present study, all genotypes exhibited a reduction in leaf water potential with more than −1 MPA difference from control to 0.3 M salinity stress (Figure 6.3). Earlier studies have shown that leaf water potential

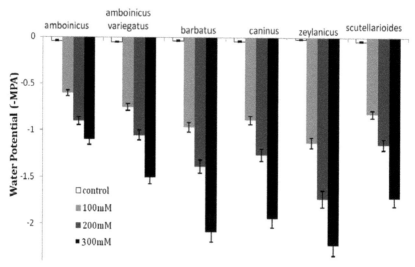

FIGURE 6.3   Water potential of six different *Plectranthus* species under salinity stress. Each point is an average of five independent determinations ± SE, (t (4) = 8.4, P < 0.05).

decreased in 8 maize genotypes subjected to 100 mM salinity [48]. Even in halophyte *Iris lacteal,* water potential reduced with increase in salt concentration [82].

Water uptake capacity (WUC) of a plant is used to assess the capability of a plant to absorb water per unit dry weight in relation to turgid weight. Many plants under salinity stress exhibited improved water uptake capacity to attain higher turgidity than plants growing in normal conditions to balance its water requirement under stressed conditions [29]. In the present study, *P. amboinicus* and *P. amboinicus variegatus* showed superior WUC compared to remaining genotypes.

Apart from the *Plectranthus* species increased WUC has been reported in plants such as wheat [3], pepper [23], green gram [31], *vigna* genoptypes [83] due to salinity stress. When calcium was applied to the pepper and wheat plants, their WUC decreased even under saline conditions emphasizing the role of Ca^+ in maintaining water relations. However, when Mg is predominant, i.e., the Mg:Ca ratio of water is high and coupled by the action of Na, the capacity of Ca will be dwarfed allowing the plant to increase its WUC as observed in *Plectranthus* genotypes. In several *vigna* genotypes [83] and in four cultivars of *Glycine max* [63], relative water content reduced while the water uptake capacity increased due to salinity stress correlating with the behavior of *Plectranthus* species.

Relative water content, water uptake capacity and EL increased steadily with increasing salinity (Figure 6.4). Relative water content (RWC) indicates the water status of plants and is an accurate methodology to measure plant water status in the context of cellular water deficit. Its reduction is due to the sodium toxicity at plant roots preventing proper water uptake by the plants. EL is a measure of ions in irrigation water or any potable water. Plasma membrane of plant cells plays a key role as a barrier to transport of molecules which is controlled *via* protein channel and phospho-lipids in the membrane [81]. Primary site of ion specific salt injury is plasma membrane [43]. Hence to identify salt tolerance in plants, EL from the plasma membrane is an important parameter to measure. In all the six *Plectranthus* genotypes RWC decreased due to increasing salinity stress which indicate the activity of toxic ions in blocking the free transport of water in the plants. Electrolytic leakage was found to increase gradually with increased treatment with saline water (Figure 6.4), which shows the

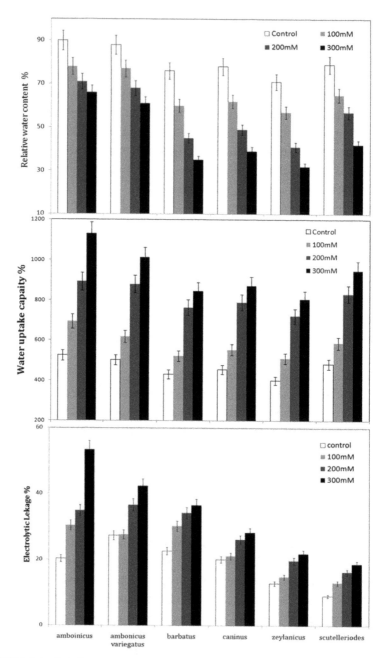

FIGURE 6.4 Relative water content, water uptake capacity and electrolytic leakage EL in six different *Plectranthus* species under salinity stress. Each point is an average of at least five independent determinations ± SE, (t (4) = 10.2, P < 0.05).

enhanced activity of ions is more in the genotypes stressed due to severe salinity. Concentration of Na^+ and Cl^- ions increased more than 10% from control to mild salinity and from there on up to severe salinity stress, which correspond to the salinity treatment concentrations and electrolytic leakage values representing activity of ions in the foliar samples.

Increase in the EL represents high membrane permeability due to increased damage of the plasma membrane by more number of ions [42]. In *Abelmoschus esculentus* plants grown under two saline conditions (2 and 4.2 dS m^{-1}), relative water content decreased and electrolytic leakage increased with increasing salinity [69]. In this study, water uptake efficiency increased with increasing salinity indicating the adaptability of *Plectranthus* species to acclimatize at higher salinity stress levels.

Salinity affects photosynthesis mainly through reduction in leaf area, chlorophyll content and stomatal conductance, and to a lesser extent through decrease in photosystem II efficiency. Total Chlorophyll content in all the *Plectranthus* species reduced due to increasing Na^+ and Cl^- ions. Reduction in the chlorophyll content was high in the *P. scutellerioides* indicating it is more sensitive to salinity when compared to other five genotypes (Table 6.3). During salinity stress the chlorophyll content was found to reduce with increasing activity of chlorophyllase enzyme [65], disruption of chloroplast, unstable pigments and protein complex due to increased concentrations of ions [18].

From the present data, it is evident that the physiological responses in *Plectranthus* species were interrelated under salinity stress. Strong relationship is found between electrolytic leakage and chlorophyll content which was also reported in previous studies [32]. Membrane permeability might have increased leading to more EL activity partly due to reduction in chlorophyll content leading to leaf senescence [17].

6.3.5 SALINITY VERSUS CAM PLANTS

Plectranthus is a Crassulacean Acid Metabolism (CAM) plant that fixes carbon nocturnally and stores as organic acids in its vacuoles. The carbon so obtained is fixed during day time substantiated by the biochemical pathways. The CO_2 concentrating mechanism in CAM plants is expected to contribute to decreased ROS generation. In many CAM plants, the regeneration of Phosphoenol pyruvate depends mainly on starch content which shows variations in starch transition [9]. When CAM mechanism is active, starch degrading enzymes are required along with the proper transport of intermediate products across membrane of chloroplast [14]. Hence degradation of transitory starch is necessary for operation of bio-chemical pathways like glycolysis as demonstrated in *Mesembryanthe-mum crystallinum* which is a facultative CAM plant. Mutant cultivars of *Mesembryanthemum crystallinum* deficient in starch synthesis were found to be incapable in maintaining the CAM mechanism due to lack of enzyme phosphoglucomutase. However, when glucose is applied the CAM mecha-nism is reported to have been recovered [14].

Most of the carbohydrates have been retained in the leaves of wild type and mutant plants for the purpose of nocturnal PEP generation. Six *Plectranthus* species under three different stress conditions exhibited a reduction in their starch contents with respect to increasing salinity and an increase in glucose, fructose and sucrose contents (Table 6.4). In *Nicotiana tabacum* under saline conditions sugar content was found to increase[68]. However when *Aloe vera* was irrigated with sea water hav-ing an EC of 23.4 dS m^{-1} there was a reduction in glucose content [87]. Extreme salt concentrations greater than the present treatment might have resulted in such variations in *Aloe vera*, which indicates that photosyn-thetic activity reduced due to loss of chlorophyll content can be correlated to accumulation of sugars in leaves. Accumulation of sugars like sucrose and glucose will not only reduce the photosynthetic rates but also act as signaling agents affecting the photosynthesis [79]. Sunflower propagated under lower nitrogen levels accumulated more sugars [52], which once again correlated to the effect of soil salinity on nutrition availability to plants affects on photosynthetic rates.

6.4 SUMMARY

Primary and secondary salinization and sodication of the soils has greatly impacted agriculture having a profound effect on the flora of the world as well as affecting growth and yield of crop plants. Genus *Plectranthus*,

TABLE 6.4 Effect of Salinity Stress on Carbohydrates of Six Different *Plectranthus* Species

Plants/treatments	Glucose (mg⁻¹g fw)	Starch (mg⁻¹g fw)	Sucrose (mg⁻¹g fw)	Fructose (mg⁻¹g fw)
P. ambonicus				
Control	81±2.1	60±3.2	98±2.5	75±1.16
100 mM	85±3.1	53±2.1	87±2.1	79±1.67
200 mM	109±3.4	49±1.5	112±2.7	87±1.54
300 mM	117±1.6	44±1.6	130±1.8	93±2.13
P. amboinicus Variegatus				
Control	74±2.3	61±2.57	95±2.6	64±1.54
100 mM	79±1.9	54±2.31	85±1.71	76±1.89
200 mM	106±2.7	48±1.91	109±2.31	103±2.08
300 mM	121±2.1	43±1.89	126±2.41	108±2.17
P. barbatus				
Control	67±2.13	36.7±1.71	84±2.3	48±1.85
100 mM	72±1.76	31±1.45	86±2.2	55±1.92
200 mM	86±2.71	27±1.53	93±2.7	84±2.03
300 mM	98±2.48	24±1.46	101±2.1	96±2.23
P. caninus				
Control	69±1.76	38±1.56	90±1.9	58±1.32
100 mM	81±2.01	29±2.12	86±2.1	74±1.38
200 mM	93±1.98	22±1.78	101±2.7	89±2.07
300 mM	99±1.79	19±1.31	106±1.6	98±2.4
P. zeylanicus				
Control	66.8±1.8	34±3.4	81.5±1.7	48±1.2
100 mM	59±2.6	29±1.24	74±2.21	55±1.51
200 mM	82±1.9	26±1.13	86±2.4	84±1.67
300 mM	91±2.3	21±1.6	99±2.56	96±1.78
P. scutelleriodes				
Control	71±1.15	44±2.18	97.8±1.87	61±1.34
100 mM	76±2.51	41±2.2	82±2.11	69±1.68
200 mM	104±1.34	36±1.87	106±1.37	77±1.91
300 mM	109±1.56	32±1.43	110±2.43	85±2.27

fw = Fresh weight; Each value is the mean ± SE of five independent determinations, ($t_{(4)}$ = 2.2, $P<0.05$).

popular for its medicinal value and for its water conservation mechanism is widely cultivated. An investigation was conducted to understand the behavior of this genus when subjected to abiotic stress such as salinity. As anticipated, *Plectranthus* species have shown negative effect on its growth and biomass with increasing salinity stress. However, *P. amboinicus* and *P. amboinicus variegatus* were quite resilient amongst the six species with comparatively better water uptake efficiency even under high salinity stress conditions. Similar studies needs to be conducted not only to identify the economically and medicinally important plants, which are capable of growing in salinized soils but also to understand the mechanisms that are responsible for their salt tolerance, if any. With such knowledge in hand, it may be possible to engineer tolerance in plants with novel genetic techniques.

ACKNOWLEDGEMENTS

Research in Plant Biotechnology laboratory of Central Research laboratory, GITAM University is funded by the grants from the Science and Engineering Research Board, Department of Science and Technology, Govt. of India. The research team is thankful to SERB for the Senior Research Fellowship.

KEYWORDS

- biochemical
- CAM metabolism
- carbohydrate
- electrical conductivity (EC)
- electrolyte leakage
- glucose
- ion uptake

- leaf area
- leaf water potential
- membrane permeability
- micro and macro nutrients
- morphology
- nitrogen
- nutrient uptake
- photosynthesis
- physiology
- plant nutrients
- *Plectranthus*
- relative water content
- salinity tolerance
- SAR
- sodicity hazard
- sodium
- soil salinization
- stress adaptation
- sucrose
- water potential
- water uptake capacity

REFERENCES

1. Abbas, G., Saqib, M., Rafique, Q., Rahman, M. A., Akhtar, J., Haq, M. A., & Nasim, M. (2013). Effect of salinity on grain yield and grain quality of wheat (*Triticum aestivum L.*). *Pakistan J. Agric. Sci.*, 50, 185–189.
2. Adriano, D. C., & Doner H. E. (1982). Bromine, chlorine and fluorine. In: *Methods of Soil Analysis Part II. Chemical and Biological Properties*. (Eds. A. L. Page, R. H. Miller, and D. R. Keeney), Madison, WI. *Soil Science Society of America and American Society of Agronomy*, 461–462.
3. Akhtar, N., Hossain, F., & Karim, A. (2013). Influence of calcium on water relation of two cultivars of wheat under salt stress. *International Journal of Environment*, 2, 1–8.

4. AOAC. (1995). *Official Methods of Analysis*. 16th edition Association of Official Analytical Chemists, Washington DC.

5. Arnon, D. I. (1949). Copper enzyme in isolated chloroplast polyphenoloxidase in *Beta vulgaris L.Plant physiology*, 24, 1–15.

6. Ashwell, G. (1957). Colorimetric analysis of sugars. In: *Methods in Enzymology*. (Eds. Cohvick and Kaplan) New York: Academic Press Inc., 3, 73–105.

7. Ayers, R. S., & Westcot, D. W. (1985). *Water Quality for Agriculture*. Irrigation and Drainage Paper, 29, Revision-1. FAO, Rome.

8. Bayer, A., Ruter, J., & van Iersel, M. (2014). Irrigation volume and fertilizer rate influence growth and leaching fraction from container-grown *Gardenia jasminoides*. *Acta Hortic*. (ISHS) 1055, 417–422.

9. Black, C., & Osmond, C. B. (2003). Crassulacean acid metabolism photosynthesis: working the night shift. *Photosynthesis Research*, 76, 329–341.

10. Bray, R. H., & Kurtz, L. T. (1945). Determination of total, organic, and available forms of phosphorus in soils. *Soil Science*. 59, 39–45.

11. Burrow, D. P., Surapaneni, A., Rogers, M. E., & Olsson, K. A. (2002). Ground water use in forage production: the effect of saline–sodic irrigation and subsequent leaching on soil sodicity. *Australian Journal of Experimental Agriculture*, 42, 237–247.

12. Carter, D. L., Bondurant J. A., & Robbins, C. W. (1971). Water soluble NO_3 nitrogen, PO_4 phosphorous and total salt balances on a large irrigation tract. *Soil Sci. Soc. Am. Proc.*, 35, 331–335.

13. Chaves, M. M., Flexas, J., & Pinheiro, C. (2009). Photosynthesis under drought and salt stress: regulation mechanisms from whole plant to cell. *Ann. Bot.*, 103, 551–560.

14. Cushman, J. C., Agarie, S., Albion, R. L., Elliot, S. M., Taybi, T., & Borland, A. M. 2008. Isolation and characterization of mutants of common ice plant deficient in Crassulacean acid metabolism. *Plant Physiology*, 147(1), 228–238.

15. Dahiya, S. S., & Singh, M. (1979). Effect of salinity alkalinity and iron sources on availability of iron. *Plant soil*, 51, 13–18.

16. Dayal, V., Anil Kumar, D., Om Prakash. A., Raghunath, P., & Anil, D. (2014). Growth, lipid peroxidation, antioxidant enzymes and nutrient accumulation in Amrapali mango (*Mangifera indica* L.) grafted on different rootstocks under NaCl stress. *Plant Knowledge Journal* 3(1):15–22 (2014).

17. De Araújo, S. A. M., Silveira, J. A. G., Almeida, T. D., Rocha, I. M. A., Morais, D. L., & Viéga, R. A. (2006). Salinity tolerance of halophyte *Atriplex nummularia* L. grown under increasing NaCl levels. *R. Bras Eng. Agric. Ambiental*, 10, 848–854.

18. Djanaguiraman, M., & Ramadass, R. (2004). Effect of salinity on chlorophyll content of rice genotypes. *Agric. Sci. Digest*, 24, 178–181.

19. Dubios, M., Gilles, K. A., Hamilton, J. K., Rebers, P. A., & Smith, F. (1956). Colorimetric method for determination of sugars in related substances. *Anal Biochem.*, 28, 350–356.

20. Eskandari, S., Mozaffari, V., & Tajabadi Pour, A. (2014). Effects of Salinity and Copper on Growth and Chemical Composition of Pistachio Seedlings. *Journal of Plant Nutrition*, 37:7, 1063–1079.

21. Fahad, S., Saddam, H., Fahad, K., Chao, Wu., Saud, S., Hassan, S., Naeem, A., Deng Gang, Abid, U., & Jianliang, H. (2015). Effects of tire rubber ash and zinc sulfate on crop productivity and cadmium accumulation in five rice cultivars under field conditions. *Environ. Sci. Pollution Res.*, 22, 12424–12434.

22. FAO—Food and Agricultural Organization of the United Nations. (2006). *Terra STAT Database*. FAO, Rome, Italy.
23. Francisco, J. C., Martinez, V., & Carvajal, M. (2004). Does calcium determine water uptake under saline conditions in pepper plants, or is it water flux which determines calcium uptake? *Plant Science,* 166, 443–450.
24. Gaines, T. P., Parker, M. B., & Gascho, G. L. (1984). Automated determination of chlorides in soil and plant tissue by sodium nitrate extraction. *Agron. J.,* 76, 371–374.
25. Grattan, S. R., & Catherine, M. G. (1993). Mineral nutrient acquisition and response by plants grown in saline environments. In: *Handbook of Plant and Crop Stress*. Marcel Dekker, Inc., New York, 203–226.
26. Grattan, S. R., & Grieve, C. M. (1999). Salinity-mineral nutrient relations in horticultural crops. *Scientia Horticulturae*, 78, 127–157.
27. Hassan, S., Muhammad, B. S., Sajjad, S., Bushra, R., Beenish, A., Mohamed, B. B., & Tayyab, H. (2014). Growth, physiological and molecular responses of cotton (*Gossypium arboreum L.*) under NaCl stress. *American Journal of Plant Sciences*, 5, 605–614.
28. Hussein, M., El-Faham, S., & Alva, A. (2012). Pepper plants growth, yield, photosynthetic pigments, and total phenols as affected by foliar application of potassium under different salinity irrigation water. *Agricultural Sciences*, 3, 241–248.
29. Islam, M. S. (2001). *Morpho-Physiology of Black gram and Mung bean as Influenced by Salinity*. An M. S. thesis. Dept. of Agronomy. Bangabandhu Sheikh Mujibur Rahman Agricultural University, Salna, Gazipur.
30. Jackson, M. L. (1980). *Soil Chemical Analysis*. New Delhi: Prentice Hall of India Pvt. Ltd.
31. Kabir, M. E., Karim, M. A., & Azad, M. A. K. (2004). Effect of potassium on salinity tolerance of Mungbean (*Vigana radiate L. Wilczek*). *Journal of Biological sciences*, 4, 103–110.
32. Kaya, C., Muhammad, A., Murat, D., & Atilla, L. T. (2013). Alleviation of salt stress-induced adverse effects on maize plants by exogenous application of indole acetic acid (IAA) and inorganic nutrients – A field trial. *AJCS*, 7, 249–254.
33. Kizilgoz, I., & Sakin, E. (2010). The effects of increased phosphorus application on shoot dry matter, shoot P and Zn concentrations in wheat (*Triticum durum* L.) and maize (*Zea mays* L.) grown in a calcareous soil. *African Journal of Biotechnology*, 9, 5893–5896.
34. Kobayashi, T., & Nishizawa, N. K. (2012). Iron uptake, translocation, and regulation in higher plants. *Ann. Rev. Plant Biol.*, 63, 131–152.
35. Lindsay, W. L. (1995). Chemical reactions in soils that affect iron availability to plants: a quantitative approach. In: *Iron Nutrition in Soils and Plants*. (Abadia, J. Ed.) Kluwer Academic Publishers, Dordrecht, 7–14.
36. Lindsay, W. L., & Norvell, W. A. (1978). Development of a DTPA soil test for zinc, iron, manganese, and copper. *Soil Sci. Soc. Am. J.*, 42, 421–448.
37. Liu, F. Y., Kuo, T. L., & Wen, J. Y. (2014). Differential responses to short-term salinity stress of heat-tolerant cherry tomato cultivars grown at high temperatures. *Hort. Environ. Biotechnol.,* 55, 79–90.
38. Lukhoba, C. W., Simmonds, M. S. J., Paton, A. J. (2006). *Plectranthus*: A review of ethnobotanical uses. *J. of Ethnopharmacology*, 103, 1–24.

39. Luo, Q., Yu, B., & Liu, Y. (2005). Differential sensitivity to chloride and sodium ions in seedlings of *Glycine maxand* and *G. sojaunder* under NaCl stress. *Journal Plant Physiology,* 162, 1003–1012.

40. Maas, E. V., & Hoffman, G. J. (1977). Crop salt tolerance – current assessment. *J. Irrig. Drain. Div. ASCE, 103,*115–134

41. Mansoori, M., Mohammad, K., & Aniseh, J. (2014). Salinity stress evaluation on nutrient uptake and chlorophyll sugarcane genotypes. *J. Bio. & Env. Sci.,* 5, 163–172.

42. Mansour, M. M. F. (1997). Cell permeability under salt stress. In: *Strategies for Improving Salt Tolerance in Higher Plants.* (Eds. Jaiwl, P. K., Singh, R. P., Gulati, A). Science Publisher, Enfield, USA. 87–110.

43. Mansour, M. M. F., & Salama, K. H. A. (2004). Cellular basis of salinity tolerance in plants. *Environ. & Exp. Bot.,* 52, 113–122.

44. Marohn, C., Distel, A., Dercon, G., Wahyunto, Tomlinson, R., Noordwijk, M. V., & Cadisch, G. (2012). Impacts of soil and ground water salinization on tree crop performance in post-tsunami Aceh Barat, Indonesia. *Nat. Hazards Earth Syst. Sci.,* 12, 2879–2891.

45. Marschner, H. (1995). Mineral Nutrition of Higher Plants. London: Academic Press.

46. Minhas, P. S., & Gupta, R. K. (1992). Quality of Irrigation Water – Assessment and Management. Indian Council of Agricultural Research, New Delhi, 123 p.

47. Munns, R., & Tester, M. (2008). Mechanisms of salinity tolerance. *Annual Rev. Plant Biol.,* 59, 651–681.

48. Neto, A. A. D., Prisco, J. T., Enéas-Filho, J., Lacerda, C. F., Silva, J. V., Costa, P. H. A., & Gomes-Filho, E. (2004). Effects of salt stress on plant growth, stomatal response and solute accumulation of different maize genotypes. *Brazilian Journal of Plant Physiology,* 16, 31–38.

49. Niu, G., Denise S. R., & Lissie, A. (2008). Effect of saline water irrigation on growth and physiological responses of three rose rootstocks. *Hort. Science,* 43, 1479.

50. Ondrasek, G., Rengel, Z., Romic, D., & Savic, R. (2010). Environmental salinisation processes in agro-ecosystem of Neretva River estuary. *Novenytermeles,* 59, 223–226.

51. Ondrasek, G., Romic, D., Rengel, Z., Romic, M., & Zovko, M. (2009). Cadmium accumulation by muskmelon under salt stress in contaminated organic soil. *Sci. Total Environ.,* 407, 2175–82.

52. Ono, K., Terashima, I., & Watanabe, A. (1996). Interaction between nitrogen deficit of a plant and nitrogen content in the old leaves. *Plant and Cell Physiology,* 37, 1083–1089.

53. Pandey, S. K., & Singh H. (2011). A simple, cost effective method for leaf area estimation. *J. Bot.,* 240, 1–6.

54. Pankova, Y., Konyushkova, M., & Luo, G. (2010). Effect of soil salinity in subboreal deserts of Asia. *19th World Congress of Soil Science, Soil Solutions for a Changing World,* 1–9.

55. Parihar, P., Singh, S., Rachana, S., Vijay Pratap, S., & Prasad, S. M. (2015). Effect of salinity stress on plants and its tolerance strategies: a review. *Environmental Science and Pollution Research,* 22, 4056–4070.

56. Paul, D. (2012). Osmotic stress adaptations in rhizobacteria. *J. Basic Microbiol.,* 52, 1–10.

57. Pitman, M. G., & Laüchli, A. (2002). Global impact of salinity and agricultural eco-systems. In *Salinity: Environment—Plants—Molecules*. (Eds. Laüchli, A., Lüttge, U.) Kluwer Acad., Netherlands, 3–20.

58. Rabhi, M., Barhoumi, Z., Ksouri, R., Abdelly, C., & Gharsalli, M. (2007). Interactive effects of salinity and iron deficiency in *Medicagociliaris*. *C. R. Biol.*, 330, 779–788.

59. Rahi, T. S., Singh, K., & Singh, B. (2013). Screening of sodicity tolerance in Aloe vera: an industrial crop for utilization of sodic lands. *Industrial Crops Products*, 44, 528–533.

60. Raja, R. R. (2012). Medicinally potential plants of *Labiatae* (Lamiaceae) family: An overview. *Research Journal of Medicinal Plant*, 6, 203–213.

61. Ram, B., Kumer, S., Sahi, B. K., & Tripthi, B. K. (1999). Traits for selecting elite sugarcane clones under water and salt stress conditions. *Proceedings International Society Sugar Cane Technology 23rd Congress*, 132- 139.

62. Ramachandra Reddy, A., Reddy, K. R., & Hodges, H. F. (1996). Mepiquat chloride (PIX)-induced changes in photosynthesis and growth of cotton. *Plant Growth Regul.*, 20, 179–183.

63. Ramana, G. V., Sweta Padma, P., & Chaitanya, K. V. (2012). Differential responses of four soybean *(Glycine max.* L) cultivars to salinity stress. *Legume Res.*, 35, 185 – 193.

64. Raven, J. A., Evans, M. C. W., & Korb, R. E. (1999). The role of trace metals in photosynthetic electron transport in O_2 evolving organisms. *Photosynthetic Research*, 60, 111–149.

65. Reddy, M. P., & Vora, A. B. (1985). Effect of salinity on protein metabolism in *bajra* leaves. *Indian J. Plant Physiol.*, 28, 190– 195.

66. Richard, L. A. (1954). Diagnosis and Improvement of Saline and Alkali Soils. *Agri. Handbook* No.60, USDA. Riverside, CA.

67. Rivero, R. M., Mestre, T. C., Mittler, R., Rubio, F., Sanchez, F. G., & Martinez, V. (2014). The combined effect of salinity and heat reveals a specific physiological, biochemical and molecular response in tomato plants. *Plant Cell Environ.*, 37, 1059–1073.

68. Ryuichi, S., Nguyen, T. N., Hirofumi, S., Reda, M., & Kounosuke, F. (2006). Effect of salinity stress on photosynthesis and vegetative sink in tobacco plant. *Soil Science and Plant Nutrition*, 52, 243–250.

69. Saeed, R., Salma, M., & Rafiq, A. (2014). Electrolyte leakage and relative water content as affected by organic mulch in okra plant (*Abelmoschus esculentus* (L.) moench) grown under salinity. *Fuuast J. Biol.*, 4, 221–227.

70. Salih, H. O., & Kia, D. R. (2013). Effect of salinity level of irrigation water on cow-pea (*Vigna unguiculata*) growth. *IOSR Journal of Agriculture and Veterinary Science (IOSR-JAVS)*, 63, 37–41.

71. Sangakkara, U. R., Hartwig U. A., & Nosberger, J. (1996). Responses of root branching and shoot water potentials of french bean (*Phaseolus vulgaris L.*) of soil moisture and fertilizer potassium. *Journal of Agronomy and Crop Sciences*, 177, 165–173.

72. Saqib, M., Akhtar, J., Abbas, G., & Nasim, M. (2013). Salinity and drought interaction in wheat (*Triticum aestivum* L.) as affected by the genotype and plant growth stage. *Acta Physiol. Plant.*, 35, 2761–2768.

73. Sharma, D. K., Dubey, A. K., Srivastav, M., Singh, A. K., Pandey, R. N., & Anil, D. (2013). Effect of paclobutrazol and putrescine on antioxidant enzymes activity and

nutrients content in salt tolerant citrus rootstock sour orange under sodium chloride stress. *Journal of Plant Nutrition*, 36, 1765–1779.

74. Sidhu, G. S., & Sharma, B. D. (2010). Diethylene triamine penta acetic acid-extractable micronutrients status in soil under a rice-wheat system and their relationship with soil properties in different agro-climatic zones of Indo-Gangetic plains of India. *Communications in Soil Science and Plant Analysis*, 41, 29 – 51.

75. Singh, M. V. (2008). Micronutrient deficiencies in crops and soils in India. In: *Micronutrient Deficiencies in Global Crop Production.* (Ed. Alloway, Brian J.) Springer, Netherlands. pp. 93–125.

76. Tester, M., & Davenport, R. (2003). Na^+ tolerance and Na^+ transport in higher plants. *Ann. Bot.,* 91, 503–507.

77. Toth, S., & Prince, A. (1949). Estimation of cation exchange capacity and exchangeable calcium, potassium and sodium contents of soils by flame photometer techniques. *Soil Sci.*, 67, 439–445.

78. Valentovic, P., Luxova, M., Kolarovic, L., & Gasparikova, O. (2006). Effect of osmotic stress on compatible solutes content, membrane stability and water relations in two maize cultivars. *Plant Soil Environment,* 52, 186–191.

79. Van Doorn, W. G. (2008). Is the onset of senescence in leaf cells or intact plants due to low or high sugar levels? *Journal of Experimental Botany*, 59, 1963–1972.

80. Walkley, A., & Black, I. A. (1934). An examination of Degtjareff method for determining soil organic matter and a proposed modification of the chromic acid titration method. *Soil Sci.*, 37, 29–37.

81. Wang, W., Kohler, B., Caoand, F., & Liu, L. (2008). Molecular and physiological aspects of urea transport in higher plants. *Plant Sci.* 175, 467–477.

82. Wang, W. Y., Yan, X. F., Jiang, Y., Qu, B., & Xu, Y. F. (2012). Effects of salt stress on water content and photosynthetic characteristics in *Iris lactea* Var. Chinensis seedlings. *Middle-East Journal of Scientific Research*, 12, 70–74.

83. Win, K. T., Aung, Z. O., Tadashii, H., Taiichiro, O., & Hirata, Y. (2011). Genetic analysis of Myanmar *Vigna* species in responses to salt stress at the seedling stage. *African Journal of Biotechnology,* 10, 1615–1624.

84. Yadav, R. L., & Meena, M. C. (2009). Available micronutrients status and relationship with soil properties of Degana soil series of Rajasthan. *Journal of the Indian Society of Soil Science*, 57, 90–92.

85. Yurela, I. (2005). Copper in Plants. *Brazilian Journal of Plant Physiology,* 17145–17156.

86. Yusuf, A. A., Mofio, B. M., & Ahmed, A. B. (2007). Proximate and mineral composition of *Tamarindus indica* Linn. 1753 seeds. *Sci. World J.*, 2, 1–4.

87. Zan, M. J., Chang, H. W., Zhao, P. L., & Wei, J. G. (2007). Physiological and ecological characters studies on Aloe vera under soil salinity and seawater irrigation. *Process Biochem.*, 42, 710–714.

88. Zhang, L., Ma, H., Chen, T., Pen, J., Yu, S., & Zhao, X. (2014). Morphological and physiological responses of cotton (*Gossypium hirsutum* L.) plants to salinity. *PLoS ONE, 9*, e112807.

89. Zong, L., Tedeschi, A., Xue, X., Wang, T., Menenti, M., & Huang, C. (2011). Effect of different irrigation water salinities on some yield and quality components of two field-grown cucurbit species. *Turk J. Agric. For.*, 35, 297–307.

CHAPTER 7

EFFECT OF SALINITY STRESS ON GROWTH PARAMETERS AND METABOLITES OF MEDICINAL PLANTS: A REVIEW

YOGITA DESHMUKH and PUJA KHARE

CONTENTS

7.1 INTRODUCTION

Since people around the world are switching to indigenous systems of medicines/herbs, there is ever increasing demand to cultivate medicinal plants for the individual uses as well as for the pharmaceutical industry. World Health Organization (WHO) has assessed that about 80% of the world populations rely on medicinal herbs. In such a scenario some of the medicinal plants need to be cultivated commercially. Since most productive soils will not be spared in most developing countries for this purpose, it is the otherwise problem soils and/or harsh climatic conditions under which these plants will have to be cultivated. Saline soils being one such group of soils, researchers are trying to test important medicinal plants for their salt tolerance besides other abiotic stresses such as chilling, drought, high and low temperatures, infrared exposures and heavy metals etc. Since the quality of the herbs is controlled by the secondary metabolites, which strongly depend on environmental conditions under which the herbs are grown, such studies assume prime importance.

Salinity, a major abiotic stress, is as old as the history of agriculture. References to *usar* lands are available in ancient Indian scriptures. The soil salinization can be traced either to natural processes or to anthropogenic interventions often referred as secondary soil salinization. Expansion of irrigation especially through canals is one of the major secondary causes of soil salinization. On the other hand, increasing food requirement for the burgeoning population has brought this problem in limelight especially in India where 6.73 million hectare (M ha) of agricultural lands are lying barren because of excess salts. Besides, large areas may also be producing much below the optimum productivity because soils may be affected by salinity to various degrees. Even at global scale, ~20% of the world's cultivated land and nearly half of all the irrigated lands are affected to some degree by salinity or related stresses. It is projected that 30% of the global land might be affected by soil salinization within the next 25 years, and it may go up to 50% by the year 2050 [82].

According to Sreenivasulu et al. [79], the adverse effects of salts on plant growth can be grouped under the three broad categories: (i) a reduction in osmotic potential of the soil solution that reduces plant available water and thus creates water stress in plants, (ii) a deterioration in the physical structure of the soil resulting in diminished water permeability and soil aeration, and (iii) increase in the concentration of the certain ions that have an inhibitory effect on plant metabolism (specific ion toxicity) and/or mineral nutrient imbalances and deficiencies. Deterioration of physical structure is a major issue in alkali soils having excess of sodium, carbonate and bicarbonate ions, which pose serious problems through specific effects caused by excess intake of sodium, poor physical conditions and sometimes excess salts as well [28]. In general, a combination of these factors often acts in unison to impact the yield and quality of the products.

Soil salinization is likely to have devastating global effects because salinity induces physiological and metabolic disturbances in plants, affecting development, growth, yield, and quality of plants. High concentration of salts in soils, cause ionic imbalance and hyper-osmotic stress in plants and as a consequence secondary such as oxidative stresses develop causing damage to crops and plants. Even saline irrigation water will result in a similar situation because salts accumulate in soils with irrigation. It has been reported that steady state salinity build-up in saline water irrigated soils may be about half the salinity of irrigation water in sandy, about the same in medium and more than 2 times the salinity of irrigation water in heavy textured soils. Because of the quantitative and qualitative characteristics changes during the plant growth, it may impact the ingredients and medicinal properties of herbs as well [51]. Other abiotic stresses may also result in oxidative stresses inducing similar cellular damage in plants. For example, drought and/or salinization are manifested primarily as osmotic stress, resulting in the disruption of homeostasis and ion distribution in the cell [72, 87].

While the effects of salt stress on traditional crops have been extensively investigated there is a general lack of information in the case of medicinal plants in spite of these being among the important crops. To some extent, adverse effect of soil salinity on the survival and the productivity of medicinal plants of industrial interest have been documented, but its effects on essential oil composition besides the plant yield are not well documented [80]. For example; the biosynthesis of volatile oils, the industrially important ingredients of the medicinal plants, is influenced by various environmental factors amongst them soil salinity being one of the major factors [77]. Similarly, alkaloids, anthocyanins, flavonoids, quinones, lignans, steroids, and terpenoids are among plant secondary metabolites which have commercial applications as drug, dye, flavor, fragrance, insecticide and antioxidants [35]. These secondary metabolites have great value but their production may not always be satisfactory as several factors such as type of medicinal plant species or genus, growth or developmental stage, specific seasonal conditions, nutrient availability and stress conditions restrict/enhance their production [81]. It has been shown that salt and/or drought stress can increase /decrease the content of relevant natural products in plants. For example, Selmar [71] observed a strong increase in the concentration of tropane alkaloids in *Datura inoxia* plants under salt stress.

To maintain growth and productivity, plants must adapt to stress conditions and exercise specific tolerance mechanisms through the activation and regulation of specific stress related genes. These genes are involved in the whole sequence of stress responses, such as signaling, transcriptional control, protection of membranes and proteins, and free-radical and toxic-compound scavenging. Plant modification for enhanced tolerance is mostly based on the manipulation of genes that protect and maintain the function and structure of cellular components. In contrast to most monogenic traits of engineered resistance to biotic stresses, the genetically complex responses to abiotic stress conditions are more difficult to control and engineer. Various aspects of salt tolerance in plants were proposed by Zhu [87]. Mechanisms of homeostasis against ionic and osmotic stress and detoxification pathway in salt tolerant plants are shown in Figure 7.1.

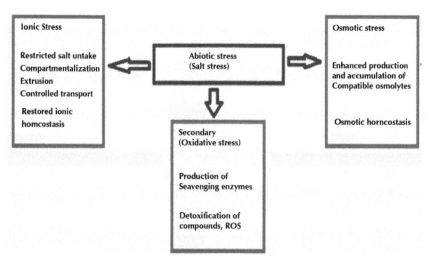

FIGURE 7.1   Mechanisms of homeostasis against ionic and osmotic stress and detoxification pathway in salt tolerant plants.

Although traditional methods of salinity management will continue to play their roles, switching over to modern methods such as genetic engineering for the salt stress management and breeding for drought and salinity stress tolerance in medicinal plants through plant biotechnology may prove useful and cost effective. Recent researches into the molecular mechanisms of stress responses has started to bear fruit and, in parallel, genetic modification of stress tolerance has also shown promising results that may ultimately apply to agriculturally and ecologically important plants as well.

This chapter summarizes the effects of salt stress on germination, growth, development and primary and secondary plant products in some medicinal plants. Chapter also deals with traditional and recent advances in elucidating stress-response mechanisms and their biotechnological applications in cultivation of medicinal plants and a brief of conventional and other techniques to improve productivity under salt stress.

7.2   EFFECTS OF SOIL SALINITY ON GERMINATION AND SURVIVAL

As a result of soil/water salinity, seed germination, survival percentage, morphological characteristics, development and yield of the plant are affected. In general, salt stress decreases germination, adversely impacts survival and growth as a result of decrease in the photosynthesis and respiration rate of plants besides other factors.

7.2.1   GERMINATION AND SEEDLING GROWTH

Germination, in general is the most salt sensitive growth stage which is severely inhibited with increasing salinity [78]. Amongst several factors, germination of seeds may be affected in the following two ways:

- Salt in the medium may increase the osmotic potential to retard or prevent the uptake of water necessary for germination and mobilization of nutrient required for germination
- The salt constituents or ions may be toxic to the embryo [38].

Germination of seeds of *Ocimum basilicum* [47], *Petroselinum hortense* ([64], S*weet marjoram* [5], and *Thymus maroccanus* [17] significantly decreased under salt stress. Such a negative response on germination has also been reported by other authors on *Ocimum basilicum, Eruca sativa, Petroselinum hortense, chamomile, sweet marjoram, and Thymus maroccanus* [17]. Besides reduced germination percentage, delay in germination with increasing salinity has also been reported in Basil. Amongst the several salts employed to create salinity stress, Na_2CO_3 and $NaHCO_3$ had the highest effect in reducing the above characters [85]. Seeds of *Trigonella foenum-graecum l.* Var. PEB (Fenugreek) subjected to NaCl salinity (0 mM to 100 mM) revealed that concentrations of NaCl (up to 40 mM) did not affect percentage germination, although it was found to be delayed. At higher salinity levels, inhibitory effect on germination was recorded to an extent that seeds did not germinate at 80 mM and above concentrations of NaCl [65].

Seedling growth is also one of the vulnerable stages in the plants' life cycle. Reports show that the typical symptom of salt injury is growth retardation. The seedling growth of *Thymus maroccanus* reduced with increasing salinity [17]. Reduced seedling growth has also been reported in basil, chamomile (*Matricaria chamomila*) and marjoram [13]. It is attributed to slow or less mobilization of reserve foods, suspending the cell division, inhibition of cell elongation and/or enlarging and injuring hypocotyls [63]. *Moringa peregrine* on the other hand showed no reduction in growth parameters and seedling emergence up to 6 dS m^{-1} but significantly decreased thereafter with increasing salinity [69].

The average plumule growth decreased gradually with increasing salt concentration. Highly significant differences were observed between species in the radicle growth [51]. However, at low concentrations, radicle and/or plumule growth either improved or had no pronounced negative effect possibly due to their nutrient like action [32]. Salinity also caused reduction in the number of flowers, the number of branches and the head diameter in Chamomile. The reduction may be attributed to suppression of growth under salinity stress during the early developmental stages (shooting stage) of the plants. Gradual decrease in root length, shoot length, fresh weight and dry weight of the seedlings was observed with increasing concentrations of NaCl (0 mM to 100 mM) in the growth medium in *Trigonella foenum – graecum l.* Var. PEB (Fenugreek) during the test period of 7 days [65]. Significant (P < 0.05) decrease in the dry biomass was observed following the application of increasing NaCl levels, the maximum reduction being 65% at 100 mM NaCl.

Hendawy and Khalid [30] observed a significant decrease in sage (*Salvia officinalis L.*) dry weight at 50 mM NaCl ranging from 34% to 48%. Several processes have been outlined for such reductions in growth rates. According to Munns [52], it may either be due to decreased availability of water or to the toxicity of sodium chloride. Inadequate photosynthesis owing to stomatal closure and consequently limited uptake of carbon dioxide is one of the causes of reduction in growth rates as stated by Zhu (2001). Also the reduction in dry weight under salinity stress may be attributed to inhibition of hydrolysis of reserved foods and their translocation to the growing shoots.

Ansari et al. [8] compared the performance of three *Cymbopogon grasses namely C. winterianus, C. flexuosus and C. martinii* at different levels of NaCl salinity. They concluded that salinity resulted in the suppression of plant growth and a decline in essential oil concentration and yield in all species. The ability to limit Na^+ transport into the shoots and the Na^+ accumulation in the rapidly growing shoot tissues is critically important for maintenance of high growth rates and protection of the metabolic process in elongating cells from the toxic effects of Na^+.

7.2.2 SURVIVAL OF PLANTS

Excess soluble salts in the soil have often proved lethal to plants. Seedlings of *Melissa officinalis* could not survive at 6 dS m^{-1} [57]. Mortality in sage plants increased with increasing salinity up to 3000 ppm, where it could not survive [30]. In thyme, survival percentage decreased significantly under salinity conditions [24]. Similar results were obtained on *Majorana hortensis* and spearmint [4]. Ashraf [10] found that increasing salt concentrations caused a significant reduction in the fresh and dry biomass of both shoots and roots and seed yield of *Ammolei majus*, while plant fresh and dry yield reduced in *Hyoscyamus niger*. In alkali soils, survival percentage of different medicinal plant species decreased with increase in soil ESP.

TABLE 7.1   Survival (%) of Medicinal and Aromatic Plant Species Under Different ESP Levels

	ESP levels			
Plant species	**25**	**35**	**45**	**55**
Babchi (Psoralea corylifolea)	83	77	41	12
Sadabahar (Catharanthus roseus)	60	52	12	Nil
Muskdana (Abelmoschas moschatus)	40	21	8	Nil

Source: Anonymous [7].

The survival of *babchi* and *sadabahar* was more than 50% up to ESP 35 whereas survival of *muskdana* was <50%. *Babchi* and *sadabahar* had 77 and 52% survival at ESP 35 which decreased to 41 and 12%, respectively at exchangeable sodium percentage (ESP) 45 [7].

7.2.3 MOISTURE CONTENT

Moisture content of seedlings/leaves plays an important role in various physiological processes including plant growth. Plant growth reduces under limited water availability for cell expansion and nutrient uptake [84]. Moisture content of seedling has been found to increase or decrease and it seems no consistent trend has emerged so far. The development of succulence in plants to minimize the adverse effect of salts has been reported. It also explains why the fresh weight might get enhanced under salt treatment?

Ibrar and Hussain [34] reported that moisture content of the roots was enhanced with increasing salinity level in *Medicago polymorpha*. This preliminary laboratory study suggested that the tested medicinal species could be grown on marginally saline habitats due to its tolerance to moderate salinity at germination and seedling stage. Leaf relative water content (LRWC), a reflection of the water status of plants under stress condition [39], was decreased in the leaves of rosemary grown under the 50 and 100 mM NaCl treatments. Salinity caused water deficit through an increase in soluble salts concentrations. Further studies are needed to test its salt tolerance under field conditions to assess the possibility of cultivation of medicinal plants on otherwise unproductive lands.

7.2.4   MORPHOLOGICAL   CHARACTERISTICS,   DEVELOPMENT AND YIELD

Morphological characteristics such as number of leaves, leaf area, and leaf biomass were found to reduce under salt stress in a number of medicinal plants. On *Mentha piperita* var. officinalis and *Lipia citriodora* var. *verbena* significant reduction in number of leaves, leaf area and leaf biomass under salt stress were noted. On milk thistle, growth parameters such as plant height, number of leaves per plant, number of capitula per plant, main shoot capitulum's diameter reduced with salinity more than 9 dS m^{-1}. Besides, plant height, number of flowers, number of branches and head diameter, flower fresh weight, flower dry weight, and peduncle length also reduced with increasing salinity.

In *Melissa officinalis*, root growth was reduced with increase in irrigation water salinity. In *Satureja hortensis*, leaf area, leaf and stem fresh weight, as well as dry weight of leaves, stems and roots decreased in plants grown at different levels of NaCl [68]. The shoot and root dry weight of rosemary was significantly ($P < 0.05$) decreased under salinity stress. Several investigators have reported reduced plant growth under salinity stress in *Foeniculum vulgare* subsp. vulgare; *Majorana hortensis*; peppermint, pennyroyal, and apple mint; *Matricaria recutita*; *Thymus maroccanus*; geranium; *Thymus vulgaris*; sweet fennel; sage. The application of 50 mM NaCl in the nutrient solution (Electrical conductivity (EC) = 7 dS m^{-1}) caused only 7% reduction in the shoot and root dry matter production.

Westervelt [83] also reported that rosemary could tolerate high levels of soil salinity (4–8 dS m^{-1}). The plant height and root length of *Salvia miltiorrhiza* remained virtually unchanged among all the salt treated treatments (0, 25, 50, 75 and 100 mM NaCl), suggesting the growth of the plant was not affected on a morphological basis. *Salvia miltiorrhiza* is a known salt tolerant crop and can bear more than 100 mM in NaCl stress. However, high salinity significantly decreased plant fresh and dry weights when compared with the control. All these findings are similar to the large volume of data available in the literature for cereal, oil seed, vegetables and other crop species.

The detrimental effects of high salinity on plants can be observed at the whole plant level through their survival and/or decrease in the productivity.

In fennel, cumin and *Ammi majus* increasing salt concentrations caused a significant reduction in the number of umbels, fruit yield/plant and weight of 1000 seeds. Umbel no./plant and seed yield in fennel decreased with the increase in salinity level of the irrigation water. In control (0 mM) umbel no./plant was 60 and seed yield was 59.56 (g) while at 40 and 80 mM NaCl, these reduced to 41.66 and 28.44 (umbel no./plant) and 26.7 and 11.5 (g seed yield) respectively [62]. Meena et al. [45] concluded that fennel is quite tolerant to salts and can withstand saline irrigation water up to 8.6 dS m^{-1} especially with organic inputs. Reduced seed yield and yield components per plant were also observed on milk thistle and *Trachypermum ammi* [68].

The threshold salinity (EC$_t$) as determined using Maas and Hoffman [43] model revealed that palmarosa and lemon grass have almost similar threshold, i.e., 4.49 and 4.44 dS m^{-1}, respectively. Among the crops tested, vetiver was relatively more salt-tolerant with the EC$_t$ value of 6.13 dS m^{-1} (2008–2009) and 5.99 dS m^{-1} (2009–2010). During 2009–2010, *kamaksturi* and *shatavari* recorded EC$_t$ values of 5.38 dS m^{-1} and 4.05 dS m^{-1} respectively (Table 7.2). The trend line indicated that oil content remains almost constant and is not much affected by salinity.

As far as percent oil content and the soil salinity relationship are concerned, as revealed by the R^2 values, there was no good relationship between palmarosa, vetiver, lemon grass, *kamakasturi* and citronella. However, Salt tolerance studies reveal that palmarosa performed well even with waters having EC of 18 dS m^{-1}, the herb and essential oil yield increasing upto 10.4 dS m^{-1} and decreasing thereafter. The herb and essential oil yield of lemon grass decreased consistently with increase in salinity of irrigation water. The yield declined by more than 50% at EC of 18.8 dS m^{-1}. Vetiver locally known as *khus* is capable to withstand high pH and water logging conditions.

German Chamomile is another aromatic crop which could be cultivated successfully with saline waters, having EC up to 8.0 dS m^{-1} and alkaline waters having RSC up to 15.0 meq L^{-1}. Citronella is relatively sensitive as it cannot withstand salinity beyond 5.5 dS m^{-1}. Periwinkle, popularly known as *Sadabahar* can be grown well with waters having EC up to 10.0 dS m^{-1}. Ergot is more tolerant as it could withstand salinity up to

16.0 dS m^{-1}. Egyptian henbane and *isabgol* could also perform well with waters having EC up to 8.0 dS m^{-1} [58].

Seed yield of *Isabgol* and *chandrasur* significantly decreased with increasing levels of ESP. The highest seed yield of Isabgol (0.88 tha^{-1}) and chandrasur (0.99 t ha^{-1}) was recorded at ESP 25, the lowest being at ESP 55 (Table 7.3). The reduction in seed yield of *isabgol* and *chandrasur* was more than 50% beyond ESP 35 [7]. Both the crops seem to be sensitive to higher levels of soil ESP.

TABLE 7.2 Economic Threshold Salinity (ECt) Level of Medicinal and Aromatic Crops [7]

Crop/Economic plant part	2008–2009		2009–2010	
	Threshold salinity EC$_t$ (dS m^{-1})	Slope beyond threshold (g/unit EC)	Threshold salinity EC$_t$ (dSm^{-1})	Slope beyond threshold (g EC^{-1})
Ashwaganda				
Seed	4.06	−222	−	−
Root	3.34	−120		
Palmarosa (foliage)	4.49	−449	−	−
Lemon grass (foliage)	4.44	−2080	−	−
Vetiver (root)	6.13	−250	5.99	−144
Kamakasturi (foliage)	−	−	5.38	−2330
Shatavar (bulb)	−	−	4.05	−5420

TABLE 7.3 Effect of Different ESP Levels on Seed Yield (t ha^{-1}) of Isabgol and *Chandrasur*

ESP Levels	Isabgol		*Chandrasur*	
	Seed yield	% reduction	Seed yield	% reduction
25	0.88	−	0.987	−
35	0.555	37	0.678	31
45	0.307	65	0.385	61
55	0.217	75	0.268	72
CD (5%)	0.183	−	0.146	−

7.3   EFFECT OF SALT STRESS ON PRIMARY PLANT PRODUCTS

Salinity plays a major role in modifying the organized and integrated complex of biochemical and biophysical reactions in plant cells. Total carbohydrate, fatty acid and protein content gets adversely affected but increased levels of amino acids, particularly proline has been well documented. The content of some secondary plant products is significantly higher in plants grown under salt stress than in those cultivated in normal conditions.

7.3.1   CARBOHYDRATES

Like all living organisms, plants require energy in chemical form so they can grow and carry out the basic life functions. Plants produce, store and burn carbohydrates in the form of sugar for energy. Since salinity induces stomatal closure, decrease in photosynthesis is inevitable. Thus, limited CO_2 can alter leaf carbohydrate content and source to sink translocation patterns.

In fennel, total carbohydrate was found adversely affected with salinity mainly attributable to the nutritional imbalance due to specific toxic effects of salinity, hyperosmotic stress and reduced photosynthesis. In contrast, total carbohydrate content pronouncedly increased with increasing salt stress levels in *Salvia officinalis*. Content of soluble sugars was higher in *Satureja hortensis* treated with NaCl than control [68].

In a study comparing the salt tolerant and salt sensitive rice cultivars, Pattanagul and Thitisaksakul [59] observed that partitioning sugars into starch may be involved in salinity tolerance by avoiding metabolic alterations. When four medicinal plants namely *Azadirachtaindica, Cassia fistula, Aloe barbadensis* and *Catharanthus roseus* were daily subjected to 1000mg of NaCl salt for 20 days, total carbohydrate was not much different after 20 days in *Azadirachta indica* showing that *Azadirachta indica* has some mechanism of salt tolerance [41].

7.3.2   LIPIDS

Lipids are ubiquitous in plants serving many important functions including storage of metabolic energy, protection against dehydration and pathogens, carrying of electrons and absorption of light, function as insulators of delicate internal organs and hormones and play an important role as the structural constituents of most of the cellular membranes. In addition, plant lipids are agricultural commodities important to the food, medical and manufacturing industries. Salinity stress induces oxidative stress, resulting in lipid peroxidation and increase in cell membrane permeability to toxic ions (e.g., Na^+ and Cl^-) which in turn reduces the plant growth as has been observed in the case of rosemary.

Oil content and fatty acid synthesis of plants is influenced by salt stress, which modifies fatty acid composition, and as such has a vital role in tolerance to several physiological stressors. Oil yield in roots was significantly lowered in *Ricinus communis* as a result of salt stress but increased in shoots. The total fatty acid content of *Coriandrum sativum* leaves decreased significantly due to salinity, and the content of α-linolenic and linoleic acids decreased significantly at higher NaCl levels [68]. This inhibition in the oil content under salt stress is similar to that reported on sage.

7.3.3 SUGARS AND PROTEINS

Soluble sugar and protein, the primary osmolytes in plants, play key roles in osmotic adjustment under salt-stressed conditions. Proteins accumulation in plants under saline conditions may provide a storage form of nitrogen that is re-utilized later and may play a significant role in osmotic adjustment [75]. Many salt-responsive proteins such as heat shock proteins, pathogen-related proteins, protein kinases, ascorbate peroxidase, osmotin, ornithine decarboxylase, and some transcription factors, have been detected in some major crops which are thought to give them the ability to withstand salt stress. Proteomic analysis of medicinal plants also revealed that alkaloid biosynthesis related proteins such as tryptophan synthase, codeinone reductase, strictosidine synthase, and 12-oxophytodienoate reductase might have major role in production of secondary metabolites. As such, proteomics has been used to study the expression of salt stress related proteins in several crops.

Plant cells may alter their gene expression in order to tolerate salt stress which results in an increase, decrease, induction or total suppression of some stress responsive proteins such as malate dehydrogenase and pyruvate phosphate dikinase [55]. The protein content in *Catharanthus roseus* significantly decreased with increasing NaCl concentrations. The level of soluble proteins decreased in salinity stressed chamomile and sweet marjoram. Also, in *Achillea fragratissima*, it was indicated that salinity concentration of 4000 ppm significantly depressed the crude protein. NaCl treatment with 80 mM NaCl lowered the protein content in leaves of *Dioscorea rotundata* plants as compared with control plants. However, a stimulation of protein synthesis has also been found depending upon the degree of soil salinization [42]. The dramatic increase in soluble sugar and to a much lesser degree of protein contents in *Salvia miltiorrhiza* leaves with the increasing salinity of the hydroponic solution suggest that antioxidant enzymes and osmolytes are partially involved in the adaptive response to salt stress in *Salvia miltiorrhiza*, thereby maintaining better plant growth under saline conditions [26].

7.3.4 AMINO ACIDS

Plants synthesize amino acids from the primary elements, the carbon and oxygen obtained from air, hydrogen from water in the soil, forming carbohydrate by means of photosynthesis. When combined with nitrogen which the plants draw from the soil, it leads to synthesis of amino acids, by collateral metabolic pathways. Only L-Amino acids are part of these proteins and have metabolic activity. Plants under salt stress may have increased levels of certain compounds such as amino acids (alanine, arginine, glycine, serine, leucine, and valine, together with the amino acid, proline, and the non-protein amino acids, citrulline and ornithine) and amides (such as glutamine and asparagines) [44]. Many investigators recorded an accumulation of amino acids especially proline in plant exposed to salt stress [21].

Higher total free amino acids in plants subjected to salt stress conditions were reported in *Catharanthus roseus* and *Matricaria chamomilla*. Free amino acids in *H. muticus* were generally elevated under salinization treatments but in *D. stramonium* these contents were generally lowered with the rise in salinization levels except in the case of stem, where they were somewhat elevated [3]. It was observed that the amino acid pool increased in shoot of plants grown in soil with salt concentration of 2000, 4000 or 6000 ppm during two seasons 2007, 2008 except 4000 ppm level of salinity at 2007 season where it has value less than control [33].

Proline is a protective agent of enzyme and membrane and as an intracellular structure or a storage compound of carbon and nitrogen for rapid recovery from stress. Because of its zwitterionic, high-hydrophilic nontoxic character, proline can act as a 'compatible solute' and its accumulation is a common response to salinity and related stresses. Proline occurs widely in higher plants and accumulates in larger amounts than other amino acids. It accumulates in leaves as a response to salt stress in *Salvia officinalis*, *Trachyspermum ammi*, spearmint, chamomile, sweet marjoram, *Catharanthus roseus*, *Achillea fragratissima*, *Matricaria chamomilla*, sweet fennel, and *Satureja hortensis*. The contents of proline increased progressively in *Hyoscyamus muticus* and *Datura stramonium* in response to salinization treatments where N/10 Pfeffer's nutrient solution containing various salinization levels (0.0, 1000, 3000, 5000 and 7000 ppm) with a mixture of NaCl and $CaCl_2$ (1:1 by weight) was used for irrigation [3]. The proline and carbohydrate content as compatible organic solutes increased with increasing salinity in *Moringa peregrine* from control to 14 dS m^{-1}[69]. The increase in proline content is attributed to a decrease in proline oxidase activity in saline conditions [53]. However, the plant ability to accumulate proline under stress conditions varies between species or even varieties [76].

7.4 EFFECT OF SALT STRESS ON SECONDARY PLANT PRODUCTS

Several environmental stresses cause drastic changes in the growth, physiology, and metabolism of plants leading to the increased accumulation of secondary metabolites. As medicinal plants are important sources of drugs, investigations are undertaken to understand the effect of stress on the physiology, biochemistry, genomic, proteomic, and metabolic levels. Some medicinal plants under salt stress accumulate higher concentrations of secondary compounds than control plants which are cultivated under standard conditions.

7.4.1 ESSENTIAL OILS

Salt stress significantly affects the production of essential oils, important secondary metabolites and the constituents of medicinal plants. Contradictory reports have emerged concerning the response of medicinal plants in terms of essential oil production under salt stress. Sangwan et al. [70] stated that the production of essential oil not only depends upon the metabolic state of the source tissues but also may be integrated with the stress factors. Razmjoo and Sabzalian [66] indicated that increased salinity significantly reduced essential oil content of chamomile.

Ashraf and Harris [12] also showed that oil content in the seeds of bishops weed (*Ammolei majus*) decreased consistently with increase in external salt levels. Essential oil yield decreased under salt stress in *Trachyspermum ammi* [12]; *Mentha piperita*; peppermint, pennyroyal, and apple mint, *Thymus maroccanus* [17] and basil [68]. Anethole percentage was also reduced with saline water.

In marjoram, the proportions of the main compounds were differently affected by salt. In *Matricaria recutita*, the main essential oil constituents (α-bisabololoxide B, α-bisabolonoxide A, chamazulene, α-bisabolol oxide A, α-bisabolol, trans-β-farnesene) increased under saline conditions [15]. On the other hand, in *Origanum vulgare* the content of carvacrol (the main essential oil constituent), p-cymene and γ-terpinene decreased under salt stress [68]. Phenolic acid content ratio increased in *Mentha pulegium* as a result of salt stress [61]. Similar results of inhibitory effects of high salinity were observed in lemon balm; *Majorana hortensis, Matricaria chamomile, Salvia officinalis* and basil [68].

On the contrary, an increase of essential oil yield due to lower levels of salinity has been reported in other plant species, e.g., *Satureja hortensis* and *Salvia officinalis*. It has been shown that essential oil yield of coriander leaves was stimulated only under low and moderate stress, while it decreased at the high salinity level. At low stress, (E)-2-decenal and dodecenal contents increased [54]. Ben-Taarit et al. [18] highlighted that essential oil compounds were sensitive to environmental changes especially salt stress, stress mediated changes being more prominently reflected in the major oil constituent, i.e., viridiflorol, 1,8-cineole, thujone and manool.

These variations could be due to the induction of the specific enzymes involved in the biosynthesis of the later compounds under salt stress [18].

The major components of the essential oil eugenol and linalool in *Ocimum basilicum* var. purpurascens showed different responses. Soil salinity at 1500 and 4500 mg kg^{-1} levels increased linalool but decreased eugenol. There are reports of an increase in essential oil percentage due to lower levels of salinity in *Satureja hortensis*; sage [30] and thyme [24]. In basil, the highest oil percentage was achieved under salinity condition. In contrast, other reports showed a significant reduction of essential oil percentage on lemon balm and sweet marjoram (*Majorana hortensis* L.) [57].

Ashraf [10] also showed that oil content in the seed of medicinal plant, bishop's weed (*Ammolei majus*) decreased consistently with increase in external salt levels. Reduced flower weight may have resulted in oil content reduction in chamomile in saline environment. It was also noticed that moderate (50 mM) to high (75 mM) concentrations of NaCl enhanced the essential oil production in *S. officinalis* fruits [30]. These contradictory results represent the different responses of medicinal plants to salt stress regarding essential oil production. The fact that some medicinal plants may increase the production of essential oil or its main constituents in response to salt stress encourages us to determine the molecular mechanisms of salt stress on the production of secondary metabolites in medicinal plants.

7.4.2 ALKALOIDS

The close relationship between plant secondary metabolism and defense response is widely recognized. Production of secondary metabolites by plants may not always be satisfactory as several factors such as type of medicinal plant species or genus, particular growth or developmental stage, specific seasonal conditions, nutrient availability, or stress conditions can restrict their production. The saline stress has significant effect on alkaloid content of some medicinal plants like *Catharanthus roseus* when grown under salt stress, concentration of the major alkaloids constituents such as reserpine and vincristine, respectively being higher [48]. The level of ricinine alkaloids in roots of *Ricinus communis* was significantly lower in plants under salt stress, but it increased in the shoots.

Also the studies published on *Solanum nigrum* indicated significant increase in the alkaloid content of these plants exposed to salt stress [19]. Selmar [71] reported a strong increase in the concentration of tropane alkaloids in *Datura inoxia* plants which might be an expression to overcome the salt stress. Total indole alkaloid content varied with different soil salinity regimes; higher alkaloid content being recorded in NaCl-treated than control plants [36]. Previously, same authors reported alterations in the alkaloid content of *C. roseus* in response to abiotic stresses like drought, as well as under growth regulator treatments [37].

7.4.3 PHENOLIC COMPOUNDS

Phenolic compounds from medicinal herbs and dietary plants include phenolic acids, flavonoids, tannins, stilbenes, curcuminoids, coumarins, lignans, quinones, and others. Natural phenolic compounds are used in cancer prevention and treatment as bioactivities of phenolic compounds are responsible for their chemo-preventive properties. Rosemary leaf contains phenolic acids (2–3% rosemarinic, chlorogenic, carnosic acid and caffeic), rosmarinic and carnosic acid being major antioxidant phenolic compounds in rosemary plants [6].

In studies on spearmint it was found that phenolic acid concentration increased in salt stressed plants [4]. In *Achillea fragratissima* the content of phenols increased significantly with increasing salinity [1]. Also in *Matricaria chamomilla* accumulation of phenolic acids (protocatechuic, chlorogenic and caffeic acids) was observed under salinity stress [20]. Two month-old rosemary plants subjected to three salt treatments (50, 100, and 150 mM NaCl) caused an accumulation of total phenolic and antioxidants to reduce oxidative stress. It can be a possible way to produce antioxidant compounds in rosemary by subjecting the plants to salt stress [40]. Variation of phenolic compounds at different growth stages of *O. majorana*, and the possible role of these changes in the response of the plant to salt has been studied by Baâtour et al. [14]. Levels of polyphenols also increase under increasing levels of salinity, which shows that, the induction of secondary metabolism is one of the defense mechanisms adapted by the plants to face saline environment.

7.5   PLANTS RESPONSE TO COMBAT THE SALINITY STRESS

Many genes have been induced under stress conditions. The products of stress-inducible genes protecting against these stresses includes the enzymes responsible for the synthesis of various osmoprotectants. Genetic engineering of tolerance to abiotic stresses help in molecular understanding of pathways induced in response to one or more of the abiotic stresses. The salt tolerance mechanism of plants is complex. The detailed salt tolerance mechanism is summarized in Figure 7.2.

7.5.1   APPLICATIONS OF COMPATIBLE SOLUTE ENGINEERING

Salt stress can reportedly result in oxidative stress through increased reactive oxygen species (ROS), such as the superoxide radical (O_2), hydrogen peroxide (H_2O_2), and the hydroxyl radical (OH). To mitigate this damage, plants have developed a complex antioxidative defense system that includes ionic and osmotic homeostasis and antioxidative enzymes. The primary function of compatible solutes is to maintain cell turgor and thus the driving gradient for water uptake. Recent studies indicate that compatible solutes can also act as free-radical scavengers or chemical chaperones by directly stabilizing membranes and/or proteins [22].

Compatible solutes are categorized under the three groups namely amino acids (e.g., proline), quaternary amines (e.g., glycine betaine, dimethylsulfoniopropionate) and polyol/sugars (e.g., mannitol, trehalose). Over expression of compatible solutes in transgenic plants can result in improved stress tolerance. Some plants also have the potential to increase cellular concentrations of osmotically compatible solutes, osmolytes and osmoprotectants.

During stress conditions plants need to maintain internal water potential below that of soil and maintain turgor and water uptake for growth. This requires an increase in osmotica, either by uptake of soil solutes or by synthesis of the compatible solutes that do not interfere with normal biochemical reactions. These compatible solutes help to maintain ion homeostasis and water relations, and they alleviate the negative effects of high ion concentrations on the enzymes, stabilizing proteins, protein complexes,

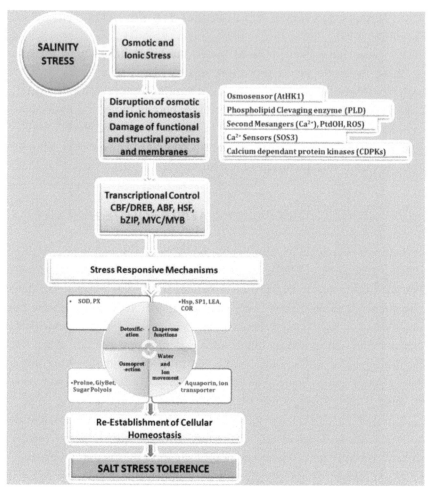

FIGURE 7.2 The complexity of the plant response to abiotic stress; the initial stress (e.g., osmotic and ionic effects) triggers the downstream signaling process and transcription controls which activate stress-responsive mechanisms to re-establish homeostasis and protect and repair damaged proteins and membranes.

membranes and cellular function under stress conditions. Moreover, salinity-induced oxidative stress could be ameliorated through the action of antioxidative enzymes such as superoxide dismutase (SOD), peroxidase (POD), ascorbate peroxidase (APX), and catalase (CAT). Compatible solutes or osmolytes accumulate in organisms in response to osmotic stress. In plants, glycine betaine, a representative member of betaine group of osmolytes, is synthesized in the chloroplast from choline by a two-step process. The first step (choline to betaine aldehyde) is mediated by choline monooxygenase (CMO), which can be induced by drought and salinity [67]. The second step (betaine aldehyde to glycine betaine) is catalyzed by betaine aldehyde dehydrogenase (BADH), and NAD-dependent dehydrogenase.

7.5.2   ENGINEERING OF ION TRANSPORT

Osmotic stress, ion toxicity and high salt content in the soil (especially Na^+ and Cl^-) and the irrigation water, significantly impair plant growth. Ion transporters selectively transport ions and maintain them at physiologically relevant concentrations. Under salt stress, tolerant plant cells must maintain high K^+ (100–200 mM) and low Na^+ (less than 1 mM) levels for normal metabolic functions.

An important strategy for achieving greater tolerance to salinity stress is to help plants to reestablish homeostasis under stressed environments, restoring both ionic and osmotic homeostasis. This strategy continues to be a major approach to improve salt tolerance in plants through genetic engineering, where the target is to achieve Na^+ excretion, or vacuolar storage. The Na^+/H^+ antiporters also play a crucial role in maintaining cellular ion homeostasis and are responsible for plant survival and growth under saline conditions. The Na^+/H^+ antiporters catalyze the exchange of Na^+ for H^+ across membranes and have a variety of functions, such as regulating cytoplasmic pH, sodium levels and cell turgor [72]. Constitutive expression of vacuolar Na^+/H^+ antiporter (NHX1) or AVP1 (A. thaliana vacuolar H+-translocating pyrophosphatase) gene energized the pumping of Na^+ into the vacuole, and increased both accumulation and Na^+ tolerance in *Arabidopsis* [25]. No specific reports are found in medicinal plants where ion transport has been employed for the salt tolerance.

7.6   CASE STUDIES OF SOME MEDICINAL PLANTS

7.6.1   SALVIA MILTIORRHIZA

Salvia miltiorrhiza (also known as red sage, Chinese sage, tan shen, or danshen) is a perennial flowering plant in the genus *Salvia*, highly valued for its roots in traditional Chinese medicine. Native to China and Japan, it grows between 90–1,200 m in elevation, preferring grassy places in forests, hillsides, and along stream banks. The outside of the tap root is used in medicine. An antioxidant called salvianolic acid or salvianolic acid B can be isolated from this material, which is still a subject of medical research although it has been widely used for hundreds of years to treat cardiovascular diseases.

A recent study has revealed that *Salvia miltiorrhiza* hydrophilic extract (SMHE) clearly reduced oxidative stress in diabetic patients with chronic heart disease [60]. Increase in soluble sugar and proteins, the primary osmolytes, has been found to increase with increasing salt concentration in this plant. While Na^+ content in salt-treated *Salvia miltiorrhiza* was significantly higher with the increasing salinity but the K^+ content was lower, implying that a high level of external Na^+ caused a decrease in the K^+ concentrations in *Salvia miltiorrhiza* tissues [26]. Spraying brassinosteroid has been found to increase the drought stress tolerance in *S. miltiorrhiza* seedling showing increase in the antioxidant enzyme activity (SOD, CAT, POD) and Pro content and a decrease in the MDA content. With the treatment of 0.1 mg L^{-1} brassinosteroid, the activity of SOD, CAT and the content of Pro raised to the highest level, while the MDA content reached its lowest point [88].

7.6.2   MENTHA PIPERITA

Peppermint (*Mentha piperita* L.) belongs to mint (*Lamiaceae* family) and is herbaceous and perennial. It is a medicinal and aromatic plant and is used extensively in medicines and food. Fundamental components of peppermint essential oil include menthol, menthone, methylacetat, menthofuran and pulegone. It has been observed that the amount of peppermint essential oil and its constituents are considerably impacted by different factors such as climate, soil type, geographical area, harvest time and fertilizer usage. The effect of salinity stress based on growth parameters, essential oil constituents and yield in peppermint revealed that essential oil yield as well as essential oil percentage reduced with increasing salinity from 0 to 100 mmol L^{-1} [61].

7.6.3   AZADIRACHTA INDICA

The salt tolerant tree species are used for medicines, fodder, fuel wood, timber, and shade. *Azadirachta indica, Salvadora persica, Acacia nilotica,* and *Prosopis juliflora* are some such plants. *Azadirachta indica* (Neem) is a tree in the mahogany family *Meliaceae*. It is native to India growing in tropical and semi-tropical regions. Its fruits and seeds are the source of neem oil, which is light and dark brown in color, bitter with a rather strong odor and is used in preparing cosmetics, soaps, hairs products, ceramics, etc.

The most frequently reported application of Neem in ancient Ayurveda are skin diseases, inflammations, malaria and tuberculosis and more recently in rheumatic disorders, insects repellent and insecticides. Amongst the four medicinal plants namely *Azadirachta indica, Cassia fistula, Aloe barbadensis, Catharanthus roseus,* the neem was found to be the most tolerant to salt stress as revealed by growth parameters like leaf area, total carbohydrates and total chlorophyll content [41]. It was observed that chlorophyll content was the highest in *Azadirachta indica* as compared to other plants. Since total chlorophyll content is an indicative of photosynthetic and metabolic activity, it has resulted in high tolerance compared to others where chlorophyll content was less [41].

In another study, proline, glycine betaine and peroxidase contents of calli and regenerated shoots of *Azadirachta indica* increased with increase in salinity up to 150 and 200 mM, respectively. The catalase activity in the calli and regenerated shoot increased up to 150 mM saline and declined thereafter. This study revealed that *A. indica* exhibits protection against the oxidative damage caused by salinity up to 250 mM, at least partly, due to constitutive and induced activities of antioxidative enzymes [73].

7.7   SALT STRESS AND ACTIVE INGREDIENTS OF MEDICINAL PLANTS

Turmeric (*Curcuma longa* L.) is a perennial rhizomatous plant from the family of Zingibraceae, native in South Asia. The main components of turmeric are curcuminoids and essential oil which are responsible for turmeric characteristic such as odor and taste. Curcumionoids, forming the yellow color of turmeric, consist of curcumin, methoxy curcumin and bismethoxy curcumin. The essential oil of turmeric consists of several sesquiterpenoids such as ar-turmerone, curlone, α-turmerone, β-turmerone and bisacumol. Other sesquiterpenoids are zingibrene, curcumenone, curcumenol, procurcumenol and dehydrocurdione and germacrone-13-al.

The analysis of curcuminoids and volatile oil in turmeric are considered as a significant character in order to evaluate the quality of plant materials. However, it was reported that curcuminoid and essential oil components of turmeric vary at different stages of growth. In the study conducted by Mostajeran et al. [50], two months old plants were exposed to salinity (0, 20, 60 and 100 mM NaCl) for two months. As per these results, the addition of curcumin in rhizomes for four months old plant versus three months were almost 5 fold for 0 mM NaCl and 2 fold for 100 mM NaCl. Low salinity has positive effect in curcumin production but higher salinity (higher than 60 mM) had adverse effect and caused 24% reduction of curcumin compared to control plants. There were more paracymene and terpineol in volatile oils of turmeric rhizome than the other components such as most of the volatile oil compounds were unchanged or varied slightly as salinity changed [50].

In *Ambrosia maritima* L. (Damsisa) which is one of the wild plants present in Egypt and different African countries of Nile valley, ambrosin and damsin are the most active ingredients in damsisa plants. Ambrosin has been described as attribute to a group of natural products known as psuedo-guaianolides. Drinking decoctions of damsisa were the most commonly used remedy for schistosomiasis in Upper Egypt. Aqueous methanol extract of damsisa plant had hepatoprotective and antioxidant actions in drug that induce hepatotoxicity in rats. In the experiments by Hussein et al. [33], *A. maritime* seeds grown at different levels of soil salinity revealed that at different levels of salinity either irradiated or un-irradiated control salinity mostly produced shoots having high ambrosin percentage than the control set. The opposite is true in percentage to damsin content, except for 40 Gy dose. The results show that shoots produced from seeds treatment by 40 Gy contain more damsin percentage than its corresponding control at different salinity levels. The damsin percentage increased by 4.9, 87.6 and 19.7% above its corresponding control at 2000, 4000 and 6000 ppm salt concentration.

Sangwan et al. [70] stated that the production of essential oil not only depends upon the metabolic state of the source tissues but also may be integrated with the stress factors. Razmjoo and Sabzalian [66] indicated that increased salinity significantly reduced essential oil content of chamomile. Ashraf and Harris [11] also showed that oil content in the seed of medicinal plant, bishops weed (*Ammolei majus*) decreased consistently with increase in external salt levels.

Madagascar periwinkle a glycophyte is a medicinal plant which contains a number of terpenoid indole alkaloids with more than 130 separated and identified compounds. The juice of the leaves is applied externally to treat wasp stings. All parts of the plant are credited with hypoglycaemic properties and are used to treat diabetes. In one of the studies conducted to investigate the effect of salinity stress (0, 2, 4, 6, 8 dS m^{-1}) on some morphological parameters and Ajmalicine alkaloid content in two cultivars of *Catharanthus roseus Don.* (Rosea and Alba) showed that shoot and root length, fresh and dry weight of shoots and roots decreased under salinity conditions. Ajmalicine alkaloid content increased under salinity conditions and Alba cultivar produced an amount less than Rosea. Highest fresh and dry weight of shoots and roots were in Alba and lowest length of shoot and root in Rosea [16].

7.8   AMELIORATION AND FUTURE PROSPECTS

There are various ways and means by which salt stress to plants can be managed. Conventional techniques include, drainage, leaching, application of chemical amendments, selection of crops and varieties, avoiding stresses at salt sensitive growth stages, irrigation water management, nutrient management, rainwater management, land forming, use of improved irrigation techniques such as drip and sprinkler irrigation, conjunctive use of saline and fresh water, bio-reclamation and several others.

Hopefully the wide experience of all these techniques either as such or with modifications [2] can be replicated in medicinal plants. Although, even the conventional methods may pose several problems in future, yet some can easily be applied without any major difficulty. For example crop selection can be made amongst several medicinal plants such as *Cassia senna, Lepidium sativum, Glycyrrhiza glabra,* and *Withania somnifera* that have performed well when irrigated with saline water up to EC 10 dSm^{-1}. Conjunctive use of saline and fresh water is also a good strategy that has shown promise in increasing the yield of Isabgol (Table 7.4).

Nutrient management also holds good promise [27]. Since high concentrations of Na$^+$ and Cl$^-$ in the soil solution depress nutrient-ion activities and disturb the nutrient ratios by producing extreme ratios of Na$^+$/Ca^{2+}, Na$^+$/K$^+$, Ca^{2+}/Mg^{2+}, and Cl$^-$/NO^{-3}, consequently, both osmotic and ionic injury may result in reduced yield or quality. Salt tolerance in plants increases by reducing the sodium uptake of plants and in this reduction, potassium (K) plays a key role. Nitrogen (N) is the main constituent of all amino acids in protein and a number of nitrogen containing compounds such as amino acids (proline and glycinebetaine), amids, imino acids and polyamines in plants that are subjected to salinity. Thus, application of N may reduce the effect of salinity and enhance the growth of plants. Similarly, externally supplied calcium (Ca^{2+}) can ease the toxic effects of NaCl apparently by facilitating a higher K$^+$ to Na$^+$ selectivity [74]. Application of Ca^{2+} also decreases permeability of plasma membranes and maintains membrane integrity and selectivity thus reducing Na+ and Cl$^-$ toxicity in plant.

TABLE 7.4 Yield of Isabgol (t ha^{-1}) with Different Quality of Water and Irrigation Frequency

Saline water/irrigation frequency	I$_1$	I$_2$	I$_3$	I$_4$	Mean
BAW	0.98	1.16	1.2	1.19	1.13
SW	0.96	1.35	1.38	1.39	1.27
Alternate	1.24	1.28	1.34	1.42	1.32
Mean	1.06	1.26	1.31	1.33	1.24

LSD ($p = 0.0\ 5$): Irrigation waters (Iw): NS Sch.: 1.99 Interaction Iw x Sch.: NS

BAW = Best available water, SW = Saline water, Alternate = Fresh-saline water applied alternatively. I$_1$, I$_2$, I$_3$ and I$_4$ show number of irrigations, i.e., 1, 2, 3 and 4 applied during crop growth.

A study on *Cichorium* plant showed that 10 mM $CaCl_2$ with 50 mM NaCl not only counteracted salinity stress inhibitory effect for the plant growth but also induced metabolism of the plant important secondary metabolites those give the plant its medicinal importance [23]. Addition of calcium may increase α-amylase activity, as α-amylase requires Ca^{2+} for its activity [37]. Since most saline soils have a mixture of salts, the studies conducted in mediums that represent the real field situation can help to translate the results to real field conditions. The exogenous application of K^+ regulates the NH^{4+} toxicity and consequently helps the salt stressed plant in reducing the overload energy of toxic NH^{4+} efflux to uphold plant in such an environment. Copper nutrition reduced lipid peroxidation and membrane permeability while it increased total phenol content of salt stressed plants. It seems that copper nutrition can effectively ameliorate salt-induced oxidative damage as in the case of rosemary [46].

Although many studies have been reported in the literature about plant growth regulators, the mechanisms of these regulators in promoting stress tolerance in plants need to be elucidated. The exogenous applications of 5-aminolevulinic acid (ALA) regulate plant growth and development, as well as enhance Chl biosynthesis and photosynthesis, resulting in increased crop yield[31]. In *Cassia obtusifolia L.* its role in alleviation of NaCl stress is achieved via improved antioxidantenzyme activities, increased Chl content and photosynthetic efficiency, strengthened capacity of scavenging ROS, increased membrane stability, decreased cell osmotic potential, as well as decreased membrane lipid peroxidation [86]. Similar exercise for other chemicals and bio-regulators for various medicinal plants needs to be undertaken.

Ardebili et al. [9] indicated that foliar application of amino acid at suitable concentrations had positive effects on the content of secondary metabolites, antioxidants and antioxidant activity. The stimulated values of biochemical constituents strengthened the role of the applied amino acids in the metabolism of *Aloe vera* plants. Future work could usefully examine the ability of exogenous proline to ameliorate the injurious effects of high concentrations of NaCl.

Genomics, proteomics, and metabolomics are the new challenging fields representing a challenging complexity in phytomedical research. Proteomics provides a promising approach for studying secondary metabolism in plants and plant cells. The natural yield of secondary metabolites in medicinal plants is generally low and the biochemistry of the biosynthesis of these compounds is also poorly understood. A further understanding of the biochemical and molecular mechanisms underlying stress should be achieved with the advent of functional genomics, transcriptomics and proteomics.

Cell suspension cultures and metabolic engineering are among strategies to increase the yield of the commercially important plant based chemicals. Over-expression of rate limiting enzymes which are involved in biosynthesis of these compounds has been used for this purpose. Therefore, it is necessary to identify proteins involved in the secondary metabolites biosynthesis. Enzyme isolation and characterization by current approaches is time consuming and troublesome, thus, the proteomic approach is a faster and more complete and using this technology it is possible to identify regulatory and transport proteins as well as enzymes. *Catharanthus roseus* used as a model system to produce secondary metabolite resulted in some effective anti-tumor drugs, vinblastine and vincristine, which are alkaloid compounds. The discovery and use of new stress-tolerance-associated genes, as well as heterologous genes, to confer plant stress tolerance (including those unique to extreme-growth-environment organisms, e.g., halophytes, thermophilic organisms) has been the subject of ongoing efforts to obtain tolerant plants.

While adaptation to stress under natural conditions has some ecological advantages, the metabolic and energy costs may sometimes mask and limit its benefits in agriculture and result in yield penalty. Therefore, the improvement of abiotic stress tolerance of agricultural/medicinal and other plants can only be achieved, practically, by combining traditional and molecular breeding. Thus, a comprehensive breeding strategy for abiotic stress tolerance should include the following steps and approaches:

- Conventional breeding and germplasm selection, especially of wild relevant species;
- Elucidation of the specific molecular control mechanisms in tolerant and sensitive genotypes;
- Biotechnology-oriented improvement of selection and breeding procedures with marker-assisted breeding programs with stress-related genes and QTLs;
- Improvement and adaptation of current agricultural practices.

There is need to identify new halophytic plant species to detect new tolerance determinants, operating pathways and genes that confer salt tolerance. It may help trigger generation of new genetically modified stress-tolerant medicinal plants. Since soil salinity is a major abiotic stress worldwide, it is imperative that more crops tolerant to this abiotic stress be designed, tested, and eventually released for application as new commercial varieties.

7.9 SUMMARY

Salt stress can have devastating effect on the growth, development, physiology and yield of plants. However, the response to salinity differs greatly among various plant species due to the levels of stress as well as the environmental condition. In recent years, the biochemical responses of plants to salt stress have been studied intensively. A comprehensive overview of the studies reveal that the impact of salt stress on medicinal plants in terms of germination, growth and other morphological characters including yield is similar to what has been widely reported for cereal, vegetable, oilseed and other crops of human interest. On the contrary, salt stress and corresponding enhancement effect for secondary metabolites remains inconclusive and needs to be studied in greater details and under varying environments masking the real field situations. Even if there are any gains in the quality with mild salt stress, it may be interesting to evaluate the yield penalty and its impact on the overall production of the industrially important constituents.

Since most plants are salt sensitive at the seedling stage and become increasingly tolerant as growth proceeds through the vegetative and reproductive stages, there is a need to extend many of these studies to other growth stages going in some cases even up to the harvest stage. The generated information on the tolerance mechanism would be useful for developing new cultivars that are adaptable in salinity environments although defining salt tolerance is quite difficult because of the complex nature of salt stress and the wide range of plant responses. Also, there is a greater need of dialog between the conventional crop scientists and the medicinal plant scientists on salinity management through conventional and other techniques in a cost effective manner.

KEYWORDS

- abiotic stress
- alkaloids
- amelioration of salt stress
- amino acids
- *Azadirachta indica*
- carbohydrate
- compatible solute engineering
- crop yield
- essential oils
- germination
- ion transport
- lipids
- *Majorana hortensis*
- medicinal plants
- *Mentha piperita*
- metabolites
- moisture content
- morphological parameters
- *Ocimum basilicum*
- osmotic stress
- oxidative stress
- phenols
- proline
- protein
- salt stress
- *Salvia miltiorrhiza*
- secondary plant products
- soil salinity
- solute engineering
- survival
- threshold salinity

REFERENCES

1. Abd EL-Azim, W. M., & Ahmed, S. T. (2009). Effect of salinity and cutting date on growth and chemical constituents of *Achillea fragratissima Forssk.*, under Ras Sudr conditions. *Res. J. Agr. Biol. Sci.,* 5, 1121–1129.

2. Aghaei, K., & Komatsu, S. (2013). Crop and medicinal plants proteomics in response to salt stress. *Frontiers in Plant Science, 4:* (http://www.ncbi.nlm.nih.gov/pubmed/23386857 opened on 24.06.2015)

3. Ahmad, A. M., Heikal, M. D., & Ali, R. M. (1989). Changes in amino acids and alkaloid contents in *Hyoscyamus muticus* and *Datura stramonium* in response to salinization. *Phyton (Austria),* 29, 137–147.

4. Al-Amier, H., & Craker, L. E. (2007). In-Vitro selection for stress tolerant spearmint. In: *Issues in New Crops and New Uses.* J. Janick and A. Whipkey (Eds.). ASHS Press, Alexandria, VA. 306–310.

5. Ali, R. M, Abbas, H. M., & Kamal R. K. (2007). The effects of treatment with polyamines on dry matter, oil and flavonoid contents in salinity stressed *chamomile* and *sweet marjoram. Plant Soil Environ.* 53, 529–543.

6. Almela, L., Sanchez-Munoz, B., Fernandez-Lopez, J. A., Roca, M. J., & Rabe, V. (2006). Liquid chromatograpic-mass spectrometric analysis of phenolics and free radical scavenging activity of rosemary extract from different raw material. *J. Chromatogr.,* 1120, 221–229.

7. Anonymous, (2011). *Biennial Report (2008–10). AICRP on Management of Salt Affected Soils and Use of Saline Water in Agriculture.* CSSRI, Karnal. 208p.

8. Ansari, S. R., Frooqi, A. H. A., & Sharma, S. (1998). Inter-specific variation in sodium and potassium ion accumulation and essential oil metabolism in three *Cymbopogon* species raised under sodium chloride stress. *J. Essen. Oil Res.,* 10, 413–418.

9. Ardebili, Z. O., Moghadam, A. R. L., Ardebili, N. O., & Pashaie, A. R. (2012). The induced physiological changes by foliar application of amino acids in *Aloevera* L. plants. *Plant Omics J.,* 5, 279–284.

10. Ashraf, M. (2004). Photosynthetic capacity and ion accumulation in a medicinal plant henbane (*Hyoscyamus niger L.*) under salt stress. *J. Appl. Bot.,* 78, 91–6.

11. Ashraf, M., & Harris, P. J. C. (2004). Potential biochemical indicators of salinity tolerance in plants. *Plant Sci.,* 166, 3–16.

12. Ashraf, M., & Orooj, A. (2006). Salt stress effects on growth, ion accumulation and seed oil concentration in an arid zone traditional medicinal plant ajwain (*Trachyspermum ammi L. Sprague*). *J. Arid Environ.,* 64, 209–220.

13. Baatour, O., Kaddour, W., Wannes, A., Lachaal, M., & Marzouk, B. (2010). Salt effects on the growth, mineral nutrition, essential oil yield and composition of marjoram (*Origanum majorana*). *Acta Physiol Plant,* 32, 45–51.

14. Baâtour, O., Tarchoun, I., Nasri, N., Kaddour, R., Harrathi, J., Drawi, E. L., Mouhiba, Ben Nasri-Ayachi, Marzouk, B., & Lachaâl, M. (2012). Effect of growth stages on phenolics content and antioxidant activities of shoots in sweet marjoram (*Origanum majorana L.*) varieties under salt stress. *African Journal of Biotechnology,* 11, 486–493.

15. Baghalian, K., Haghiry, A., Naghavi, M. R., & Mohammadi, A. (2008). Effect of saline irrigation water on agronomical and phytochemical characters of chamomile (*Matricaria recutita L.*). *Scientia Hortic.*, 116, 437–441.

16. Behzadifar, M. C., & Abdolhossein A. (2013). Effect of salt stress by using unconventional water on some morphological characters and ajmalicine alkaloid amount in the roots of *Catharanthus roseus* Cvs. *Rosea and Alba* Mandana. *Annals of Biological Research*, 4, 229–231.

17. Belaqziz, R., Romane, A., & Abbad, A. (2009). Salt stress effects on germination, growth and essential oil content of an endemic thyme species in Morocco (*Thymus maroccanus Ball.*). *J. Appl. Sci. Res.*, 5, 858–863.

18. Ben-Taarit, M. K., Msaada, K., Hosni, K., Hammami, M., Kchouk, E., & Marzouk, B. (2009). Plant growth, essential oil yield and composition of sage (*Salvia officinalis L.*) fruits cultivated under salt stress conditions. *Ind. Crops Prod.*, 30, 333–337.

19. Bhat, M. A., Ahmad, S., Aslam, J., Mujib. A., & Uzzfar, M. (2008). Salinity stress enhances production of solasodine in *Solanum nigrum L. Chem Pharm Bulletin*, 56, 17–21.

20. Cik, J. K., Klejdus, B., Hedbavny, J., & Bačkor, M. (2009). Salicylic acid alleviates NaCl-induced changes in the metabolism of *Matricaria chamomilla* plants. *Ecotoxicology*, 185, 544–554.

21. Dhingra, H. B., & Varghese, T. M. (1985). Effect of salt stress on viability, germination and endogenous levels of some metabolites and ions in maize (*Zea mays L.*) pollen. *Ann. Bot.*, 55, 415–420.

22. Diamant, S., Eliahu, N., Rosenthal, D., & Goloubinoff, P. (2001). Chemical chaperones regulate molecular chaperones in vitro and in cells under combined salt and heat stresses. *J. Biol. Chem.*, 276, 39586–39591.

23. Elhaak, M. A., Abo-Kassem, E. M., & Saad-Allah, K. M. (2014). Effect of the combined treatment with sodium and calcium chlorides on the growth and medicinal compounds of Cichorium intybus. *Int. Journal of Current Microbiology and Applied Sciences*, 3, 613–630.

24. Ezz, E. A. A., Aziz, E. E., Hendawy, S. F., & Omer, E. A. (2009). Response of *Thymus vulgaris L.* to salt stress and alar (B9) in newly reclaimed soil. *J. Appl. Sci. Res.*, 5, 2165–2170.

25. Gaxiola, R. A., Li, J., Unurraga, S., Dang, L. M., Allen, G. J., Alper, S. L., & Fink, G. R. (2001). Drought and salt tolerant plants result from over expression of the AVP1 H+-pump. *Proc. National Acad. Sci. USA*, 98, 11444–11449.

26. Gengmao, Z., Quanmei, S., Yu, H., Shihui, L., & Changhai, W. (2014). The physiological and biochemical responses of a medicinal plant (*Salvia miltiorrhiza* **L.**) to stress caused by various concentrations of NaCl. http://www.ncbi.nlm.nih.gov/pmc/articles/PMC3934908/ opened on 20.06.2015.

27. Grattan, S. R., & Grieve, C. M. (1999). *Salinity-mineral nutrient relations in horticultural crops. Scientia Horticulturae*, 78, 127–157.

28. Gupta, S. K., & Gupta, I. C. (2014). *Management of Saline and Waste Water in Agriculture.* Scientific Publishers, Jodhpur. 321 pages.

29. Hanselin, M. H., & Eggen, T. (2005). Salinity tolerance during germination of seashore halophytes and salt tolerant grass cultivars. *Seed Sci. Research*, 15, 43–50.

30. Hendawy, S. F., & Khalid, K. A. (2005). Response of sage (*Salvia officinalis L.*) plants to zinc application under different salinity levels. *J. Appl. Sci. Res.*, 1, 147–155.

31. Hotta, Y., Tanaka, H., Takaoka, Y., Takeuchi, Y., & Konnai, M. (1997). Promotive effects of 5-aminolevulinic acid on the yield of several crops. *Plant Growth Regul.*, 22, 109–114.

32. Hussain, F., & Ilahi, I. (1992). Effect of Magnesium sulfate, sodium sulfate and mixture of both salts on germination and seedling growth of three cultivars of *Brassica campestris* L. *Sarhad J. Agric.*, 3, 175–183.

33. Hussein, O. S., Hanafy, A. H., Ahmed, G. A. R., & El-Hefny, A. M. (2012). Some active ingredients, total protein and amino acids in plants produced from irradiated *Ambrosia maritima* seeds growing under different soil salinity levels. *American Journal of Plant Physiology*, 7, 70–83.

34. Ibrar, M., & Hussain, F. (2003). The effect of salinity on the growth of *Medicago polymorpha Linn*. *J. Sci. & Tech. Univ. Peshawar*, 27, 35–38.

35. Jacobs, D. I., van der Heijden, R., & Verpoorte, R. (2000). Proteomics in plant biotechnology and secondary metabolism research. *Phytochem. Anal.*, 11, 277–287.

36. Jaleel, C.A, Sankar, B., Sridharan, R., & Panneerselvam, R. (2008). Soil salinity alters growth, chlorophyll content, and secondary metabolite accumulation in *Catharanthus roseus*. *Turk J. Biol.*, 32, 79–83

37. Jaleel, C. A., Manivannan, P., & Sankar, B. (2007). Calcium chloride effects on salinity-induced oxidative stress, proline metabolism and indole alkaloid accumulation in *Catharanthus roseus*. *C R Biologies*, 330, 674–683.

38. Jamil, M., Lee, D. B., Jung, K. Y., Lee, S. C., & Rha, E. S. (2006). Effect of salt (NaCl) stress on germination and early seedling growth of four vegetables species. *JCEA*, 7, 273–282.

39. Kaya, C., Tuna, A. L., Ashraf, M., & Altunlu, H. (2007). Improved salt tolerance of melon (*Cucumismelo L.*) by the addition of proline and potassium nitrate. *Environ. Exp. Bot.*, 60, 397–403.

40. Kiarostami, K., Ohseni, R., & Saboora, A. (2010). Biochemical changes of *Rosmarinus officinalis* under salt stress. *Journal of Stress Physiology & Biochemistry*, 6, 114–122.

41. Kiran kumari, S. P., Sridevi, V., & Chandana Lakshmi, M. V. V. (2012). Studies on effect of salt stress on some medicinal plants. *International Journal of Computational Engineering Research*, 2, 143–149.

42. Levitt, J. (1980). *Responses of Plants to Environmental Stresses*. Volume II, 2nd Ed. Academic Press, New York.

43. Maas, E. V., & Hoffman G. J. (1977). Crop salt tolerance: Current assessment. *J. Irrigation Drainage Div. ASCE*, 103, 115–134.

44. Mansour, M. M. F. (2000). Nitrogen containing compounds and adaptation of plants to salinity stress. *Biol Plant.*, 43, 491–500.

45. Meena, R. L., Ambast, S. K., Gupta, S. K., Chinchmalatpure, A. R., & Sharma, D. K. (2014). Performance of fennel (*Foeniculun vulgare Mill.)* as influenced by saline water irrigation and organic input management in semi-arid conditions. *Journal of Soil Salinity and water Quality*, 6, 52–58.

46. Mehrizi, M. H., Shariatmadari, H., Khoshgoftarmanesh, A. H., & Dehghani, F. (2012). Copper effects on growth, lipid peroxidation, and total phenolic content of rosemary leaves under salinity stress. *J. Agr. Sci. Tech.*, 14, 205–212.

47. Miceli, A., Moncada, A., & D'Anna, F. (2003). Effect of water salinity on seeds-germination of *Ocimum basilicum* L. *Eruca sativa* L. *Petroselinum hortense Hoffm. Acta Hortic.*, 609, 365–370.

48. Misra, N., & Gupta, A. K. (2006). Effect of salinity and different nitrogen sources on the activity of antioxidant enzymes and indole alkaloid content in *Catharanthus roseus* seedlings. *J. Plant Physiol.*, 163, 11–18.

49. Mogahdam, A., Fathi A., Lotfi, M., Asadi, K., & Amouzadeh, A. (2014). The analysis of the effects of salinity and ascorbic acid on growth properties of German Chamomile (*Chamomilla recutita* L). *International Research Journal of Applied and Basic Sciences*, 8, 1982–1987.

50. Mostajeran, A., Gholaminejad, A., & Asghari G. (2014). Salinity alters curcumin, essential oil and chlorophyll of turmeric (*Curcuma longa L.*). *Res. Pharm Sci.*, 9, 49–57.

51. Muhammad, Z., & Hussain, F. (2010). Effect of NaCl salinity on the germination and seedling growth of some medicinal plants. *Pakistan Botany J.*, 42, 889–897.

52. Munns, R. (2003). Comparative physiology of salt and water stress. *Plant, Cell and Environment*, 25, 239–250.

53. Muthukumarasamy, M., Gupta, D. S., & Panneerselvam, R. (2000). Influence of triadimefon on the metabolism of NaCl stressed radish. *Biol. Plant*, 43, 67–72.

54. Neffati, M., & Marzouk, B. (2008). Changes in essential oil and fatty acid composition in coriander (*Coriandrum sativum L.*) leaves under saline conditions. *Ind. Crops Prod.*, 28, 137–142.

55. Ngara, R., Ndimba, R., Jensen, J. B., Jensen, O. N., & Ndimb, B. (2012). Identification and profiling of salinity stress-responsive proteins in *Sorghum bicolor* seedlings.*J. Proteomics,* 75, 4139–4150.

56. Omami, E. N., Hammes, P. S., & Robbertse, P. J. (2006). Differences in salinity tolerance for growth and water-use efficiency in some amaranth (*Amaranthus* spp.) genotypes. *New Zeal J. Crop Hort. Sci.*, 34, 11–22.

57. Ozturk, A., Unlukara, A., Ipekl, A., Gurbuz, B. (2004). Effect of salt stress and water deficit on plant growth and essential oil content of lemon balm (*Melissa officinalis L.*). *Pak J. Bot.*, 36, 787–792.

58. Patra, D., Anwar, D. M., Singh, S., Prasad, A., & Singh D. V. (1999). Aromatic and medicinal plants for salt and moisture stress conditions. In: *Recent advances in Management of Arid Ecosystem.* Proceeding Symposium, India, March 1997, 347–350.

59. Pattanagul, W., & Thitisaksakul, M. (2008). Effect of salinity stress on growth and carbohydrate metabolism in three rice (*Oryza sativa L.*) cultivars differing in salinity tolerance. *Indian J Exp Biol.*, 46, 736–742.

60. Qian, Q., Qian, S., Fan, P., Huo, D., & Wang, S. (2012). Effect of *Salvia miltiorrhiza* hydrophilic extract on antioxidant enzymes in diabetic patients with chronic heart disease: a randomized controlled trial. *Phytother Res.* http://www.ncbi.nlm.nih.gov/pubmed/21544882 (Opened on 23.06.2015).

61. Queslati, S., Karray-Bouraoui, N., Attia, H., Rabhi, M., Ksouri, R., & Lachaal, M. (2010). Physiological and antioxidant responses of *Mentha pulegium* (Pennyroyal) to salt stress. *Acta Physiol. Plant*, 32, 289–296.

62. Rahimi, R., Mohammakhani, A., Roohi, V., & Armand, N. (2102). Effects of salt stress and silicon nutrition on chlorophyll content, yield and yield components in fennel (*Foeniculum vulgar Mill.*). *International Journal of Agriculture and Crop Sciences*, 4, 1591–1595.

63. Rahman, M., Soomro, U., Zahoor-Ul-Hag, M., & Gul, S. H. (2008). Effects of NaCl salinity on wheat (*Triticum aestivum L.*) cultivars. *World J Agri. Sci.*, 4, 398–403.

64. Ramin, A. A. (2005). Effects of salinity and temperature on germination and seedling establishment of sweet basil (*Ocimum basilicum L.*). *J. Herbs Spices Med. Plants*, 11, 81–90.

65. Ratnakar, A., & Rai, A. (2013). Effect of sodium chloride salinity on seed germination and early seedling growth of *Trigonella foenum-graecum* l. var. PEB. *Octa Jour. Env. Res.*, 1, 304–309.

66. Razmjoo, K. P. H., & Sabzalian, M. R. (2008). Effect of salinity and drought stresses on growth parameters and essential oil content of *Matriaria chamomile*. Int. J. Agric. Biol. 10, 451–454.

67. Russell, B. L., Rathinasabapathi, B., Hanson, A. D. (1998). Osmotic stress induces expression of choline monooxygenase in sugar beet and amaranth. *Plant Physiol.*, 116, 859–865.

68. Said-Al Ahl, H. A. H., & Omer, E. A. (2011). Medicinal and aromatic plants production under salt stress. *Herba pol.,* 57, 72–87.

69. Salehi, M., Hoseini, M. N., NaghdiBadi, H., & Mazaheri, D. (2012). Biochemical and growth responses of *Moringa peregrine* (Forssk.) Fiori to different sources and levels of salinity. *Journal of Medicinal Plants*, 43, 54–61.

70. Sangwan, N. S., Farooqi, A. H. A., Shabih, F., & Sangwan, R. S. (2001). Regulation of essential oil production in plants. *Plant Growth Regul.*, 34, 3–21.

71. Selmar, D. (2008). Potential of salt and drought stress to increase pharmaceutical significant secondary compounds in plants. *Agric. For. Res.*, 58, 139–144.

72. Serrano, R., Mulet, J. M., Rios, G., & Marquez, J. A. (1999). A glimpse of the mechanisms of ion homeostasis during salt stress. *J. Experi. Botany*, 50, 1023–1036.

73. Shivanna, M. B., Nagashree, B. R., & Gurumurthy, B. R. (2013). In vitro response of *Azadirachta indica* to salinity stress and its effect of certain osmoprotectants and antioxidative enzymes. http://citeseerx.ist.psu.edu/viewdoc/summary?doi=10.1.1.302.3632.

74. Siddiqui, M. H., Mohammad, F., Khan, M. N., Al-Whaibi, M. H., & Bahkali, A. H. A. (2010). Nitrogen in relation to photosynthetic capacity and accumulation of osmoprotectant and nutrients in *Brassica* genotypes grown under salt stress. *Agricultural Sciences in China*, 5, 671–680.

75. Singh, N. K., Bracker, C. A., Hasegawa, P. M., Handa, A. K., Buckel, S., Hermodson, M. A., Pfankock, E., Regnier, F. E., & Bressan, R. A. (1987). Characterization of osmotin, athumatin-like protein associated with osmotic adaptation in plant cells. *Plant Physiol.*, 85, 126–137.

76. Singh, T. N., Aspinnall, D., Patel, L. G., & Boggess, S. E. (1973). Stress metabolism, II Changes in proline concentration in excised plant tissues. *Aust. J. Biol. Sci.*, 26, 57.

77. Solinas, V., & Deiana, S. (1996). Effect of water and nutritional conditions on the *Rosmarinus officinalis* L., phenolic fraction and essential oil yields. *Riv. Ital. EPPOS*, 19, 189–198.

78. Sosa, L., Llanes, A., Reinoso, H., Reginato, M., & Luna, V. (2005). Osmotic and specific ion effect on the germination of *Prosopis strombulifera. Ann. Bot.*, 96, 261–267.

79. Sreenivasulu, N., Sopory S. K., & Kavikishor, P. B. (2007). Deciphering the regulatory mechanisms of abiotic stress tolerance in plants by genomic approaches. *Gene*, 388, 1–13.

80. Tabatabaie, S. J., & Nazari, J. (2007). Influence of nutrient concentrations and NaCl salinity on the growth, photosynthesis, and essential oil content of peppermint and lemon verbena. *Turk. J. Agric. Forest*, 31, 245–253.

81. Verpoorte, R., Contin, A., & Memelink, J. (2002). Biotechnology for the production of plant secondary metabolites. *Phytochemistry*, 1, 13–25

82. Wang, W., Vinocur, B., & Altman A. (2003). Plant responses to drought, salinity and extreme temperatures: towards genetic engineering for stress tolerance. *Planta*, 218, 1–14.

83. Westervelt, P. (2003). *Greenhouse Production of Rosemarinus officinalis.* Master Thesis, Virginia Polytechnic Institute, Blacksburg, Virginia.

84. Yang, F., Xiao, X., Zhang, S., Korpelainen, H., & Li, C. (2009). Salt stress responses in *Populus cathayana Rehder. Plant Sci.*, 176, 669–677.

85. Zahedi, S. M., Nabipour, M., Azizi, M., Gheisary, H., Jalali, M., & Amini, Z. (2011). Effect of kinds of salt and its different levels on seed germination and growth of basil plant. *World Applied Sciences Journal*, 15, 1039–1045.

86. Zhang, C. P., Li, Y. C., Yuan, F. G., Hu, S. J., Liu, H. Y., & He, P. (2103). Role of 5-aminolevulinic acid in the salinity stress response of the seeds and seedlings of the medicinal plant *Cassia obtusifolia L. Botanical Studies*, 54, 1–13.

87. Zhu, J. K. (2001). Plant salt tolerance. *Trends Plant Sci.*, 6, 66–71.

88. Zhu, J., Lu, P., Jiang, Y., Wang, M., & Zhang, L. (2014). Effects of *Brassinosteroid* on antioxidant system in *Salvia miltiorrhiza* under drought stress. *Journal of Research in Agriculture and Animal Science*, 2, 1–6.

APPENDIX

List of plants that have been mentioned in this chapter

Scientific Name	Common Name
Abelmoschus moschatus	Muskdana, Ambrette
Achillea fragrantissima (Forssk) Sch. Bip	Lavender cotton
Aloe barbadensis	Aloe vera
Lipia citriodora var. verbena (syn *Aloysia citrodora*)	Lemon verbena and lemon bee brush
Amaranthus tricolor	Amaranth
Ammolei majus	Bishop's Flower, Bishop's weed
Arabidopsis thaliana	Thale cress, mouse-ear cress, arabidopsis
Asparagus racemosus	Satavar, shatavari, shatamull
Azadirachta indica	Neem
Beta vulgaris L.	Sugar beet
Cassia fistula	Golden shower tree
Cassia senna	Senna
Catharanthus roseus	Madagascar periwinkle or rosy periwinkle,
Chrysopogon zizanioides	Vetiver
Cichorium intybus	Common chicory
Coriandrum sativum	Coriander
Cucumis melo	Muskmelon
Cymbopogon flexuosus	Lemongrass
Cymbopogon martinii	Palmarosa, Indian geranium, ginger grass, rosha, rosha grass
Cymbopogon winterianus Jowitt ex Bor	Citronella grass, Java citronella grass, Citronella
Datura inoxia	Datura
Datura stramonium	Iimson weed, Devil's snare
Dioscorea cayennensis	Guinea Yam.
Dioscorea cayennensis subsp. Rotundata	White yam
Eruca sativa	Rucola, rucoli, rugula, colewort, roquette, arugula
Foeniculum vulgare subsp. vulgare;	Fennel
Glycyrrhiza glabra L.	Liquorice, common liquorice, licorice, licorice-root
Hyoscyamus niger	Henbane, nightshade, black henbane

Scientific Name	Common Name
Hyoscyamus muticus L.	Henbane, sakaran, egyptiskbolmört
Lepidium sativum	Chandrasur seeds, Garden Cress, Pepper Grass
*Matricaria chamomilla (*syn. *Matricaria recutita)*	Chamomile, German chamomile
Medicago polymorpha	California burclover, toothed bur clover, toothed medick
Melissa officinalis	Lemon balm, balm, common balm, balm mint
Mentha piperita var. officinalis	Peppermint
Mentha pulegium	Pennyroyal, squaw mint, mosquito plant
Mentha spicata	Spearmint
Moringa peregrine (Forssk.)	Ben tree, wispy-needled yasar tree,
Ocimum basilicum	Basil, Sweet basil, kamaksturi
Origanum vulgare	Oregano
*Majorana hortensis Moench (*syn. *Origanum majorana)*	Marjoram, Sweet marjoram
Petroselinum hortense	Parsley
Plantago ovate	Isabgol
Populus cathayana Rehder	Mantsurianpoppeli
Psoralea corylifolia	Babchi
Raphanus sativus	Radish
Ricinus communis	Castor oil plant
Rosmarinus officinalis	Rosemary
Salvia miltiorrhiza	Red sage, Chinese sage, tan shen, or danshen
Salvia officinalis L.	Sage, garden sage, common sage
Satureja montana	Summer savory
Satureja thymbra	Leaved savory, pink savory
Silybum marianum	Milk thistle, blessed milk thistle, Marian thistle
Solanum nigrum	Black nightshade, wonder berry
Sorghum bicolor	Jawar
Thymus maroccanus	Zeeter
Trachyspermum ammi	Ajwain
Trigonella foenum-graecum	Fenugreek
Triticum aestivum L.	Common wheat
Withania somnifera	Ashwagandha, Indian ginseng, winter cherry

PART III

SOIL SALINITY MANAGEMENT IN CROP PRODUCTION

CHAPTER 8

BACTERIAL-MEDIATED AMELIORATION PROCESSES TO PLANTS UNDER SALT STRESS: A REVIEW

AMRITA KASOTIA, AJIT VARMA, and DEVENDRA KUMAR CHOUDHARY

CONTENTS

8.1 INTRODUCTION

Based on report issued by the United Nations in the year 2013, global human population is likely to reach 11 billion by the end of this century wherein about 8.7 billion individual would be undernourished; abundantly from developing countries like India [81]. Dawson et al. [64] reported that 31% (2.5 billion people by 2050) of the global population is at risk of undernourishment if no adaptation or agricultural innovation is made in the intervening years. It imposes great pressure to ecosystem especially from the expansion of agriculture to feed the billions [73, 97]. To increase global food supply by 70–110% by 2050 the key priorities are to improve technologies and policies that promote more ecological efficient food production while optimizing the allocation of lands to conservation and agriculture [34, 161].

Dietary protein is required to overcome malnutrition and the needs can be met through intakes of plant and animal source foods [301]. Among crops, legumes are richest source of protein and commonly called poor man's meat [234]. Table 8.1 shows the protein content of many important pulse crops. On an average, 6 kg of plant protein is required to yield 1 kg of meat protein which can prove to be uneconomical for the developing countries [220, 262].

Change in climatic conditions and anthropogenic pressure accounts most in reduced crop production. Abiotic factors affect crop yield and the available records show estimated crop yield loss globally including legumes due to drought 17%, salinity 20%, high temperature40%, low temperature 15% and other factors 8% [10, 235]. For example, pulse production increased marginally (from 12.70 million tons in 1960–1961 to 17.79 million tons covering an area of 27.94 million hectare (Mha) with productivity of 636.5 kg ha^{-1} in 2011 [82]. Ku et al. [154] reported remarkable reduction in the potential yield of soybean seed (24–50%) under abiotic prone soil conditions through greenhouse and field studies operated in different regions at different times. For India, being a developing country, it is a challenge to overcome crop loss to feed 1.252 billion people. Hence, a climate-smart and sustainably productive agriculture is a must for assured livelihood security in an agriculturally important country like India where over 600 million people are directly dependent on agriculture [261].

This chapter presents exhaustive review on bacterial-mediated amelioration to plants under salt stress.

TABLE 8.1 Kilo Calories and Protein Content of Major Pulses, Soybean and Cereals Crops

Crop category	Scientific name	Common name	Value per 100 grams	
			Kcal	Protein
Pulses	*Phaseolus vulgaris*	Black bean	333	23.58
		Kidney bean	341	21.60
		Pinto bean	347	21.42
	Vigna angularis	Adzuki bean	329	19.87
	Vigna radiata	Mung bean	347	23.86
	Vigna mungo	Black gram (Urad beans)	341	25.21
	Phaseolus lunatus	Lima bean	335	20.62
	Vigna unguiculata	Cowpea	336	23.52
	Vicia faba	Faba bean	341	26.12
	Cicer arietinum	Chickpea (Garbanzo)	364	19.30
	Lens culinaris	Lentil	353	25.80
	Cajanus cajan	Pigeon pea	343	21.70
	*Cyamopsis tetragonoloba**	Guar bean	347	3.0
Legume oil crop	*Glycine soja*	Soybean	446	36.49
Cereals	*Triticum durum*	Wheat, durum	339	13.68
	Triticum aestivum	Wheat, bread	340	10.69
	Zea mays	Maize	365	9.42
	Oryza sativa	Rice, medium grain	360	6.61
	Pennisetum glaucum	Millet	378	11.02
	Sorghum	Sorghum	339	11.30
	Hordeum vulgare	Barley	352	9.91

Source: USDA National Nutrient databases (http://www.usda.gov/foodcomp/search/).
**Source*: [128, 129].

8.2 PLANT SIGNALING UNDER SALT AND DROUGHT STRESSES

Plants regulate their growth and development to counter numerous external stimuli and ever-changing environment upon induction of abiotic stress namely: drought, salinity, cold, heat, freezing chilling, nutrient, high light intensity, ozone (O_3) and anaerobic stresses (less oxygen) [123, 182, 195, 213, 265, 293, 299]. There is a kind of consensus for signaling mechanisms under most abiotic stresses; only difference may be in cell/tissues/parts affected. Responses to salt and drought stress are analogous as they lead to osmotic imbalance to plants. Osmotic stress arises immediately after the roots come in contact with highly salt concentrated solution in hydroponic system or in soil [258]. Plants adapt and achieve homeostasis within no more than one day after the occurrence of osmotic stress [191]. Salt stress/shock also shows ionic stress in addition to osmotic stress and occurs only when Na^+ concentration inside the cell reaches toxic levels [258], which may take time to respond after 1–3 days of NaCl application [191, 192, 241].

After experiencing abiotic stress, signaling cascade starts in cells that involves transcription factors and stress responsive elements. Cascade starts with signal perception leading to generation of secondary messengers, viz, calcium, Reactive Oxygen Species (ROS) and inositol phosphates. Subsequently calcium interacting proteins such as Ca^{2+}-dependent protein kinases (CDPKs), calmodulin and calcineurin B-like proteins (CBLs), the proteins with structural 'EF-hand' calcium-binding motif gets activated [145]. In eukaryotes, CDPKs together with Mitogen-Activated Protein Kinases (MAPKs) are two signaling cascades widely activated in response to changing environmental abiotic and biotic stress which are then induced by MAPKs. They regulate various cellular activities, such as gene expression, mitosis, differentiation, proliferation and cell survival/apoptosis [89].

ABA-dependant signaling in osmotic imbalance lead by drought and salinity stress involves transcriptional activators: AERB/ABF (ABA responsive element-binding protein/ABA-binding factor) (bZIP (basic leucine zipper)) and MYC/MYB (myelocytomatosis oncogene/ myeloblastosis oncogene) [246]. ABA-independent signaling involve expression of DREB2 (Dehydration Responsive Element-binding factor); and NAC (NAC was derived from the names of the three described TFs containing NAC domain, namely NAM (no apical meristem), ATAF (Arabdopsis transcription activator factor) and CUC (cup-shaped cotyledons) and ZF-HD (zinc-finger homeodomain) regulon [194, 246]. Besides ABA dependent and independent signaling pathways, a third element in describing the salt tolerance works independently to take care of ionic stress. These three pathways that imparts tolerance to salinity stress have been compiled and shown in Figure 8.1

A study employing the model plant *Arabidopsis thaliana* to salt stress describes elevated level of essential Na^+ transporters, e.g., AtHKT1, AtNHX1, AtAVP1 and AtSOS [7, 8, 25, 126, 134, 173, 281, 303]. AtHKT1 functions in selective transport of Na^+ by over-accumulation in aerial parts that implies Na^+ loading into phloem and Na^+ exclusion from xylem [23, 63, 125, 268]. AtNHX1 maintains K^+/Na^+ ratio in the cytosol by sequestration into the vacuole during salinity stress [7, 42, 120]. AtAVP1 increases the difference in the electrochemical potential of H^+ between the vacuole and cytoplasm by encoding a vacuolar H^+-translocating pyrophosphatase, thereby energizing the movement of Na^+ into vacuole through Na^+/H^+ antiporters such as AtNHX1 [95, 193, 273].

Salt Overly Sensitive (SOS) pathway is regarded as the key mechanism for Na^+ exclusion and ion homeostasis control at the cellular level [323]. The SOS1, SOS2 and SOS3 loci were first identified through forward genetic screens for salt-hypersensitive growth [324]. The SOS1

SALT STRESS				
Osmotic effect				Specific ions effects
ABA dependent pathway		ABA independent pathway		Signaling pathway that comprises SOS3/SCaBP ⬇
MYC/MYB	AREB/ABF (bZIP)	NAC/ZF-HD	DREB2	SOS2/PKS ⬇
MYC/MYBR	ABRE	NACR/HDZFR	DRE	SOS1plasma membrane Na^+/H^- antiporter ⬇
Activation of genes				Ionic Homeostasis
Salt tolerance				

FIGURE 8.1   Three operational pathways that determine salt tolerance of crops exposed to salt stress.

(plasma membrane Na^+/H^+ exchanger) forms a Ca^{2+}-dependent signaling cascade along with SOS3 and SOS2 [259, 303]. SOS1 is regulated through phosphorylation of its autoinhibitory C-terminal domain by SOS2 protein kinase, a member of the sucrose non-fermenting 1-related protein family [226]. SOS2 is regulated through interaction of the Ca^{2+}-sensing protein SOS3 [300]. SOS1 also shows expression in the cells surrounding the vasculature and its transport from the root to shoot via xylem suggests its role in the long distance Na^+ transport in plants [204].

8.3 PLANT GROWTH PROMOTING BACTERIA (PGPB)

PGPB are the group of bacteria colonizing spermosphere of seed and are directly or indirectly involved in plant growth promotion via production and secretion of various regulatory chemicals in the vicinity of rhizosphere. Plant growth promoting rhizobacteria (PGPR) are most widely studied group of PGPB and based on their degree of association with plant root cells are now classified into extracellular-PGPR (ePGPR) and intracellular-PGPR (iPGPR). The ePGPRs may exist in the rhizosphere or in the spaces between the cells of root cortex; iPGPRs locates generally inside the specialized structure of root cells. PGPB include those that are free living, colonize plant's interior tissue [239], those that form specific relationships with plants (eg. *Rhizobia* spp. and *Frankia* spp.) and cynobacteria (formerly called blue-green algae) [102]. Plant growth is facilitated directly by either assisting resources acquisition (nitrogen, phosphorous and essential minerals) or modulating plant hormone levels or by decreasing the inhibitory effects of biotic and abiotic stresses [35, 54, 269].

  PGPB-mediated reports have been published on amelioration of salinity stress, e.g., *Achromobacter* in tomato [175], *Azospirillum* in Lettuce [13], *Klebsiella* and *Pseudomonas* in cotton [309, 314], *Azotobacter* in *Zea mays* [240], *Pseudomonas* in canola [45], *Pseudomonas* in radish [184], *Azospirillum* and *Pantoea* in sweet pepper [68], *Xanthobacter*, *Enterobacter* and *Bacillus* in Eggplant [1], *Bacillus* in rice [199], *Bacillus*, *Staphylococcus* and *Kocuria* in strawberry [138], *Pseudomonas* in barley and oats [41], *Burkholderia*, *Promicromonospora* and *Acinetobacter* in *Cucumis sativus*[136]. Similarly PGPB-mediated alleviation of drought stress has been reported, viz., *Achromobacter* in tomato and pepper [176], *Pseudomonas* in *Catharanthus roseus* [133], *Pseudomonas* in *Lactuca sativa* [151], *Azospirillum* in rice [243], *Bacillus* in tomato [49], *Phyllobacterium* in *Arabidopsis* [32], *Pseudomonas* in mung bean [252]. Among them several PGPB have been commercialized, namely: *Agrobacterium radiobacter*, *A. lipoferum*, *A. brasilense*, *B. fimus*, *B. pumilus*, *B. subtilis* var. *amyloliquefaciens*, *Burkholderia cepacia*, *P. fluorescens*, *P. macerans*, *P. syringae*, *Serratia*, *Streptomyces lydicus*, various *Rhizobia* spp. etc. [102].

8.3.1 DIRECT MECHANISMS

PGPB-mediated direct methods make plants to grow well with good yield and biomass. Some of these methods improve availability of plant nutrients and include phosphate solubilization, iron sequestration and nitrogen fixation. The compounds made up of inorganic-phosphate (Pi), iron and nitrogen is least available to plants. These are generally bound to cations and anions to form a stable compound which further makes them less available in soil. PGPB solubilizes these nutrients and makes them available to plants.

8.3.1.1 Phosphate Solubilization

Phosphorus (P) is one of the major essential macronutrients for plant growth and development. P exists in two forms in soil, as organic and inorganic phosphates. It is present at the concentration of 400–1,200 mg kg^{-1} of soil. Most P in soil is part of insoluble compounds, which makes P unavailable for plant nutrition. The concentration of soluble P in soils is usually very low, only 5% or less of the total amount of P in soils is available for plant nutrition. The fixation and precipitation of P in soils is highly dependent on soil type and pH. In acid soils, free oxides and hydroxides of aluminum and iron fix P while in alkaline soils calcium fixes it. Organic acid metabolite production and decrease in medium pH appear to be the major mechanisms for P solubilization [53, 76, 79, 212, 302]

Phosphate solubilizing bacteria (PSB) are capable of hydrolyzing organic and inorganic phosphorus from insoluble compounds. PSBs bring out a number of transformations of phosphorus including: altering the solubility of inorganic compounds of phosphorus, mineralization of organic phosphate compounds into inorganic phosphates, conversion of inorganic available anion into cell components, i.e., an immobilization process and oxidation or reduction of inorganic phosphorus compounds are the most important reactions/processes in phosphorus cycle [238, 239]. PGPB excretes low molecular weight organic acids such as gluconic acid, citric acid, succinic acid, propionic acid, lactic acids that dissolves phosphate minerals and/or chelate phosphorous ion, i.e., PO_4^{3-} directly whereby solubilize inorganic phosphate [44, 69, 106, 313]. In addition to secretion of organic acids, PGPB also decrease the rhizospheric pH with production of proton/bicarbonate release (anion/cation balance) or gaseous (O_2/CO_2) exchanges [143]. Mineralization of most organic phosphorus compounds in most soils is carried out by means of phosphatase enzymes [17, 293]. These catalyze dephosphorylating reactions involving the hydrolysis of phosphoester or phosphoanhydride bonds [111, 263].

8.3.1.2 Iron Sequestration

Iron is required by most living organisms. It is a necessary cofactor for nearly 140 enzymes catalyzing specific biochemical reactions and processes. A minimum of 10^{-8} M iron is needed by plants. In soil iron exist in form of ferric state (Fe^{3+}) and produces insoluble hydroxides, oxyhydroxides that are largely unavailable to plants and microorganisms [167].

In neutral and or alkaline soils, the availability of ferric is further reduced due to decreased solubility at higher pH [121, 171, 275]. PGPB secrete siderophore to provide sufficient amount of iron to plants. Bacterial siderophores are mostly catecholates, with some polycarbonates, e.g., azotobactin [256]. Siderophore ligand shows strong affinity for the higher oxidation state of iron, i.e., Fe-III than Fe-II. The weak complexing of Fe (II) thus affords an efficient means of release, via reduction, inside the cell [200]. The plant roots then uptake iron from siderophore-iron by either chelate degradation, the direct uptake for the complex or ligand exchange reactions [231]. Siderophore also enhance bacterialIndole-3-Acetic Acid

(IAA) synthesis through chelating of heavy metals that caused detrimental effects in plant growth [71].

8.3.1.3 Nitrogen Fixation

Nitrogen, being an essential component of proteins, nucleic acids and other nitrogenous compounds, is considered as one of the vital component of the living system. PGPB have the capacity to fix atmospheric dinitrogen by forming nodule in roots and provide nitrogen to plant in the form of ammonia. The bacteria responsible for nitrogen fixation are called diazotrophs. The members of the alpha and beta subgroup of the phylum proteobacteria are the main rhizobial bacteria associated with legumes [31, 298]. PGPB can fix nitrogen symbiotically or non-symbiotically. Symbiotic N_2 fixation accounts nearly 65% of the total biologically fixed nitrogen [232]. Symbiotic N_2 fixation occurs in *Azotobacter* spp. *Bacillus* spp. *Beijerinckia* spp. etc. [26] whereas non-symbiotic nitrogen fixation occurs in free living diazotrophs, *Azospirillum* [15], *Pseudomonas* [181] and *Burkholderia* [80].

Diazotrophs contains nitrogenase enzyme complex that is mainly involved in nitrogen fixation. This enzyme system consists of three subunits and it is regulated by a complex system of multiple genes namely Nitrogenase 1 (classic), encoded by *nif* gene dependent on iron and molybdenum, Nitrogenase 2, encoded by *vnf* gene and dependent on vanadium and Nitrogenase 3, encoded by *anf* gene and dependent on iron [36, 90, 308]. Some novel nitrogenases have been discovered in *Streptomyces thermoautotrophicus* [322] and in *Rhodopseudomonas palustris* [158, 163]. PGPB accelerate nodulation and increase nitrogen fixation activity in soybean [62], in *Phaseolus* [86] and many other legumes [72].

8.3.2 PHYTOHORMONE PRODUCTION

PGPB are known to synthesize plant hormone which directly influence plant growth via production of IAA, gibberellin, cytokinin, nitric oxide and Abscisic acid (ABA). Physiological and agronomical aspects of phytohormone production in model PGPB *Azospirillum* has been recently reviewed by [40].

8.3.2.1 Indole-3-Acetic Acid (IAA)

IAA is one of the best studied naturally occurring auxin (generic name) that represent a group of chemically characterized compounds known by their capability to induce plant growth at their subapical zone of the stem [40]. These are allied to different plant processes like stimulation of cell division; vascular tissue differentiation; stem and root elongation; lateral and adventitious root initiation; apical dominance; gravitropism and phototropism [271]. Hence, maintaining appropriate cellular levels of active auxin is important for regulating all aspects of plant growth and development [152]. A high degree of similarity is found between biosynthesis of IAA in plants and microbes [264]. Mainly there are five different pathways found in PGPB for biosynthesis of IAA in microbes, i.e., indole-3-acetamide (IAM), indole-3-pyruvate pathway (IPyA), trypamine pathway (TAM), tryptophan side-chain oxidase pathway (TSO), indole-3-acetonitrile pathway (IAN) and tryptophan independent pathway [263].

Except tryptophan independent pathway all other pathways use tryptophan as precursor for the biosynthesis [264]. Although several pathways are there for IAA biosynthesis but IAM and IPyA pathways are adapted by most of the PGPB [156]. The precursor tryptophan is synthesized from chorismate in plant. The formed tryptophan oozes out from loosely bound root cells and available for uptake by microbes which then synthesize IAA (Figure 8.2). Apart from providing IAA to plants, it also acts as signaling molecule in bacteria.

IAA moreover crosstalk synergistically with ethylene for controlling root elongation and root hair formation; and antagonize lateral root formation and hypocotyls elongation [189]. IAA maintains high cytokinin levels in the center of meristem by acting synergistically with cytokinin in shoot apical meristem, eventually, allowing cell differentiation [75]. IAA producing *Azospirillum brasilense* Az39 and *Bradyrhizobium japonicum* E109 enhanced plant growth in corn and soybean [39].

8.3.2.2 Gibberellin

Gibberellins (GAs) are ubiquitous plant hormones that are involved in several metabolic functions during plant growth. PGPB-secreted GAs are

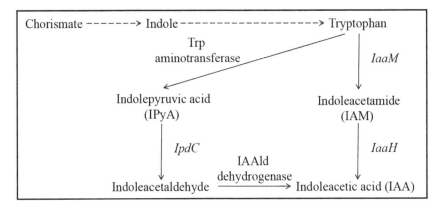

FIGURE 8.2   IAM and IPyA metabolic routes of IAA biosynthesis pathways. IaaM, tryptophan-2-monooxygenase; IaaH, indole-3-acetamide hydrolase; IpdC, indole-3-pyruvate decarboxylase.

beneficial to plant growth and development [135]. GAs synthesized by PGPB are regulated in three ways, either by de-conjugation of glucozyl gibberellins, direct synthesis of gibberellins, or change inactive gibberellins to active gibberellins [37]. The activation from inactive to active 3-deoxy GAs in roots is achieved by β-hydroxylation to give GA_3, GA_1 and GA_4 [37, 38]. Inoculation of GA secreting *Azospirillum* increased germination rate and rapid growth of seedling is attributed atleast partially to GA production [40]. The PGPB, *Promicromonospora* sp. SE188 inoculated tomato plants showed a significant higher shoot length and biomass [135].

8.3.2.3   Cytokinin

Cytokinins are a group of natural compounds that regulate cell division and differentiation processes in meristematic tissues of higher plants [40], besides being involved in many physiological and cellular processes, including chlorophyll accumulation. It leads to formation of organs, stem initiation, leaf expansion, root elongation, root development and root hair formation and delayed senescence, indicative of beneficial effect on plant growth and yield [249, 296]. Cytokinin is necessary for inducing root nodule organogenesis for nitrogen fixation [147]. The inoculation of *Bacillus megaterium* (cytokinin producing PGPB) in *A. thaliana* increased the level of cytokinin in roots which predicted the role of intact cytokinin-signaling pathway in the plant [206].

8.3.2.4   Nitric Oxide

Nitric oxide (NO) is a recently discovered phytohormone involved in induction of lateral root and adventitious root formation [40, 186] as well as increasing the root hair number and length of roots [15]. Here, NO plays main role in IAA pathways as intermediate in IAA induced root development [58]. Moreover, it regulates biofilm formation [70], which can be directly related to bacterial density. It has been shown as key signaling molecule for wide range of functions in plants [122, 186]. Certain bacteria also produce NO as a metabolic product and some produce specifically for cell signaling and for nitration of important secondary metabolite in one group [280]. NO produced by these bacteria regulate their mobility and gene expression and can elicit responses in neighboring organisms, including plants [52]. In plants, NO is a versatile signal molecule that control and synchronizes a complex network of auxin, cytokinin, ethylene and abscisic acid-stimulated process [52]. NO is produced under aerobic and anaerobic denitrification, and heterotrophic nitrification [185]. Inoculation with *Azospirillum* induced lateral root and adventitious root formation in tomato plants traced to phytohormone NO [59, 186].

8.3.2.5   Abscisic Acid

ABA phytohormone is mainly associated with biotic and abiotic stresses. It induces stomata closure, inhibits seed germination and flowering [231]. ABA is produced by several bacteria such as *B. japonicum* [29], *Azospirillum brasilense* [50, 218] and *Azospirillum lipoferum* [51]. It is proposed that bacteria synthesize ABA as product of carotenoid metabolism and its biosynthesis increases with increasing osmotic stress [50, 88, 132]. It has been shown that inoculation of ABA producing strain *Azospirillum lipoferum* USA 59b stimulated growth of plants in dry soil

[51]. Similarly, inoculation of *B. licheniformis* and *P. fluorescens* in grapevine acted as stress alleviators by inducing ABA synthesis thereby diminishing water losses [250]. Recently it was observed that ABA metabolizing rhizobacteria decreased ABA concentration *in planta* and altered plant growth [19].

8.3.3 ENZYME SECRETION

Apart from nutrient acquisition and phytohormone productions, PGPB secrete enzyme, viz; 1-aminocyclopropane-1-carboxylate, ACC deaminase and protease that help stimulate plant growth.

8.3.3.1 ACC Deaminase

ACC deaminase enzyme is the most exploited enzyme in PGPB produced enzymes. It is known to cleave ACC, the ultimate precursor of ethylene in plants, which is a stress induced hormone tending to increase with stress and is responsible for various physiological changes *in planta*. Ethylene is also an inhibitor of rhizobial nodulation of legumes [112, 166] and produced by following yang cycle [308], as shown in Figure 8.3. ACC synthase (ACS) catalyzing SAM to ACC and ACC oxidase (ACO) catalyzing ACC to ethylene, are the two key enzymes for ethylene biosynthetic pathway, are tightly regulated both transcriptionally and post-transcriptionally to modulate ethylene biosynthesis [130]. In yang cycle, 5'-methylthioadenosine is released as byproduct when SAM is converted to ACC by ACS, is subsequently recycled to methionine [183]. In some conditions ACC can be converted to a conjugated form, N-malonyl ACC (MACC) by ACC n-malonyltransferase or 1-(γ-L-glutamylamino) cyclopropane-1-carboxylic acid (GACC) [172] or jasmonyl-ACC (JA-ACC) [266]. These conjugations control ethylene biosynthesis, in a manner analogous to the conjugation of auxin and cytokinin [87, 169, 219, 285]. ACC can be transported to particular stressed or senescent organ resulting in synthesis of ethylene in the affected tissue [311].

ACC deaminase cleaves ACC to α-ketobutyrate and ammonia; and thereby lowers the ethylene level in developing or stressed plants [99,

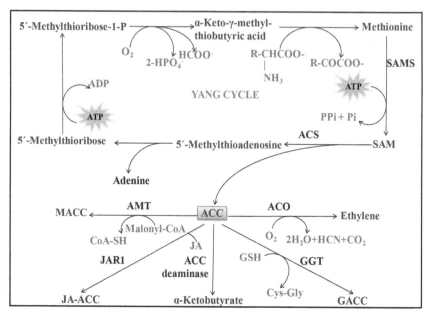

FIGURE 8.3 Ethylene biosynthetic pathway and the Yang cycle (ACC = 1-aminocyclopropane-1-carboxylic acid; SAM = S-adenosyl-methionine; SAMS = SAM-synthetase; ACS = ACC-synthase; ACO = ACC-oxidase; GGT = γ-glutamyl-transpeptidase; GACC = γ-glutamyl-ACC; JA = jasmonic acid; JAR1 = jasmonic acid resistance 1; AMT = ACC-N-malonyl transferase; MACC = 1-malonyl-ACC. Source: [270, 285]).

100, 104, 205] According to a previously proposed model [105], it was predicted that ACC deaminase-producing bacteria first bind to the either seed or root surface as below ground surrounding surface is rich in plant exudates (photosynthetically fixed carbon, organic acids and amino acids). Later on, it reduces the deleterious levels of ethylene by using ACC as nitrogen source [124] and carbon source [20]. Another model suggests that plant exude some ACC from roots or seeds, which are taken up by ACC deaminase secreting bacteria [19, 216]. However, it is not clear from these studies that the effects are local (confined to the roots) or systemic (occurring throughout the plant) via mediation of long-distance ACC signaling [18]. ACC deaminase induction and regulation in bacterium is a relatively complex process. ACC deaminase structural gene (*acd*S) had been studied in *Pseudomonas* UW4 by that are all transcriptionally regulated [108, 164].

the receptor on bacterium is specific to iron siderophore complex located in the outer membrane of the bacterium [5]. Siderophore producing PGPB have shown promise in the biocontrol of several plant diseases, like root rot of wheat, damping off of cotton, potato seed piece decay, stem rot of peanut, vascular wilts, charcoal rot in ground nut, *Fusarium* wilt of pepper, anthracnose (*Colletotrichum achutatum*) in strawberry and dry rot of potato etc. [9, 260, 278, 312].

8.3.4.3   ACC Deaminase

Under biotic stress, plant responds to stress by raising the levels of ethylene known as stress ethylene which worsen the plants health. For this an effective measure is to lower ethylene level by ACC deaminase producing bacteria [101, 103]. ACC deaminase producing bacteria limits the phytopathogen infection in plants by reducing the level of ethylene in plants and provides resistance to both fungal and bacterial pathogen [277, 291] as well as nematodes [197]. Diverse reports have shown disease reduction for example in soybean from *Sclerotium rolfsii* and *Rhizoctonia solani* [131], tomatoes from *Ralstonia solanacearum* [310]. In addition, limiting the crown gall formation in tomato plants when infected with *Agrobacterium tumefaciens* [117, 119, 277] or *A. vitis* [277] has been documented. But when the *acd*S gene is introduced in *Agrobacterium tumefaciens* GV3101:pM90 it promoted transformation efficiency in canola cultivars: *Brassica napus* cv. Westar, *B. napus*cv Hyola 401 and *B. napus* cv. 4414RR by reducing ethylene level in host plant [118].

8.3.4.4   Induced Systemic Resistance

To combat the invasion of phytopathogen, PGPB helps the plant to acquire resistance by a mechanism called induced systemic resistance (ISR) [286–288] by means of modulating host immunity [315]. ISR is different from systemic acquired resistance (SAR) which is induced by virulent pathogens, avirulent form of pathogen, incompatible race of pathogens, or certain chemicals [177]. Both ISR and SAR provide resistance in systemic manner, i.e., protection is seen throughout the plant as they just reduce the severity of disease, not prevent the disease [66, 120]. The major difference between ISR and SAR is necrosis [hypersensitive reaction (HR)], which is absent in ISR [116]. Unlike SAR, ISR is not associated with necrotic lesion formation. ISR also differs in that it depends on the perception of ethylene and jasmonic acid; and is not associated with expression of the pathogenesis-related genes. Hence, this enhanced capacity to express basal defense makes the plants 'primed' [56, 92]. ISR is induced in plants mainly by induction of jasmonic acid and ethylene signaling [315] but in some PGPB via SA signaling [14]. While some of the bacteria produce SA itself (174, 225, 247], others reported to produce jasmonic acid and ABA [88]. Elicitation of ISR by plant associated bacteria is demonstrated by gram positive and gram negative bacteria that provided reduction in the incidence or severity of various diseases in tomoto, bell, pepper, muskmelon, water melon, sugarbeet, tobacco, *Arabidopsis* sp., cucumber, loblolly pine, cayenne pepper and green kuang futsoi [48].

8.4 ALLEVIATION OF ABIOTIC STRESS BY INDUCED SYSTEMIC TOLERANCE

Plants continuously face abiotic stress due to ever-changing climate or depletion of nutrients in soil. Apart from providing ISR, PGPB have been reported to provide induced systemic tolerance (IST) to plants under abiotic stress [307]. The main components of IST are reduction in ethylene, induction of HKT gene, scavenging of ROS, diminution of ABA and exopolysaccharide production along additional beneficial PGPB properties described in earlier section with special reference to salt and drought stress.

8.4.1 REDUCTION IN ETHYLENE

Reduction in the level of ethylene allows the plant to grow in stress. If ethylene levels can be prevented to excessively rise as a consequence of stress, significant portion of the damage to the plant might be avoided [96]. It has been projected that ethylene is produced in two peaks in response to environmental stresses. The first peak occurs few hours after the stress of small strength. The second peak is of greater strength which occurs 1 to 3 days after the stress. The first peak triggers synthesis of plant defensive/protective proteins. The second peak initiates processes such as senescence, chlorosis and abscission, exacerbating the effects of the stress [102].

Plant exudates contain tryptophan, amino acid along with several other organic molecules. In response to tryptophan and other small molecules in the plant root exudates, the associated bacteria synthesize and secrete IAA. Some of this is taken up by plant and some activates ACC synthase that catalyze synthesis of ACC. As a result ethylene production stimulates in plants. Increased IAA concentration stimulates plant cell proliferation, cell elongation and loosen plant cell walls, thereby increasing exudation of ACC (non ribosomal) [215] that may be taken up by ACC deaminase enzyme of bacteria [214]. As a result ACC outside the plant cell is reduced. To maintain this equilibrium more ACC is exuded from seed or roots, as a consequence ACC become less available for oxidation to ethylene in plant cell (Figure 8.4). There is a cross talk between IAA and ACC deaminase, i.e., by lowering plant ethylene levels, ACC deaminase facilitates the stimulation of plant growth by IAA [103] and changes root architecture by forming long roots and large root hair number. ACC deaminase producing bacteria also reduces second ethylene peak.

Moreover, cross talk between cytokinin and ethylene increases ethylene biosynthesis through stabilization of ACC synthase involved in ethylene biosynthesis [77, 203].

8.4.2 INDUCTION OF HIGH AFFINITY K^+ TRANSPORTER (HKT)

Under high salinity Na^+ ion concentration increases in plants which can be regulated by either Na^+ exclusion, Na^+ tolerance in the tissues or osmotic tolerance [188]. This is regulated by members of integral membrane protein, HKT that facilitate cation transport across the plasma membranes of plant cells [295]. HKT1 was first identified as HKT and sequenced from wheat (*Triticum aestivum*) roots [255]. In heterologous system (in yeast) and in *Xenopus laevis* oocytes, it was shown to mediate high-affinity Na^+-K^+ co-transport and also preferred Na^+-selective low-affinity Na^+ transport in the presence of a detrimental millimolar [Na^+] [94, 242]. In *Arabidopsis*, K^+ nutrition is negatively regulated by HKT1 in spite of K^+ uptake [244],

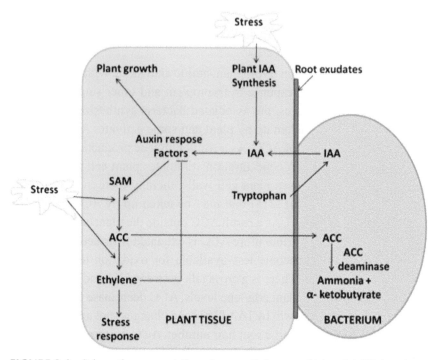

FIGURE 8.4   Schematic representation of cross talk between IAA and ACC deaminase adapted from Glick [103] with slight modifications.

whereas AtHKT1 expressed in *Xenopus oomycetes* selectively transports Na$^+$ but not K$^+$ [281]. Further studies of *Arabidopsis* plants carrying loss of function mutations in HKT1 revealed that Na$^+$ accumulation increased in shoots and rendered the plant hypersensitivity to Na$^+$ [23, 187]. HKT isoforms have been involved in long-distance Na$^+$ movement in several other plant species, especially cereals [168]. In rice, OsHKT 1:5 retrieved Na$^+$ from the xylem sap through xylem parenchyma cells [236]. This resulted in low sodium load in shoot tissue. Na$^+$ selective transport via AtHKT1;1 has an essential role in Na$^+$ exclusion from leaves and K$^+$ homeostasis in leaves during salinity stress thereby facilitating shoot to root recirculation of Na$^+$ and also decrease uptake of Na$^+$ from root [23, 63, 125, 127, 173, 268, 320].

To regulate the Na$^+$: K$^+$ ratio in plants, volatile compounds are secreted by bacteria which down-regulates the expression of HKT1 in root and up-regulates in shoot [307]. Blend of volatile organic chemicals (VOCs) are secreted by PGPB which facilitates interaction with plants [52]. Bacterial volatiles are able to regulate expression of HKT1 gene in maintaining low Na^+:K^+ ratio in plant [317]. Bacterial volatiles from *B. subtilis* GB03 helps plants to regulate whole-plant auxin redistribution, leaf cell expansion, photosynthetic efficiency, chlorophyll content, root branching, underlie plant growth promotion [316, 320]. The two well known volatiles emitted by *B. subtilis* (GB03) and *B. amyloliquefaciens* (IN937a) are 2, 3-butanediol and acetoin which positively affected growth of *Arabidopsis thaliana* [244, 245] *B. subtilis* GB03 treated *Arabidopsis* plants also showed increased iron uptake [319]. Apart from this GB03 VOCs modulated altered sodium homeostasis and provided salinity tolerance in *Arabidopsis* [317]. VOCs activates HKT1 which controls Na^+ import in roots and adjust Na^+ and K^+ levels differentially, depending on the plant tissues [307]. Furthermore, increased biosynthesis of choline and glycine betaine, compatible solutes in GB03 treated plants has been reported [318] and induced SA signaling in *Pseudomonas chlororaphis* O6 offered drought tolerance in model plant *Arabidopsis* [47]. Likewise maintaining higher stomatal conductance and photosynthetic activities could alleviate osmotic stress [68] which in turn improves leaf K^+:Na^+ and lowers accumulation of toxic (Na^+ and Cl^-) [217]. Hence, these studies provide insight into the biological and ecological plant self-immunity and/or adaptation to biotic and abiotic stresses in modern agriculture [83].

8.4.3 SCAVENGING OF REACTIVE OXYGEN SPECIES

Under optimal growth conditions, ROS are mainly produced at low level in organelles such as chloroplast, mitochondria and peroxisomes [6]. The enhanced production of ROS during stress can pose a threat to cells but it is thought that ROS also act as signal for the activation of stress-response and defense pathways [221]. ROS damage on biological membranes can be modulated by regulation of membrane structures by adaptive mechanism such as alteration of composition and organization of lipids inside the bilayer in a way that prevents lipid peroxidation, modification of degree of polyunsaturated fatty acids (PUFA) unsaturation, mobility of lipids within the bilayer and preventive antioxidant systems [28]. ROS also attack to other macromolecules such as proteins and DNA, with the formation of nucleotide peroxides especially at the level of thymine [60]. Oxidative attack on proteins results in site-specific amino-acid modifications, fragmentation of the peptide chain, aggregation of cross-linked reaction products and altered electrical charge. After oxidative modification, proteins become sensitive to proteolysis and/or may be inactivated, or may show reduced activity. ROS-induced DNA damage includes single- and double-strand breaks, abasic sites and base damages. Furthermore, mitochondrial DNA is more sensitive to oxidative damage than nuclear DNA, in particular because of the absence of chromatin organization and lower mitochondrial DNA repair activities [305].

As major site of ROS production both in animal and plant cells is mitochondrial electron transport chain (ETC), the importance of ROS dependent damage on mitochondrial proteins such as ETC proteins and mitochondrial DNA becomes clearer. Because ROS are toxic but also participate in signaling events, plant cells require at least two different mechanisms to regulate their intracellular ROS concentrations by scavenging of ROS, one that will enable the fine modulation of low levels of ROS for signaling purposes, and one that will enable the detoxification of excess ROS, especially during stress [178].

Plants have evolved both non-enzymatic and enzymatic mechanisms to cope with deleterious effects of ROS in the cells. To control the level of ROS and to protect cells under stress conditions, plant tissues contain several enzymes scavenging ROS (superoxide dismutase [SOD; E.C. 1.15.1.1], ascorbate peroxidase [APX; E.C. 1.1.1.11], catalase (CAT; E.C. 1.11.1.6], glutathione reductase [GR; E.C. 1.6.4.2], monodehydroascorbate reductase [MDHAR; E.C. 1.6.5.4], glutathione peroxidase, alternative oxidase and dehydroascorbate reductase [DHAR; E.C. 1.8.5.1]) and detoxifying lipid peroxidation products (glutathione S-transferase and phospholipid-hydroperoxide glutathione peroxidase) [28].

Peroxidase (POD, guaiacol peroxidase; E.C. 1.11.1.7) located in apoplastic space and vacuole also plays an important role in catalyzing the reduction of H_2O_2 and O_2 [321]. Apart from enzymatic defenses, plants also possess a non-enzymatic defense system, including phenolic compounds, ascorbate, glutathione, carotenoid and α-tocopherol [98]. Recent studies showed that polyphenol oxidase (PPO; catechol oxidase; E.C. 1.10.3.2) and peroxidase (POD) catalyzing the oxidation of phenolics into quinones is involved in the plant defense mechanisms against environmental stresses [57]. Plants with high levels of antioxidants, either constitutive or induced, have been reported to have greater resistance to this oxidative damage [3, 150].

8.4.4 DIMINUTION OF ABSCISIC ACID

In salinity and drought stress, an increased ABA level induces the expression of many dehydration responses and tolerance in both vegetative tissues and seeds [306]. ABA is mainly involved in cellular dehydration processes in the seed maturation and vegetative growth stages [93]. It is also a key mediator in drought response as it induces stomatal closing to minimize transpirational water loss [61, 146, 230]. Moreover, ABA interacts with other hormones such as auxin, cytokinin, JA and ethylene in stress challenged plants [165]. When plants are exposed to stress, the accumulation of ABA results in stomatal closure to minimize water losses, accelerates leaf senescence, retards plant growth and induces the biosynthesis of protective substances (e.g., late embryogenesis-abundant proteins). Whereas, cytokinin triggers responses to delay leaf senescence and stomatal closure [109, 202, 224], counteraction of cytokinin is linked to the interruption of ABA signal transduction which leads to reduced ABA level in plants [294].

8.4.5 EXOPOLYSACCHARIDE (EPS)

A large number of microorganisms produce variety of exopolysaccharides [22, 78, 304], consisting of polysaccharides, as well as proteins, nucleic acid, lipids and humic substances [289]. They have high moisture holding capacity and serve to maintain minimum moisture in their surrounding environment which allows the microorganisms to survive at low water content [254]. It protects the bacteria from stress [290]. EPS production by bacteria helps to overcome osmotic stress caused by salt and matric stress caused by drought; and can improve the crop productivity and physiochemical properties of soil by binding the soil particles to roots and making the nutrients available to plants [16, 228]. It also serves as potential energy reserve as it can be catabolized under nutrient deficient conditions [254]. Qurashi and Sabri [227] found that EPS by PGPB increases under salt stress and allows the plant to grow in stress by increasing the plant growth characteristics.

8.5   COMPATIBLE SOLUTES IN BACTERIA

Bacteria are able to adapt to change in osmolarity. This adaptation results by accumulation of low molecular weight hydrophilic molecules, which do not interfere with cell metabolism and hence called compatible solutes [33]. These compatible solutes include sugars, amino acids and their derivatives [157]. Sugars include sucrose, maltose, cellobiose, gentibiose, terhalose and palatinose; amino acids include proline [24], aspartic acid, glutamic acid, serine [209], threonine, glycine, alanine [208]; derivatives include glycine betaine and carnitine [24]. These compounds exclude from the polypeptide backbone and stabilize native protein structure [30].

Change in osmolarity occurring during drought and salinity stress causes change in turgor pressure. This can be maintained by accumulation of proline, glycine betaine, mannitol and sorbitol in transgenic plants that over produce such solutes to show tolerance to drought and salt stress [46]. Salt tolerance in *Zea mays* co-inoculated with *Rhizobium* and *Pseudomonas* was also experimented with increased production of proline along with decreased electrolyte leakage, maintenance of relative water content of leaves and selective uptake of K^+ ions [12]. Over expression of terhalose-6-phosphate synthase gene revealed upregulation of genes involved in stress tolerance and carbon and nitrogen metabolism. Thus, trehalose metabolism in *rhizobia* can be considered as a key for signaling plant growth, yield and adaptation to abiotic stress and its manipulation can have a major agronomical impact on leguminous plants [267].

Chen et al. [43] correlated proline accumulation with drought and salt tolerance in plants. Introduction of *proBA* gene derived from *Bacillus subtilis* into *Arabidopsis thaliana* resulted in production of higher levels of free proline resulting in increased tolerance to osmotic stress in the transgenic plants. Grover et al. [110] and Sandhya et al. [251] also described role of microorganisms in adaptation of agriculture crops to abiotic stress including mechanisms triggering osmotic responses and induction of novel genes in plants.

8.6 MUTAGENESIS IN BACTERIA

Strain improvement through induced mutagenesis became a successful method for commercially significant microbial metabolites [148]. Strain enhancement can be done by classical mutagenesis, which involves exposing the microbes to physical mutagens such as X-rays, UV rays, etc., and chemical mutagens such as N-methyl-N'-nitro-N-nitrosoguanidine (MNNG), Ethyl methanesulfonate (EMS), etc. [207]. MNNG has been proved best mutagen after mid-twentieth century which increased production of many industrially important microorganisms till now.

Development of over-producing PGPB mutants could be a more beneficial way to increase their chances of survival in stress. There are several reports of mutational methods to improve the PGPB for their properties. Achal et al. [2] developed phosphate solubilizing UV induced mutants of *Aspergillus tubingensis*. Kumar et al. [155] used 8-hydroxyquinoline to produce siderophore overproducing properties. Similarly, mutants of *Pseudomonas corrugata* (NRRL B-30409) based on their phosphate solubilization ability, production of organic acids, and subsequent effect on plant growth at lower temperature under *in vitro* and *in situ* conditions were developed using MNNG as mutagen [140, 279]. Enhancement of cellulolytic nitrogen fixing activity of *Alcaligenes* sp. was created by MNNG mutagenesis [160].

8.7 PLANT GROWTH PROMOTING BACTERIA AS BIOFERTILIZER

With the current pace of increase in population, there is an immediate need to increase agriculture productivity. To meet this demand, developing country like India sought effective measures to fulfill the needs and hence, use of chemical fertilizers gained momentum in agriculture. As a result consumption, production and import of chemical fertilizers tremendously increased from 1981 to 2012 (Figure 8.5: One lakh = 10^5).

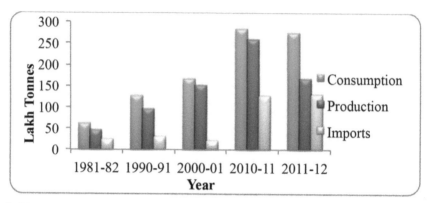

FIGURE 8.5   Trends in consumption, production and import of fertilizers in India [180].

Chemical fertilizers have low use efficiency [113]. This is so because some of the nutrients are less available to plants for example P, which gets precipitated when added to soil [113] and nitrogenis lost through nitrate leaching which in return contaminate ground water [27]. Savci [253] described chemical fertilizers as an agricultural pollutant, excessive fertilization causing soil acidity [284], soil salinity, heavy metal accumulation (cadmium and chromium), water eutrophication and accumulation of nitrates resulting in changes in soil's ecological environment [162, 253].

The use of PGPB as biofertilizers is eco-friendly approach for enhancing plant growth and productivity in stressed environment (lack of nutrients, abiotic and biotic stress). Many scientists have made good reviews of biofertilizing properties of PGPB (discussed briefly in earlier sections of this chapter). This has led to increase in the development and use of microbial-based fertilizers worldwide. Enhanced knowledge of plant microbe interaction occurring in rhizosphere will further boost their application to improve productivity in stressed environments [170].

8.8   SUMMARY

The major limiting factors impacting plant growth and agricultural productivity worldwide are environmental stresses. Ecosystem of arid and semi-arid regions is mainly affected by high temperature, lack of moisture, soil salinity/alkalinity, high pH and metal toxicity. Out of this, soil degradation through salinization accounts the most. Besides the conventional techniques of avoiding these stresses through land reclamation, a number of other options are being tried and found useful. For example; PGPB have been known to play an essential role in the metabolism and growth of the plants. PGPB that colonize the roots of plants in the spermosphere (sphere that surround the seed) enhance plant growth, after attaching to root surface. In the last two decades or so various PGPB strains have been reported to alleviate abiotic stress. These are *Bacillus, Burkholderia, Acinetobacter, Enterobacter, Azospirillum, Beijierinckia, Rhizobium, Serratia, Erwinia, Flavobacterium, Alcaligenes,* etc. These microbes secrete bacterial AAC deaminase, volatiles, antioxidants, cytokinin, IAA and unknown metabolites in response to plant's ethylene, HKT1, ROS and ABA under salt and drought stress. This results in Induced Systemic Tolerance (IST), and brings in physical and chemical changes in plants that result in enhanced tolerance to abiotic stresses.

Moreover PGPB influence soil fertility by mineralizing organic phosphorous in soil and by solubilizing precipitated phosphates. PGPB excretes low molecular weight organic acids such as gluconic acid, citric acid etc. that dissolve phosphatic minerals and/or chelate cationic partners of the phosphorous ion, i.e., PO_4^{3-} (othophosphate) directly, solubilize inorganic phosphate available largely in soil to bioavailable phosphorous. Besides many phosphatase and cellulolytic enzymes are released for enzyme-labile soil organic phosphorous in favor of plant availability. PGPB directly facilitate the proliferation of plant by producing siderophore (Fe-III chelating agent), which can solubilize and sequester iron and provide it to the plants. These are low molecular weight (<10,000 Da), ferric-specific ligands produced by microbes as scavenging agent to combat low iron stress.

Proteases secreted by PGPB breaks down complex proteins available in soil into plant usable amino acids. They catalyze total hydrolysis of proteins to peptides; thereby function as degradative enzymes. In response to osmotic stress in soil, PGPB secretes compatible solutes that include sugars, amino acids and their derivatives which help them to adapt in external osmolarity. PGPB colonizes plant roots and alleviates the debilitating effects of salt stress. Production of exoplolysaccharide (EPS) protects these bacteria from hydric stress and fluctuations in water potential by enhancing water retention and regulation of carbon sources.

In addition EPS binds to cations, including sodium (Na^+) and, therefore, decrease the content of Na^+ available for plant uptake.

An understanding of controls over the below-ground function constitutes an important challenge as natural and agro-ecosystem around the globe are exposed to anthropogenic pressures. In addition, the physico-chemical and structural properties of soils, including their development, have been greatly affected by the action of the rhizosphere. PGPB have been proved to be best eco-friendly remedy to accelerate the growth of plant in nutrient deficient soil with respect to chemical fertilizers which are least available to plant. There are some legumes which are incapable of growing in drought and salt stresses as they may be devoid of mechanisms to survive in stressed conditions or unavailability of nutrient or increased secretion of ethylene hormone or decreased secretion of plant growth promoting hormones. In such cases, rhizospheric bacteria interact with plant tissues and cells with differing need. The use of PGPB is eco-friendly approach for enhancing plant growth and productivity in stressed environments.

ACKNOWLEDGEMENTS

Part of this research has been supported by DBT and SERB grant no. BT/PR1231/AGR/021/340/2011 and SR/FT/LS-129/2012 respectively to Dr. Devendra Kumar Choudhary. Authors acknowledge this financial support. Authors also acknowledge UGC-RGNF grant no. F1-17.1/RGNF 2012–2013-SC-RAJ-19482.

KEYWORDS

- abiotic stress
- abscisic acid
- amino acids
- antibiotics
- auxin response factors

- bacteria
- bacterial mediated amelioration
- bio-chemical reactions
- bio-control agents
- bio-fertilizer
- biotic stress
- blue-green algae
- compatible solutes
- cytokinin
- drought stress
- enzymes
- enzymes secretion
- ethylene
- ethylene biosynthesis
- exopolysaccharide (EPS)
- gene expression
- genetic pathways
- gibberellin
- ground water
- growth promoting bacteria
- indole-3-acetic acid (IAA)
- induced resistance
- induced tolerance
- iron sequestration
- K^+ transporters
- leaching
- leguminous plants
- lipids
- lytic enzymes
- metabolic activities
- microorganisms
- mutagenesis
- $Na^+:K^+$ ratio

- nitric oxide
- nitrogen
- nitrogen fixation
- non-symbiotic nitrogen fixation
- osmolarity
- osmotic stress
- P fixation
- phosphate solubilizing bacteria
- physiological processes
- phytohormone
- plant growth promoting bacteria
- plant hormones
- plant tissues
- protease
- protein
- reactive oxygen species (ROS)
- salt overly sensitive (SOS)
- salt stress
- sequestration
- siderophore
- signaling cascades
- signal perception receptors
- sodium transport
- stomata
- sugars
- symbiotic N_2 fixation
- toxic

REFERENCES

1. Abd El-Azeem, S. A., Elwan, M. W., Sung, J. K., & Ok, Y. S. (2012). Alleviation of salt stress in eggplant (*Solanum melongena* L.) by plant-growth-promoting Rhizo-bacteria. *Communications in Soil Science and Plant Analysis, 43,* 1303–1315.

2. Achal, V., Savant, V. V., & Reddy, M. S. (2007). Phosphate solubilization by a wild type strain and UV-induced mutants of *Aspergillus tubingensis*. *Soil Biology and Biochemistry*, 39, 695–699.

3. Ahmad, P., Sarwat, M., & Sharma, S. (2008). Reactive oxygen species, antioxidants and signaling in plants. *Plant Biology*, 51, 167–173.

4. Akhtar, M. S., & Siddiqui, Z. A. (2008). Bio-control of a root-rot disease complex of chickpea by *Glomus intraradices*, *Rhizobium* sp. and *Pseudomonas straita*. *Crop Protection*, 27, 410–417.

5. Akhtar, M. S., & Siddiqui, Z. A. (2011). Role of Plant growth promoting *rhizobacteria* in bio-control of plant diseases and sustainable agriculture. In: *Plant Growth and Health Promoting Bacteria*. Maheshwari, D. K. (Ed.). Springer, Berlin Heidelberg, 157–195.

6. Apel, K., & Hirt, H. (2004). Reactive oxygen species: metabolism, oxidative stress, and signal transduction. *Annual Reviews of Plant Biology*, 55, 373–399.

7. Apse, M. P., Aharon, G. S., Snedden, W. A., & Blumwald, E. (1999). Salt tolerance conferred by overexpression of a vacuolar Na^+/H^+ antiport in *Arabidopsis*. *Science*, 285, 1256–1258.

8. Apse, M. P., & Blumwald, E. (2007). Na^+ transport in plants. *FEBS Letters*, 581, 2247–2254.

9. Arora, N. K., Kang, S. C., & Maheshwari, D. K. (2001). Isolation of siderophore producing strains of *Rhizobium meliloti* and their bio-control potential against *Macrophomina phaseolina* that causes charcoal rot of groundnut. *Current Opinion in Plant Biology Science*, 81, 673–677.

10. Ashraf, M., Athar, H. R., Harris, P. J. C., & Kwon, T. R. (2008). Some prospective strategies for improving crop salt tolerance. *Advance in Agronomy*, 97, 45–110.

11. Babalola, O. O. (2010). Beneficial bacteria of agricultural importance. *Biotechnology Letters*, 32, 1559–1570.

12. Bano, A., & Fatima, M. (2009). Salt tolerance in *Zea mays* (L.) following inoculation with *Rhizobium* and *Pseudomonas*. *Biology and Fertility of Soils*, 45, 405–413.

13. Barassi, C. A., Ayrault, G., Creus, C. M., Sueldo, R. J., & Sobrero, M. T. (2006). Seed inoculation with *Azospirillum* mitigates NaCl effects on lettuce. *Scientia Horticulturae.*, 109, 8–14.

14. Barriuso, J., Ramos Solano, B., & Gutierrez Mañero, F. J. (2008). Protection against pathogen and salt stress by four PGPR isolated from *Pinus* sp. on *Arabidopsis thaliana*. *Phytopathology*, 98, 666–672.

15. Bashan, Y., & de-Bashan, L. E. (2010). How the plant growth-promoting bacterium *Azospirillum* promotes plant growth-A critical assessment. In: *Advances in Agronomy* vol-108. Sparks, D. L. (Ed.). Elsevier, Academic Press, 77–136.

16. Batool, R., & Hasnain, S. (2005). Growth stimulatory effects of *Enterobacter* and *Serratia* located from biofilms on plant growth and soil aggregation. *Biotechnology*, 4, 347–353.

17. Behera, B. C., Singdevsachan, S. K., Mishra, R. R., Dutta, S. K., & Thatoi, H. N. (2014). Diversity, mechanism and biotechnology of phosphate solubilizing microorganism in mangrove—A review. *Biocatalysis and Agricultural Biotechnology*, 3, 97–110.

18. Belimov, A. A., Dodd, I. C., Hontzeas, N., Theobald, J. C., Safronova, V. I., & Davies, W. J. (2009). Rhizosphere bacteria containing 1-aminocyclopropane-1-carboxylate

deaminase increase yield of plants grow in drying soil via both local and systemic hormone signaling. *New Phytologist*, 181, 413–423.

19. Belimov, A. A., Dodd, I. C., Safronova, V. I., Dumova, V. A., Shaposhnikov, A. I., Ladatko, A. G., & Davies, W. J. (2014). Abscisic acid metabolizing rhizobacteria decrease ABA concentrations *in planta* and alter plant growth. *Plant Physiology and Biochemistry*, 74, 84–91.

20. Belimov, A. A., Hontzeas, N., Safronova, V. I., Demchinskaya, S. V., Piluzza. G., Bullitta, S., & Glick, B. R. (2005). Cadmium-tolerant plant growth-promoting bacteria associated with the roots of Indian mustard (*Brassica juncea* L. Czern.). *Soil Biology and Biochemistry*, 37, 241–250.

21. Beneduzi, A., Ambrosini, A., & Passaglia, L. M. (2012). Plant growth-promoting rhizobacteria (PGPR): Their potential as antagonists and biocontrol agents. *Genetics and Molecular Biology*, 35, 1044–1051.

22. Bergmaier, D., Champagne, C. P., & Lacroix, C. (2003). Exopolysaccharide production during batch cultures with free and immobilized *Lactobacillus rhamnosus* RW-9595 M. *Journal of Applied Microbiology*, 95, 1049–1057.

23. Berthomieu, P., Conejero, G., Nublat, A., Brackenbury, W. J., Lambert, C., Savio, C., Uozumi, N., Oiki, S., Yamada, K., Cellier, F., Gosti, F., Simonneau, T., Essah, P. A., Tester, M., Véry, A. A., Sentenac, H., & Casse, F. (2003). Functional analysis of *AtHKT1* in *Arabidopsis* shows that Na⁺ recirculation by the phloem is crucial for salt tolerance. *The EMBO Journal*, 22, 2004–2014.

24. Beumer, R. R., Te Giffel, M. C., Cox, L. J., Rombouts, F. M., & Abee, T. (1994). Effect of exogenous proline, betaine, and carnitine on growth of Listeria monocytogenes in a minimal medium. *Applied and Environmental Microbiology*, 60, 1359–1363.

25. Bhattacharyya, D., Yu, S. M., & Lee, Y. H. (2015). Volatile compounds from *Alcaligenes fecalis* JBCS1294 confer salt tolerance in *Arabidopsis thaliana* through the auxin and gibberellin pathways and differential modulation of gene expression in root and shoot tissues. *Journal of Plant Growth Regulation*, 75, 297–306.

26. Bhattacharyya, P. N., & Jha, D. K. (2012). Plant growth-promoting rhizobacteria (PGPR): emergence in agriculture. *World Journal of Microbiology and Biotechnology*, 28, 1327–1350.

27. Biswas, J. C., Ladha, J. K., & Dazzo, F. B. (2000). Rhizobia inoculation improves nutrient uptake and growth of lowland rice. *Soil Science Society of America Journal*, 64, 1644–1650.

28. Blokhina, O., Virolainen, E., & Fagerstedt, K. V. (2003). Antioxidants, oxidative damage and oxygen deprivation stress: a review. *Annals of Botany*, 91, 179–194.

29. Boiero, L., Perrig, D., Masciarelli, O., Penna, C., Cassan, F., & Luna, V. (2007). Phytohormone production by three strains of *Bradyrhizobium japonicum* and possible physiological and technological implications. *Applied Microbiology and Biotechnology*, 74, 874–880.

30. Bolen, D. W., & Baskakov, I. V. (2001). The osmophobic effect: natural selection of a thermodynamic force in protein folding. *Journal of Molecular Biology*, 310, 955–63.

31. Bomfeti, C. A., Florentino, L. A., Guimarães, A. P., Cardoso, P. G., Guerreiro, M. C., & Moreira, F. M. S. (2011). Exopolysaccharides produced by the symbiotic nitrogen-fixing bacteria of Leguminosae. *Revista Brasileira de Ciência do Solo*, 35, 657–671.

32. Bresson, J., Varoquaux, F., Bontpart, T., Touraine, B., & Vile, D. (2013). The PGPR strain *Phyllobacterium brassicacearum* STM196 induces a reproductive delay and physiological changes that result in improved drought tolerance in *Arabidopsis*. *New Phytologist*, 200, 558–569.

33. Brown, A. D. (1976). Microbial water stress. *Bacteriological Reviews*, 40, 803–846.

34. Bruinsma, J. (2009). *The Resource Outlook to 2050: Expert Meeting on How to Feed the World*, United Nations Food and Agricultural Organization, Rome.

35. Canbolat, M. Y., Bilen, S., Çakmakçı, R., Şahin, F., & Aydın, A. (2006). Effect of plant growth-promoting bacteria and soil compaction on barley seedling growth, nutrient uptake, soil properties and rhizosphere microflora. *Biology and Fertility of Soils*, 42, 350–357.

36. Canfield, D. E., Glazer, A. N., & Falkowski, P. G. (2010). The evolution and future of Earth's nitrogen cycle. *Science*, 330, 192–196.

37. Cassán, F., Bottini, R., Schneider, G., & Piccoli, P. (2001a). *Azospirillum brasilense* and *Azospirillum lipoferum* hydrolyze conjugates of GA20 and metabolize the resultant aglycones to GA1 in seedlings of rice dwarf mutants. *Plant Physiology*, 125, 2053–2058.

38. Cassán, F., Lucangeli, C., Bottini, R., & Piccoli, P. (2001b). *Azospirillum* spp. metabolize $[17,17-^2H_2]$ gibberellin A_{20} to $[17,17-^2H_2]$ gibberellin A_1 in vivo in *dy* rice mutant seedlings. *Plant and Cell Physiology*, 42, 763–767.

39. Cassán, F., Perrig, D., Sgroy, V., Masciarelli, O., Penna, C., & Luna, V. (2009). *Azospirillum brasilense* Az39 and *Bradyrhizobium japonicum* E109, inoculated singly or in combination, promote seed germination and early seedling growth in corn (*Zea mays* L.) and soybean (*Glycine max* L.). *European Journal of Soil Biology*, 45, 28–35.

40. Cassán, F., Vanderleyden, J., & Spaepen, S. (2014). Physiological and agronomical aspects of phytohormone production by model plant-growth-promoting rhizobacteria (PGPR) belonging to the genus *Azospirillum*. *Journal of Plant Growth Regulation*, 33, 440–459.

41. Chang, P., Gerhardt, K. E., Huang, X. D., Yu, X. M., Glick, B. R., Gerwing, P. D., & Greenberg, B. M. (2014). Plant Growth-Promoting Bacteria Facilitate the Growth of Barley and Oats in Salt-Impacted Soil: Implications for Phytoremediation of Saline Soils. *International Journal of Phytoremediation*, 16, 1133–1147.

42. Chen, H., An, R., Tang, J. H., Cui, X. H., Hao, F. S., Chen, J., & Wang, X. C. (2007a). Over-expression of a vacuolar Na^+/H^+ antiporter gene improves salt tolerance in upland rice. *Molecular Breeding*, 19, 215–225.

43. Chen, M., Wei, H., Cao, J., Liu, R., Wang, Y., & Zheng, C. (2007b). Expression of *Bacillus subtilis* proAB genes and reduction of feedback inhibition of proline synthesis increases proline production and confers osmotolerance in transgenic *Arabdopsis*. *Journal of Biochemistry and Molecular Biology*, 40, 396–403.

44. Chen, Y. P., Rekha, P. D., Arun, A. B., Shen, F. T., Lai, W. A., & Young, C. C. (2006). Phosphate solubilizing bacteria from subtropical soil and their tricalcium phosphate solubilizing abilities. *Applied Soil Ecology*, 34, 33–41.

45. Cheng, Z., Woody, O. Z., McConkey, B. J., & Glick, B. R. (2012). Combined effects of the plant growth-promoting bacterium *Pseudomonas putida* UW4 and salinity stress on the *Brassica napus* proteome. *Applied Soil Ecology*, 61, 255–263.

46. Chinnusamy, V., Jagendorf, A., & Zhu, J. K. (2005). Understanding and improving salt tolerance in plants. *Crop Science*, 45, 437–448.

47. Cho, S. M., Kang, B. R., Han, S. H., Anderson, A. J., Park, J. Y., Lee, Y. H., Cho, B. H., Yang, K. Y., Ryu, C. M., & Kim, Y. C. (2008). 2R, 3R-butanediol, a bacterial volatile produced by *Pseudomonas chlororaphis* O6, is involved in induction of systemic tolerance to drought in *Arabidopsis thaliana*. *Molecular Plant-Microbe Interaction*, 21, 1067–1075.

48. Choudhary, D. K., & Johri, B. N. (2009). Interactions of *Bacillus* spp. and plants-With special reference to induced systemic resistance (ISR). *Microbiological Research*, 164, 493–513.

49. Chun Juan, W., Ya Hui, G., Chao, W., Hong Xia, L., Dong Dong, N., Yun Peng, W., & Jian Hua, G. (2012). Enhancement of tomato (*Lycopersicon esculentum*) tolerance to drought stress by plant-growth-promoting rhizobacterium (PGPR) *Bacillus cereus* AR156. *Journal of Agricultural Biotechnology*, 20, 1097–1105.

50. Cohen, A. C., Bottini, R., & Piccoli, P. N. (2008). *Azospirillum brasilense* SP245 produces ABA in chemically-defined culture medium and increases ABA content in *Arabidopsis* plants. *Journal of Plant Growth Regulation*, 54, 97–103.

51. Cohen, A. C., Travaglia, C. N., Bottini, R., & Piccoli, P. N. (2009). Participation of abscisic acid and gibberellins produced by endophytic *Azospirillum* in the alleviation of drought effects in maize. *Botany*, 87, 455–462.

52. Cohen, M. F., Lamattina, L., & Yamasaki, H. (2010). Nitric oxide signaling by plant-associated bacteria. In: *Nitric Oxide in Plant Physiology*. Hayat, S., Mori, M., Pichtel, J., & Ahmad, A. (Eds.). Wiley-VCH, Weinheim, 161–172.

53. Collavino, M. M., Sansberro, P. A., Mroginski, L. A., & Aguilar, O. M. (2010). Comparison of in vitro solubilization activity of diverse phosphate-solubilizing bacteria native to acid soil and their ability to promote *Phaseolus vulgaris* growth. *Biology and Fertility of Soils*, 46, 727–738.

54. Compant, S., Clément, C., & Sessitsch, A. (2010). Plant growth-promoting bacteria in the rhizo- and endosphere of plants: their role, colonization, mechanisms involved and prospects for utilization. *Soil Biology and Biochemistry*, 42, 669–678.

55. Compant, S., Duffy, B., Nowak, J., Clement, C., & Barka, E. A. (2005). Use of plant growth promoting bacteria for biocontrol of plant diseases: principles, mechanisms of action and future prospects. *Applied and Environmental Microbiology*, 71, 4951–4959.

56. Conrath, U. (2011). Molecular aspects of defense priming. *Trends in Plant Science*, 16, 524–531.

57. Constabel, C. P., & Barbehenn, R. (2008). Defensive roles of polyphenol oxidase in plants. In: *Induced Plant Resistance to Herbivory*. Schaller, A. (Ed.). Springer, Netherlands, 253–270.

58. Correa-Aragunde, N., Graziano, M., Chevalier, C., & Lamattina, L. (2006). Nitric oxide modulates the expression of cell cycle regulatory genes during lateral root formation in tomato. *Journal of Experimental Botany*, 57, 581–588.

59. Creus, C. M., Graziano, M., Casanovas, E. M., Pereyra, M. A., Simontacchi, M., Puntarulo, S., Barassi, C. A., & Lamattina, L. (2005). Nitric oxide is involved in the *Azospirillum brasilense*-induced lateral root formation in tomato. *Planta*, 221, 297–303.

60. Cullis, P. M., Jones, G. D., Symons, M. C., & Lea, J. S. (1987). Electron transfer from protein to DNA in irradiated chromatin. *Nature*, 330, 773–774.
61. Cutler, S. R., Rodriguez, P. L., Finkelstein, R. R., & Abrams, S. R. (2010). Abscisic acid: emergence of a core signaling network. *Annual Review of Plant Biology*, 61, 651–679.
62. Dashti, N., Zhang, F., Hynes, R., & Smith, D. L. (1998). Plant growth promoting rhizobacteria accelerate nodulation and increase nitrogen fixation activity by field grown soybean [*Glycine max* (L.) Merr.] under short season conditions. *Plant and Soil*, 200, 205–213.
63. Davenport, R. J., Munoz-Mayor, A., Jha, D., Essah, P. A., Rus, A. N. A., & Tester, M. (2007). The Na+ transporter *AtHKT1:1* controls retrieval of Na+ from the xylem in *Arabidopsis. Plant, Cell and Environment*, 30, 497–507.
64. Dawson, T. P., Perryman, A. H., & Osborne, T. M. (2014). Modeling impacts of climate change on global food security. *Climatic Change*, 1–12.
65. de Souza, J. T., de Boer, M., de Waard, P., van Beek, T. A., & Raaijmakers, J. M. (2003). Biochemical, genetic, and zoosporicidal properties of cyclic lipopeptide surfactants produced by *Pseudomonas fluorescens. Applied and Environmental Microbiology*, 69, 7161–7172.
66. de Vleesschauwer, D., & Höfte, M. (2009). Rhizobacteria-induced systemic resistance. *Advances in Botanical Research*, 51, 223–281.
67. Defago, G. (1993). 2, 4-Diacetylphloroglucinol, a promising compound in biocontrol. *Plant Pathology*, 42, 311–312.
68. del Amor, F., & Cuadra-Crespo, P. (2012). Plant growth-promoting bacteria as a tool to improve salinity tolerance in sweet pepper. *Functional Plant Biology*, 39, 82–90.
69. Deubel, A., Gransee, A., & Merbach, W. (2000). Transformation of organic rhizodepositions by rhizosphere bacteria and its influence on the availability of tertiary calcium phosphate. *Journal of Plant Nutrition and Soil Science*, 163, 387–392.
70. Di Palma, A. A., Pereyra, M. C., Ramirez, L. M., Vázquez, M. L. X., María, L. Baca, B. E., Pereyra, M. A., Lamattina, L., & Creus, C. M. (2013). Denitrification-derived nitric oxide modulates biofilm formation in *Azospirillum brasilense. FEMS Microbiology Letters*, 338, 77–85.
71. Dimkpa, C. O., Svatoš, A., Dabrowska, P., Schmidt, A., Boland, W., & Kothe, E. (2008). Involvement of siderophores in the reduction of metal-induced inhibition f auxin synthesis in *Streptomyces* spp. *Chemosphere*, 74, 19–25.
72. Divito, G. A., & Sadras, V. O. (2014). How do phosphorus, potassium and sulfur affect plant growth and biological nitrogen fixation in crop and pasture legumes? A meta-analysis. *Field Crops Research*, 156, 161–171.
73. Dobrovolski, R., Diniz-Filho, J. A. F., Loyola, R. D., & Júnior, P. D. M. (2011). Agricultural expansion and the fate of global conservation priorities. *Biodiversity Conservation*, 20, 2445–2459.
74. Dowling, D. N., Sexton, R., Fenton, A., Delany, I., Fedi, S., McHugh, B., Callanan, M., Mognne-Loccoz, Y., & O'Gara, F. (1996). Iron regulation in plant-associated *Pseudomonas fluorescens* M114: implications for biological control. In: *Molecular Biology of Pseudomonads*. Nakazawa, T., Furukawa, K., Haas D., & Silver, S. (Eds.). American Society for Microbiology Press, Washington, DC, USA, 502–511.

75. Durbak, A., Yao, H., & McSteen, P. (2012). Hormone signaling in plant development. *Current Opinion in Plant Biology*, 15, 92–96.
76. Ehrlich, H. L. (1990). *Geomicrobiology*. Dekker, New York.
77. El-Showk, S., Ruonala, R., & Helariutta, Y. (2013). Crossing paths: cytokinin signaling and crosstalk. *Development*, 140, 1373–1383.
78. El-Tayeb, T. S., & Khodair, T. A. (2006). Enhanced production of some microbial exo-polysaccharides by various stimulating agents in batch culture. *Research Journal of Agriculture and Biological Sciences*, 2, 483–492.
79. Estrada, G. A., Baldani, V. L. D., de Oliveira, D. M., Urquiaga, S., & Baldani, J. I. (2013). Selection of phosphate-solubilizing diazotrophic *Herbaspirillum* and *Burkholderia* strains and their effect on rice crop yield and nutrient uptake. *Plant and Soil*, 369, 115–129.
80. Estrada-De Los Santos, P., Bustillos-Cristales, R., & Caballero-Mellado, J. (2001). *Burkholderia*, a genus rich in plant-associated nitrogen fixers with wide environmental and geographic distribution. *Applied and Environmental Microbiology*, 67, 2790–2798.
81. FAO, (2012). World Agriculture towards 2030/2050: The 2012 Revision. ESA Working Paper No. 12–03. Food and Agriculture Organization of the United Nations, Rome. 147 p.
82. FAOSTAT. (2013). http://faostat.fao.org/site/452/default.aspx.
83. Farag, M. A., Zhang, H., & Ryu, C. M. (2013). Dynamic chemical communication between plants and bacteria through airborne signals: induced resistance by bacterial volatiles. *Journal of Chemical Ecology*, 39, 1007–1018.
84. Farrell, M., Prendergast-Miller, M., Jones, D. L., Hill, P. W., & Condron, L. M. (2014). Soil microbial organic nitrogen uptake is regulated by carbon availability. *Soil Biology and Biochemistry*, 77, 261–267.
85. Fernando, W. D., Nakkeeran, S., & Zhang, Y. (2006). Biosynthesis of antibiotics by PGPR and its relation in biocontrol of plant diseases. In: *PGPR: Biocontrol and Biofertilization*. Siddiqui, Z. A. (Ed.). Springer, Netherlands, 67–109.
86. Figueiredo, M. V. B., Martinez, C. R., Burity, H. A., & Chanway, C. P. (2008). Plant growth-promoting rhizobacteria for improving nodulation and nitrogen fixation in the common bean (*Phaseolus vulgaris* L.). *World Journal of Microbiology and Biotechnology*, 24, 1187–1193.
87. Finlayson, S. A., Foster, K. R., & Reid, D. M. (1991). Transport and metabolism of 1-aminocyclopropane- 1-carboxylic acid in sunflower (*Helianthus annuus* L.) seedlings. *Plant Physiology*, 96, 1360–1367.
88. Forchetti, G., Masciarelli, O., Alemano, S., Alvarez, D., & Abdala, G. (2007). Endophytic bacteria in sunflower (*Helianthus annuus* L.): isolation, characterization, and production of jasmonates and abscisic acid in culture medium. *Applied Microbiology and Biotechnology*, 76, 1145–1152.
89. Fraire-Velázquez, S., Rodríguez-Guerra, R., & Sánchez-Calderón, L. (2011). Abiotic and biotic stress response crosstalk in plants. In: *Abiotic Stress Response in Plants— Physiological, Biochemical and Genetic Perspectives*. Shanker, A. (Ed.). InTech, New York, doi:10.5772/23217.
90. Franche, C., Lindström, K., & Elmerich, C. (2009). Nitrogen-fixing bacteria associated with leguminous and non-leguminous plants. *Plant and Soil*, 321, 35–59.

91. Fravel, D. R. (1988). Role of antibiosis in the biocontrol of plant diseases. *Annual Review of Phytopathology*, 26, 75–91.

92. Frost, C. J., Mescher, M. C., Carlson, J. E., & De Moraes, C. M. (2008). Plant defense priming against herbivores: getting ready for a different battle. *Plant Physiology*, 146, 818–824.

93. Fujita, Y., Fujita, M., Shinozaki, K., & Yamaguchi-Shinozaki, K. (2011). ABA-mediated transcriptional regulation in response to osmotic stress in plants. *Journal of Plant Research*, 124, 509–525.

94. Gassmann, W., Rubio, F., & Schroeder, J. I. (1996). Alkali cation selectivity of the wheat root high-affinity potassium transporter HKT1. *Plant Journal*, 10, 869–882.

95. Gaxiola, R. A., Rao, R., Sherman, A., Grisafi, P., Alper, S. L., & Fink, G. R. (1999). The *Arabidopsisthaliana* proton transporters, AtNhx1 and Avp1, can function in cation detoxification in yeast. *Proceedings of the National Academy of Sciences USA*, 96, 1480–1485.

96. Gepstein, S., & Glick, B. R. (2013). Strategies to ameliorate abiotic stress-induced plant senescence. *Plant Molecular Biology*, 82, 623–633.

97. Gibbs, H. K., Ruesch, A. S., Achard, F., Clayton, M. K., Holmgren, P., Ramankutty, N., & Foley, J. A. (2011). Tropical forests were the primary sources of new agricultural land in the 1980s and 1990s. *Proceedings of the National Academy of Sciences USA*, 107, 16732–16737.

98. Gill, S. S., & Tuteja, N. (2010). Reactive oxygen species and antioxidant machinery in abiotic stress tolerance in crop plants. *Plant Physiology and Biochemistry*, 48, 909–930.

99. Glick, B. R. (1995). The enhancement of plant growth by free-living bacteria. *Canadian Journal of Microbiology*, 41, 109–117.

100. Glick, B. R. (2005). Modulation of plant ethylene levels by the bacterial enzyme ACC deaminase. *FEMS Microbiology Letters*, 251, 1–7.

101. Glick, B. R. (2010). Using soil bacteria to facilitate phytoremediation. *Biotechnology Advances*, 28, 367–374.

102. Glick, B. R. (2012). Plant growth-promoting bacteria: mechanisms and applications. *Scientifica*. doi: 10.6064/2012/963401.

103. Glick, B. R. (2014). Bacteria with ACC deaminase can promote plant growth and help to feed the world. *Microbiological Research*, 169, 30–39.

104. Glick, B. R. (2015). Stress Control and ACC Deaminase. In: *Principles of Plant-Microbe Interactions*. Lugtenberg, B. (Ed.). Springer International Publishing, Switzerland, 257–264.

105. Glick, B. R., Penrose, D. M., & Li, J. (1998). A model for the lowering of plant ethylene concentrations by plant growth promoting bacteria. *Journal of Theoretical Biology*, 190, 63–68.

106. Goldstein, A. H. (1995). Recent progress in understanding the molecular genetics and biochemistry of calcium phosphate solubilization by gram negative bacteria. *Biological Agriculture and Horticulture*, 12, 185–193.

107. Gray, E. J., & Smith, D. L. (2005). Intracellular and extracellular PGPR: Commonalities and distinctions in the plant-bacterium signaling processes. *Soil Biology and Biochemistry*, 37, 395–412.

108. Grichko, V. P., Filby, B., & Glick, B. R. (2000). Increased ability of transgenic plants expressing the bacterial enzyme ACC deaminase to accumulate Cd, Co, Cu, Ni, Pb, and Zn. *Journal of Biotechnology*, 81, 45–53.

109. Grobkinsky, D. K., van der Graaff, E., & Roitsch, T. (2014). Abscisic acid-cytokinin antagonism modulates resistance against *Pseudomonas syringae* in tobacco. *Phytopathology*, 104, 1283–1288.

110. Grover, M., Ali, S. Z., Sandhya, V., Rasul, A., & Venkateswarlu, B. (2011). Role of microorganisms in adaptation of agriculture crops to abiotic stresses. *World Journal of Microbiology and Biotechnology*, 27, 1231–1240.

111. Grover, M. R. (2003). Rock Phosphate and Phosphate Solubilizing Microbes as a Source of Nutrients for Crops. MSc. Thesis. Thapar Institute of Engineering and Technology, Patiala.

112. Guinel, F. C., & Geil, R. D. (2002). A model for the development of the rhizobial and arbuscular mycorrhizal symbioses in legumes and its use to understand the roles of ethylene in the establishment of these two symbioses. *Canadian Journal of Botany*, 80, 695–720.

113. Gyaneshwar, P., Kumar, G. N., Parekh, L. J., & Poole, P. S. (2002). Role of soil microorganisms in improving P nutrition of plants. *Plant and Soil*, 245, 83–93.

114. Haas, D., & Défago, G. (2005). Biological control of soil-borne pathogens by fluorescent pseudomonads. *Nature Reviews Microbiology*, 3, 307–319.

115. Haggag, W. M. (2008). Isolation of bioactive antibiotic peptides from *Bacillus brevis* and *Bacillus polymyxa* against botrytis gray mold in strawberry. *Archives of Phytopathology and Plant Protection*, 41, 477–491.

116. Hammerschmidt, R. (2009). Systemic acquired resistance. *Advances in Botanical Research*, 51, 173–222.

117. Hao, Y., Charles, T. C., & Glick, B. R. (2007). ACC deaminase from plant growth-promoting bacteria affects crown gall development. *Canadian Journal of Microbiology*, 53, 1291–1299.

118. Hao, Y., Charles, T. C., & Glick, B. R. (2010). ACC deaminase increases the *Agrobacterium tumefaciens*-mediated transformation frequency of commercial canola cultivars. *FEMS Microbiology Letters*, 307, 185–190.

119. Hao, Y., Charles, T. C., & Glick, B. R. (2011). An ACC deaminase containing *A. tumefaciens* strain D3 shows biocontrol activity to crown gall disease. *Canadian Journal of Microbiology*, 57, 278–286.

120. He, C., Yan, J., Shen, G., Fu, L., Holaday, A, S., Auld, D., Blumwald, E., & Zhang, H. (2005). Expression of an *Arabidopsis* vacuolar sodium/proton antiporter gene in cotton improves photosynthetic performance under salt conditions and increases fiber yield in the field. *Plant and Cell Physiology*, 46, 1848–1854.

121. Hell, R., & Stephan, U. W. (2003). Iron uptake, trafficking and homeostasis in plants. *Planta*, 216, 541–551.

122. Helman, Y., Burdman, S., & Okon, Y. (2011). Plant growth promotion by rhizosphere bacteria through direct effects. In: *Beneficial Microorganisms in Multicellular Life Forms*. Rosenberg, E., & Gophna, U. (Eds.). Springer, Berlin, Heidelberg, 89–103.

123. Hirayama, T., & Shinozaki, K. (2010). Research on plant abiotic stress responses in the post-genome era: past, present and future. *Plant Journal*, 61, 1041–1052.

124. Honma, M., & Shimomura, T. (1978). Metabolism of 1-aminocyclopropane-1-carboxylic acid. *Agric Biology Chem.*, 42, 1825–1831.
125. Horie, T., Horie, R., Chan, W. Y., Leung, H. Y., & Schroeder, J. I. (2006). Calcium regulation of sodium hypersensitivities ofsos3 and *athkt1* mutants. *Plant and Cell Physiology*, 47, 622–633.
126. Horie, T., & Schroeder, J. I. (2004). Sodium transporters in plants. Diverse genes and physiological functions. *Plant Physiology*, 136, 2457–2462.
127. Horie, T., Sugawara, M., Okunou, K., Nakayama, H., Schroeder, J. I., Shinmyo, A., & Yoshida, K. (2008). Functions of HKT transporters in sodium transport in roots and in protecting leaves from salinity stress. *Plant Biotechnology*, 25, 233–239.
128. http://www.usda.gov/foodcomp/search/).
129. http://nutritiondata.self.com/facts/custom/2697353/2.
130. Hua, J. (2015). Isolation of Components Involved in Ethylene Signaling. In: *Ethylene in Plants*. Wen, C. K. (Ed.). Springer, Netherlands, 27–44.
131. Husen, E., Wahyudi, A. T., & Suwanto, A. (2011). Growth Enhancement and Disease Reduction of Soybean by 1-Aminocyclopropane-1-Carboxylate Deaminase-Producing *Pseudomonas*. *American Journal of Applied Sciences*, 8, 1073–1080.
132. Ilyas, N., & Bano, A. (2010). *Azospirillum* strains isolated from roots and rhizosphere soil of wheat (*Triticum aestivum* L.) grown under different soil moisture conditions. *Biology and Fertility of Soils*, 46, 393–406.
133. Jaleel, C. A., Manivannan, P., Sankar, B., Kishorekumar, A., Gopi, R., Somasundaram, R., & Panneerselvam, R. (2007). *Pseudomonas fluorescens* enhances biomass yield and ajmalicine production in *Catharanthus roseus* under water deficit stress. *Colloids and Surfaces B: Biointerfaces*, 60, 7–11.
134. Jha, D., Shirley, N., Tester, M., & Roy, S. J. (2010). Variation in salinity tolerance and shoot sodium accumulation in *Arabidopsis* ecotypes linked to differences in the natural expression levels of transporters involved in sodium transport. Plant, *Cell and Environment*, 33, 793–804.
135. Kang, S. M., Khan, A. L., Hamayun, M., Hussain, J., Joo, G. J., You, Y. H., Kim, J. K., & Lee, I. J. (2012). Gibberellin-producing *Promicromonospora* sp. SE188 improves *Solanum lycopersicum* plant growth and influences endogenous plant hormones. *Journal of Microbiology*, 50, 902–909.
136. Kang, S. M., Khan, A. L., Waqas, M., You, Y. H., Kim, J. H., Kim, J. G., Hamayun, M., & Lee, I. J. (2014). Plant growth-promoting rhizobacteria reduce adverse effects of salinity and osmotic stress by regulating phytohormones and antioxidants in *Cucumis sativus*. *Journal of Plant Interactions*, 9, 673–682.
137. Kang, Y., Carlson, R., Tharpe, W., & Schell, M. A. (1998). Characterization of genes involved in biosynthesis of a novel antibiotic from *Burkholderia cepacia* BC11 and their role in biological control of *Rhizoctonia solani*. *Applied and Environmental Microbiology*, 64, 3939–3947.
138. Karlidag, H., Yildirim, E., Turan, M., Pehluvan, M., & Donmez, F. (2013). Plant growth-promoting rhizobacteria mitigate deleterious effects of salt stress on strawberry plants (*Fragaria × ananassa*). *HortScience*, 48, 563–567.
139. Kasotia, A., & Choudhary, D. K. (2014a). Role of endophytic microbes in mitigation of abiotic stress for plants. In: *Emerging Technologies and Management of Crop Stress Tolerance*. Ahmad P., & Rasool, S. (Eds.). Vol-2. Elsevier, USA, 97–103.

140. Kasotia, A., & Choudhary, D. K. (2014b). Induced Inorganic Phosphate Solubilization Through N-Methyl-N'-Nitro-N-Nitrosoguanidine Treated Mutants of *Pseudomonas koreensis* Strain AK-1 (MTCC Number 12058) Under Polyethylene Glycol. *Proceedings of the National Academy of Sciences, India Section B: Biological Sciences*, doi: 10.1007/s40011–014–0416–6.

141. Kaye, J. P., & Hart, S. C. (1997). Competition for nitrogen between plants and soil microorganisms. *Trends in Ecology and Evolution*, 12, 139–143.

142. Keel, C., & Defago, G. (1997). Interactions between beneficial soil bacteria and root pathogens: mechanisms and ecological impact. In: *Multitrophic Interactions in Terrestrial System*. Gange, A. C., & Brown, V. K. (Eds.). Oxford, Blackwell Science, 27–47.

143. Khan, A. A., Jilani, G., Akhtar, M. S., Naqvi, S. S., & Rasheed, M. (2009). Phosphorus solubilizing bacteria: occurrence, mechanisms and their role in crop production. *Journal of Agricultural and Biological Science*, 1, 48–58.

144. Kim, B. S., Moon, S. S., & Hwang, B. K. (1999). Isolation, identification and antifungal activity of a macrolide antibiotic, oligomycin A, produced by *Streptomyces libani*. *Canadian Journal of Botany*, 77, 850–858.

145. Kim, M. C., Chung, W. S., Yun, D. J., & Cho, M. J. (2009) Calcium and Calmodulin-Mediated Regulation of Gene Expression in Plants. *Molecular Plant*, 2, 13–21.

146. Kim, T. H., Böhmer, M., Hu, H., Nishimura, N., & Schroeder, J. I. (2010). Guard cell signal transduction network: Advances in understanding abscisic acid, CO_2, and Ca^{2+} signaling. *Annual Review of Plant Biology*, 61, 561–591.

147. Kisiala, A., Laffont, C., Emery, R. N., & Frugier, F. (2013). Bioactive cytokinins are selectively secreted by *Sinorhizobium meliloti* nodulating and nonnodulating strains. *Molecular Plant-Microbe Interactions*, 26, 1225–1231.

148. Klein-Marcuschamer, D., & Stephanopoulos, G. (2008). Assessing the potential of mutational strategies to elicit new phenotypes in industrial strains. *Proceedings of the National Academy of Sciences USA*, 105, 2319–2324.

149. Kloepper, J. W., Leong, J., Teintze, M., & Schroth, M. N. (1980). Enhanced plant growth by siderophores produced by plant growth-promoting rhizobacteria. *Nature.*, 286, 885–886.

150. Kohler, J., Hernandez, J. A., Caravaca, F., & Roldàn, A. (2008). Plant-growth-promoting rhizobacteria and abuscular mycorrhizal fungi modify alleviation biochemical mechanisms in water-stressed plants. *Functional Plant Biology*, 35, 141–151.

151. Kohler, J., Hernandez, J. A., Caravaca, F., & Roldàn, A. (2009). Induction of antioxidant enzymes is involved in the greater effectiveness of a PGPR versus AM fungi with respect to increasing the tolerance of lettuce to severe salt stress. *Environmental and Experimental Botany*, 65, 245–252.

152. Korasick, D. A., Enders, T. A., & Strader, L. C. (2013). Auxin biosynthesis and storage forms. *Journal of Experimental Botany*, 64, 2541–2555.

153. Kraiser, T., Gras, D. E., Gutiérrez, A. G., González, B., & Gutiérrez, R. A. (2011). A holistic view of nitrogen acquisition in plants. *Journal of Experimental Botany*, 62, 1455–1466.

154. Ku, Y. S., Au-Yeung, W. K., Yung, K. L., Li, M. H., Wen, C. Q., Liu, X., & Yung, Y. L. (2013). Drought stress and tolerance in soybean. In: *A Comprehensive Survey*

of International Soybean Research- Genetics, Physiology, Agronomy and Nitrogen Relationship. J. E. Board (Ed.). InTech, doi: 10.5772/52945.

155. Kumar, K. V., Singh, N., Behl, H. M., & Srivastava, S. (2008). Influence of plant growth promoting bacteria and its mutant on heavy metal toxicity in *Brassica juncea* grown in fly ash amended soil. *Chemosphere*, 72, 678–683.

156. Lambrecht, M., Okon, Y., Vande Broek, A., & Vanderleyden, J. (2000). Indole-3-acetic acid: a reciprocal signaling molecule in bacteria plant interactions. *Trends in Microbiology*, 8, 298–300.

157. Lamosa, P., Martins, L. O., da Costa, M. S., & Santos, H. (1998). Effects of temperature, salinity and medium composition on compatible solute accumulation by *Thermococcus* spp. *Applied and Environmental Microbiology*, 6410, 3591–3598.

158. Larimer, F. W., Chain, P., Hauser, L., Lamerdin, J., Malfatti, S., Do, L., Land, M. L., Pelletier, D. A., Beatty, J. T., Lang, A. S., Tabita, F. R., Gibson, J. L., Hanson, T. E., Bobst, C., Torres, J. L. T. Y., Peres, C., Harrison, F. H., Gibson, J., & Harwood, C. S. (2004). Complete genome sequence of the metabolically versatile photosynthetic bacterium *Rhodopseudomonas palustris*. *Nature Biotechnology*, 22, 55–61.

159. Latha, P., Anand, T., Rappathi, N., Prakasam, V., & Samiyappan, R. (2009). Antimicrobial activity of plant extracts and induction of systemic resistance in tomato plants by mixtures of PGPR strains and Zimmu leaf extract against *Alternaria solani*. *Biological Control*, 50, 85–93.

160. Latt, Z. K., Yu, S. S., & Lynn, T. M. (2013). Enhancement of Cellulolytic Nitrogen Fixing Activity of *Alcaligenes* sp. by MNNG Mutagenesis. *International Journal of Innovation and Applied Studies*, 3, 979–986.

161. Laurance, W. F., Sayer, J., & Cassman, K. G. (2014). Agricultural expansion and its impacts on tropical nature. *Trends in Ecology and Evolution*, 29, 107–116.

162. Li, D. P., & Wu, Z. J. (2008). Impact of chemical fertilizers application on soil ecological environment. *Journal of Applied Ecology*, 19, 1158–1165.

163. Li, H., Pellegrini, M., & Eisenberg, D. (2005). Detection of parallel functional modules by comparative analysis of genome sequences. *Nature Biotechnology*, 23, 253–260.

164. Li, J., & Glick, B. R. (2001). Transcriptional regulation of the *Enterobacter cloacae* UW4 1-aminocyclopropane-1-carboxylate (ACC) deaminase gene (*acd*S). *Canadian Journal of Microbiology*, 47, 359–367.

165. Luo, Z. B., Janz, D., Jiang, X., Göbel, C., Wildhagen, H., Tan, Y., Rennenberg, H., Feussner, I., & Polle, A. (2009). Upgrading root physiology for stress tolerance by ectomycorrhizas: insights from metabolite and transcriptional profiling into reprogramming for stress anticipation. *Plant Physiology*, 151, 1902–1917.

166. Ma, W., Guinel, F. C., & Glick, B. R. (2003). *Rhizobium leguminosarum* biovar viciae 1-aminocyclopropane-1-carboxylate deaminase promotes nodulation of pea plants. *Applied and Environmental Microbiology*, 69, 4396–4402.

167. Ma, Y., Prasad, M. N. V., Rajkumar, M., & Freitas, H. (2011). Plant growth promoting rhizobacteria and endophytes accelerate phytoremediation of metalliferous soils. *Biotechnology Advances*, 29, 248–258.

168. Maathuis, F. J. (2014). Sodium in plants: perception, signaling, and regulation of sodium fluxes. *Journal of Experimental Botany*, 65, 849–858.

169. Machackova, I., Chauvaux, N., Dewitte, W., & VanOnckelen, H. (1997). Diurnal fluctuations in ethylene formation in *Chenopodium rubrum*. *Plant Physiology*, 113, 981–985.

170. Malusá, E., & Vassilev, N. (2014). A contribution to set a legal framework for biofertilizers. *Applied Microbiology and Biotechnology*, 98, 6599–6607.

171. Marschner, H. (1997). *Mineral Nutrition of Higher Plants*. 2nd Edn. Academic Press, London.

172. Martin, M. N., & Saftner, R. A. (1995). Purification and characterization of 1-aminocyclopropane-1-carboxylic acid N-malonyltransferase from tomato fruit. *Plant Physiology*, 108, 1241–1249.

173. Mäser, P., Eckelman, B., Vaidyanathan, R., Horie, T., Fairbairn, D. J., Kubo, M., Yamagami, M., Yamaguchi, K., Nishimura, M., Uozumi, N., Robertson, W., Sussman, M. R., & Schroeder, J. I. (2002). Altered shoot/root Na$^+$ distribution and bifurcating salt sensitivity in *Arabidopsis* by genetic disruption of the Na$^+$ transporter *AtHKT1*. *FEBS Letters*, 531, 157–161.

174. Maurhofer, M., Reimmann, C., Schmidli-Sacherer, P., Heeb, S., Haas, D., & Défago, G. (1998). Salicylic acid biosynthetic genes expressed in *Pseudomonas fluorescens* strain P3 improve the induction of systemic resistance in tobacco against tobacco necrosis virus. *Phytopathology*, 88, 678–684.

175. Mayak, S., Tirosh, T., & Glick, B. R. (2004a). Plant growth-promoting bacteria confer resistance in tomato plants to salt stress. *Plant Physiology and Biochemistry*, 42, 565–572.

176. Mayak, S., Tirosh, T., & Glick, B. R. (2004b). Plant growth-promoting bacteria that confer resistance to water stress in tomatoes and peppers. *Plant Science*, 166, 525–530.

177. Metraux, J. P. (2001). Systemic acquired resistance and salicylic acid: current state of knowledge. *European Journal of Plant Pathology*, 107, 19–28.

178. Miller, G. A. D., Susuki, N., Ciftci-Yilmaz, S., & Mittler, R. O. N. (2010). Reactive oxygen species homeostasis and signaling during drought and salinity stresses. *Plant, Cell and Environment*, 33, 453–467.

179. Milner, J. L., Silo-Suh, L., Lee, J. C., He, H., Clardy, J., & Handelsman, J. (1996). Production of kanosamine by *Bacillus cereus* UW85. *Applied and Environmental Microbiology*, 62, 3061–3065.

180. Ministry of Agriculture. (2013). Pocket Bok on Agricultural Statistics. Govt of India, Ministry of Agriculture, Dept. of Agriculture and cooperation, Directorate of Economics and Statistics, New Delhi.

181. Mirza, M. S., Mehnaz, S., Normand, P., Prigent-Combaret, C., Moënne-Loccoz, Y., Bally, R., & Malik, K. A. (2006). Molecular characterization and PCR detection of a nitrogen fixing *Pseudomonas* strain promoting rice growth. *Biology and Fertility of Soils*, 43, 163–170.

182. Mittler, R. (2006). Abiotic stress, the field environment and stress combination. *Trends in Plant Science*, 11, 15–19.

183. Miyazaki, J. H., & Yang, S. F. (1987). The methionine salvage pathway in relation to ethylene and polyamine biosynthesis. *Physiologia Plantarum*, 69, 366–70.

184. Mohamed, H. I., & Gomaa, E. Z. (2012). Effect of plant growth promoting *Bacillus subtilis* and *Pseudomonas fluorescens* on growth and pigment composition of radish plants (*Raphanus sativus*) under NaCl stress. *Photosynthetica*, 50, 263–272.

185. Molina-Favero, C., Creus, C. M., Lanteri, M. L., Correa-Aragunde, N., Lombardo, M. C., Barassi, C. A., & Lamattina, L. (2007). Nitric oxide and plant growth promoting rhizobacteria: Common features influencing root growth and development. *Advances in Botanical Research*, 46, 1–33.

186. Molina-Favero, C., Creus, C. M., Simontacchi, M., Puntarulo, S., & Lamattina, L. (2008). Aerobic nitric oxide production by *Azospirillum brasilense* Sp245 and its influence on root architecture in tomato. *Molecular Plant-Microbe Interaction*, 21, 1001–1009.

187. Møller, I. S., Gilliham, M., Deepa, J., Mayo, G. M., Roy, S. J., Coates, J. C., Haseloff, J., & Tester, M. (2009). Shoot Na$^+$ exclusion and increased salinity tolerance engineered by cell type-specific alteration of Na$^+$ transport in *Arabidopsis*. *Plant Cell*, 21, 2163–78.

188. Mondini, L., Nachit, M., Porceddu, E., & Pagnotta, M. A. (2012). Identification of SNP mutations in *DREB1, HKT1*, and *WRKY1* genes involved in drought and salt stress tolerance in durum wheat (*Triticum turgidum* L. var durum). *OMICS: A Journal of Integrative Biology*, 16, 178–187.

189. Muday, G. K., Rahman, A., & Binder, B. M. (2012). Auxin and ethylene: collaborators or competitors? *Trends in Plant Science*, 17, 181–195.

190. Muleta, D., Assefa, F., & Granhall, U. (2007). In vitro antagonism of rhizobacteria isolated from *Coffea arabica* L. against emerging fungal coffee pathogens. *Engineering in Life Sciences*, 7, 577–586.

191. Munns, R. (2002). Comparative physiology of salt and water stress. *Plant, Cell and Environment*, 25, 239–250.

192. Munns, R. (2005). Genes and salt tolerance: bringing them together. *New Phytologist*, 167, 645–663.

193. Munns, R., & Tester, M. (2008). Mechanisms of salinity tolerance. *Annual Review of Plant Biology*, 59, 651–681.

194. Nakashima, K., Ito, Y., & Yamaguchi-Shinozaki, K. (2009). Transcriptional Regulatory Networks in Response to Abiotic Stresses in *Arabidopsis* and Grasses. *Plant Physiology*, 149, 88–95.

195. Nakashima, K., Yamaguchi-Shinozaki, K., & Shinozaki, K. (2014). The transcriptional regulatory network in the drought response and its crosstalk in abiotic stress responses including drought, cold, and heat. *Frontiers in Plant Science*, doi: 10.3389/fpls.2014.00170.

196. Nakayama, T., Homma, Y., Hashidoko, Y., Mizutani, J., & Tahara, S. (1999). Possible role of xanthobaccins produced by *Stenotrophomonas* sp. strain SB-K88 in suppression of sugar beet damping-off disease. *Applied and Environmental Microbiology*, 65, 4334–4339.

197. Nascimento, F. X., Vicente, C. S., Barbosa, P., Espada, M., Glick, B. R., Mota, M., & Oliveira, S. (2013). Evidence for the involvement of ACC deaminase from *Pseudomonas putida* UW4 in the biocontrol of pine wilt disease caused by *Bursaphelenchus xylophilus*. *BioControl*, 58, 427–433.

198. Näsholm, T., Kielland, K., & Ganeteg, U. (2009). Uptake of organic nitrogen by plants. *New Phytologist*, 182, 31–48.

199. Nautiyal, C. S., Srivastava, S., Chauhan, P. S., Seem, K., Mishra, A., & Sopory, S. K. (2013). Plant growth-promoting bacteria *Bacillus amyloliquefaciens* NBRISN13

modulates gene expression profile of leaf and rhizosphere community in rice during salt stress. *Plant Physiology and Biochemistry*, 66, 1–9.

200. Neilands, J. B. (1995). Siderophores: structure and function of microbial iron transport compounds. *The Journal of Biological Chemistry*, 270, 26723–26726.

201. Nielsen, T. H., Sørensen, D., Tobiasen, C., Andersen, J. B., Christeophersen, C., Givskov, M., & Sorensen, J. (2002). Antibiotic and biosurfactant properties of cyclic lipopeptides produced by fluorescent *Pseudomonas* spp. from the sugar beet rhizosphere. *Applied and Environmental Microbiology*, 68, 3416–3423.

202. Nishiyama, R., Watanabe, Y., Fujita, Y., Le, D. T., Kojima, M., Werner, T., Vankovad, R., Yamaguchi-Shinozakib, K., Shinozakia, K., Kakimoto, T., Sakakibara, H., Schmülling T., & Tran, L. S. P. (2011). Analysis of cytokinin mutants and regulation of cytokinin metabolic genes reveals important regulatory roles of cytokinins in drought, salt and abscisic acid responses, and abscisic acid biosynthesis. *Plant Cell*, 23, 2169–2183.

203. O'Brien, J. A., & Benková, E. (2013). Cytokinin cross-talking during biotic and abiotic stress responses. *Frontiers in Plant Science*, 4, 451.

204. Olias, R., Eljakaoui, Z., Li, J. U. N., De Morales, P. A., Marín-Manzano, M. C., Pardo, J. M., & Belver, A. (2009). The plasma membrane Na^+/H^+ antiporter SOS1 is essential for salt tolerance in tomato and affects the partitioning of Na^+ between plant organs. *Plant, Cell and Environment*, 32, 904–916.

205. Onofre-Lemus, J., Hernández-Lucas, I., Girard, L., & Caballero-Mellado, J. (2009). ACC (1-aminocyclopropane-1-carboxylate) deaminase activity, a widespread trait in *Burkholderia* species, and its growth-promoting effect on tomato plants. *Applied and Environmental Microbiology*, 75, 6581–6590.

206. Ortíz-Castro, R., Valencia-Cantero, E., & López-Bucio, J. (2008). Plant growth promotion by *Bacillus megaterium* involves cytokinin signaling. *Plant Signaling and Behavior*, 3, 263–265.

207. Parekh, S., Vinci, V. A., & Strobel, R. J. (2004). Improvement of microbial strains and fermentation processes. *Applied Microbiology and Biotechnology*, 54, 287–301.

208. Paul, D. (2013). Osmotic stress adaptations in rhizobacteria. *Journal of Basic Microbiology*, 53, 101–110.

209. Paul, D., & Nair, S. (2008). Stress adaptations in a plant growth promoting rhizobacterium (PGPR) with increasing salinity in the coastal agricultural soils. *Journal of Basic Microbiology*, 48, 378–384.

210. Paungfoo-Lonhienne, C., Lonhienne, T. G., Rentsch, D., Robinson, N., Christie, M., Webb, R. I., Gamage, H. K., Carroll, B. J., Schenk, P. M., & Schmidt, S. (2008). Plants can use protein as a nitrogen source without assistance from other organisms. *Proceedings of the National Academy of Sciences USA*, 105, 4524–4529.

211. Paungfoo-Lonhienne, C., Visser, J., Lonhienne, T. G., & Schmidt, S. (2012). Past, present and future of organic nutrients. *Plant and Soil*, 359, 1–18.

212. Peix, A., Rivas-Boyero, A. A., Mateos, P. F., Rodriguez-Barrueco, C., Martınez-Molina, E., & Velazquez, E. (2001). Growth promotion of chickpea and barley by a phosphate solubilizing strain of *Mesorhizobium mediterraneum* under growth chamber conditions. *Soil Biology and Biochemistry*, 33, 103–110.

213. Peleg, Z., & Blumwald, E. (2011). Hormone balance and abiotic stress tolerance in crop plants. *Current Opinion in Plant Biology*, 14, 290–295.

214. Penrose, D. M., & Glick, B. R. (2001). Levels of 1-aminocyclopropane-1-carboxylic acid (ACC) in exudates and extracts of canola seeds treated with plant growth-promoting bacteria. *Canadian Journal of Microbiology*, 47, 368–372.
215. Penrose, D. M., Moffatt, B. A., & Glick, B. R. (2001). Determination of 1-aminocycopropane-1-carboxylic acid (ACC) to assess the effects of ACC deaminase-containing bacteria on roots of canola seedlings. *Canadian Journal of Microbiology*, 47, 77–80.
216. Penrose, D. M., & Glick, B. R. (2003). Methods for isolating and characterizing ACC deaminase-containing plant growth-promoting rhizobacteria. *Physiologia Plantarum*, 118, 10–15.
217. Pérez-Alfocea, F., Albacete, A., Ghanem, M. E., & Dodd, I. C. (2010). Hormonal regulation of source-sink relations to maintain crop productivity under salinity: a case study of root-to-shoot signaling in tomato. *Functional Plant Biology*, 37, 592–603.
218. Perrig, D., Boiero, M. L., Masciarelli, O. A., Penna, C., Ruiz, O. A., Cassán, F. D., & Luna, M. V. (2007). Plant-growth-promoting compounds produced by two agronomically important strains of *Azospirillum brasilense*, and implications for inoculant formulation. *Applied Microbiology and Biotechnology*, 75, 1143–1150.
219. Philosoph-Hadas, S., Meir, S., & Aharoni, N. (1985). Autoinhibition of ethylene production in tobacco leaf discs: enhancement of 1-aminocyclopropane-1-carboxilic acid conjugation. *Physiologia Plantarum*, 63, 431–437.
220. Pimentel, D., & Pimentel, M. (2003). Sustainability of meat-based and plant-based diets and the environment. *The American Journal of Clinical Nutrition*, 78, 660S–663S.
221. Pitzschke, A. M., Forzani, C., & Hirt, H. (2006). Reactive oxygen species signaling in plants. *Antioxidants and Redox Signaling*, 8, 1757–1764.
222. Pliego, C., Kamilova, F., & Lugtenberg, B. (2011). Plant growth-promoting bacteria: fundamentals and exploitation. In: *Bacteria in Agrobiology: Crop Ecosystems*. Maheshwari, D. K. (Ed.). Springer, Berlin, Heidelberg, 295–343.
223. Podile, A. R., & Kishore, G. K. (2006). Plant growth-promoting rhizobacteria. In: *Plant-Associated Bacteria*. Gnanamanickam, S. S. (Ed.). Springer, Netherlands, 195–230.
224. Pospisilova, J., Vagner, M., Malbeck, J., Travnickova, A., & Batkova, P. (2005). Interactions between abscisic acid and cytokinins during water stress and subsequent rehydration. *Biologia Plantarum*, 49, 533–540.
225. Press, C. M., Wilson, M., Tuzun, S., & Kloepper, J. W. (1997). Salicylic acid produced by *Serratia marcescens* 90–166 is not the primary determinant of induced systemic resistance in cucumber or tobacco. *Molecular Plant-Microbe Interaction*, 10, 761–768.
226. Quintero, F. J., Martinez-Atienza, J., Villalta, I., Jiang, X., Kim, W. Y., Ali, Z., Fujii, H., Mendoza, I., Yun, D. J., Zhu, J. K., & Pardo, J. M. (2011). Activation of the plasma membrane Na/H antiporter Salt-Overly-Sensitive 1 (SOS1) by phosphorylation of an auto-inhibitory C-terminal domain. *Proceedings of the National Academy of Sciences USA*, 108, 2611–2616.
227. Qurashi, A. W., & Sabri, A. N. (2011). Osmoadaptation and plant growth promotion by salt tolerant bacteria under salt stress. *African Journal of Microbiology Research*, 5, 3546–3554.

228. Qurashi, A. W., & Sabri, A. N. (2012). Bacterial exopolysaccharide and biofilm formation stimulate chickpea growth and soil aggregation under salt stress. *Brazilian Journal of Microbiology*, 43, 1183–1191.
229. Raaijmakers, J. M., Vlami, M., & de Souza, J. T. (2002). Antibiotic production by bacterial biocontrol agents. *Antonie Leeuwenhoek*, 81, 537–547.
230. Raghavendra, A. S., Gonugunta, V. K., Christmann, A., & Grill, E. (2010). ABA perception and signaling. *Trends in Plant Science*, 15, 395–401.
231. Rajkumar, M., Ae, N., Prasad, M. N. V., & Freitas, H. (2010). Potential of siderophore-producing bacteria for improving heavy metal phytoextraction. *Trends in Biotechnology*, 28, 142–149.
232. Rajwar, A., Sahgal, M., & Johri, B. N. (2013). Legume-Rhizobia Symbiosis and Interactions in Agroecosystems. In: *Plant Microbe Symbiosis: Fundamentals and Advances*. Arora, N. K. (Ed.). Springer, India, 233–265.
233. Recep, K., Fikrettin, S., Erkol, D., & Cafer, E. (2009). Biological control of the potato dry rot caused by *Fusarium* species using PGPR strains. *Biological Control*, 50, 194–198.
234. Reddy, A. A. (2010). Regional disparities in food habits and nutritional intake in Andhra Pradesh, India. *Regional and Sectoral EconomicStuidies,* 10, 125–134.
235. Rehman, S., Harris, P. J. C., & Ashraf, M. (2005). Stress environments and their impact on crop production. In: *Abiotic Stresses: Plant Resistance through Breeding and Molecular Approaches*. Ashraf, M., & Harris, P. J. C. (Eds.). Haworth Press, New York, 3–18.
236. Ren, Z. H., Gao, J. P., Li, L. G., Cai, X. L., Huang, W., Chao, D. Y., Zhu, M. Z., Wang, Z. Y., Luan, S., & Lin, H. X. (2005). A rice quantitative trait locus for salt tolerance encodes a sodium transporter. *Nature Genetics*, 37, 1141–1146.
237. Rentsch, D., Schmidt, S., & Tegeder, M. (2007). Transporters for uptake and allocation of organic nitrogen compounds in plants. *FEBS Letters*, 581, 2281–2289.
238. Richardson, A. E. (2001). Prospects for using soil microorganisms to improve the acquisition of phosphorus by plants. *Functional Plant Biology*, 28, 897–906.
239. Rodríguez, H., & Fraga, R. (1999). Phosphate solubilizing bacteria and their role in plant growth promotion. *Biotechnology Advances*, 17, 319–339.
240. Rojas-Tapias D., Moreno-Galván, A., Pardo-Díaz, S., Obando, M., Rivera, D., & Bonilla, R. (2012). Effect of inoculation with plant growth-promoting bacteria (PGPB) on amelioration of saline stress in maize (*Zea mays*). *Applied Soil Ecology*, 61, 264–272.
241. Roshandel, P., & Flowers, T. (2009). The ionic effects of NaCl on physiology and gene expression in rice genotypes differing in salt tolerance. *Plant and Soil*, 315, 35–147.
242. Rubio, F., Gassmann. W., & Schroeder, J. I. (1995). Sodium-driven potassium uptake by the plant potassium transporter *HKT1* and mutations conferring salt tolerance. *Science*, 270, 1660–1663.
243. Ruíz-Sánchez, M., Armada, E., Muñoz, Y., García de Salamone, I. E., Aroca, R., Ruíz-Lozano, J. M., & Azcón, R. (2011). *Azospirillum* and arbuscular mycorrhizal colonization enhance rice growth and physiological traits under well-watered and drought conditions. *Journal of Plant Physiology*, 168, 1031–1037.

244. Rus, A., Lee, B. H., Munoz-Mayor, A., Sharkhuu, A., Miura, K., Zhu, J. K., Bressan, R. A., & Hasegawa, P. M. (2004). *AtHKT1* facilitates Na$^+$ homeostasis and K$^+$ nutrition in planta. *Plant Physiology*, 136, 2500–2511.

245. Ryu, C. M., Farag, M. A., Hu, C. H., Reddy, M. S., Wie, H. X., Pare, P. W., & Kloepper, J. W. (2003). Bacterial volatiles promote the growth of *Arabidopsis*. *Proceedings of the National Academy of Sciences USA*, 100, 4927–4932.

246. Saibo, N. J. M., Lourenço, T., & Oliveira, M. M. (2009). Transcription factors and regulation of photosynthetic and related metabolism under environmental stresses. *Annals of Botany*, 103, 609–623.

247. Saikia, R., Kumar, R., Arora, D. K., Gogoi, D. K., & Azad, P. (2006). *Pseudomonas aeruginosa* inducing rice resistance against *Rhizoctonia solani*: Production of salicylic acid and peroxidases. *Folia Microbiologica*, 51, 375–380.

248. Sailaja, P. R., Podile, A. R., & Reddanna, P. (1998). Biocontrol strain of *Bacillus subtilis* AF 1 rapidly induces lipoxygenase in groundnut (*Arachis hypogaea* L.). compared to crown rot pathogen *Aspergillus niger*. *European Journal of Plant Pathology*, 104, 125–132.

249. Sakakibara, H. (2006). Cytokinins: activity, biosynthesis, and translocation. *Annual Review of Plant Biology*, 57, 431–449.

250. Salomon, M. V., Bottini, R., de Souza Filho, G. A., Cohen, A. C., Moreno, D., Gil, M., & Piccoli, P. (2013). Bacteria isolated from roots and rhizosphere of *Vitis vinifera* retard water losses, induce abscisic acid accumulation and synthesis of defense-related terpenes in in vitro cultured grapevine. *Physiologia Plantarum*, 151, 359–374.

251. Sandhya, V., Ali, S. Z., Grover, M., Reddy, G., & Venkateswarlu, B. (2010). Effect of plant growth promoting *Pseudomonas* spp. on compatible solutes, antioxidant status and plant growth of maize under drought stress. *Plant Growth Regulation*, 62, 21–30.

252. Sarma, R. K., & Saikia, R. (2014). Alleviation of drought stress in mung bean by strain *Pseudomonas aeruginosa* GGRJ21. *Plant and Soil*, 377, 111–126.

253. Savci, S. (2012). An agricultural pollutant: chemical fertilizer. *International Journal of Environmental Science and Development*, 3, 77–80.

254. Sayyed, R. Z., Jamadar, D. D., & Patel, P. R. (2011). Production of Exo-polysaccharide by *Rhizobium* sp. *Indian Journal of Microbiology*, 51, 294–300.

255. Schachtman, D. P., & Schroeder, J. I. (1994). Structure and transport mechanism of a high-affinity potassium transporter from higher plants. *Nature*, 370, 655–658.

256. Schalk, I. J., Hannauer, M., & Braud, A. (2011). New roles for bacterial siderophores in metal transport and tolerance. *Environmental Microbiology*, 13, 2844–2854.

257. Schippers, B., Bakker, A. W., & Bakker, A. H. M. (1987). Interactions of deleterious and benefcial rhizosphere microorganisms and the effect of cropping practice. *Annual Review of Phytopathology*, 25, 339–358.

258. Shavrukov, Y. (2013). Salt stress or salt shock: which genes are we studying? *Journal of Experimental Botany*, 64, 119–127.

259. Shi, H., Lee, B. H., Wu, S. J., & Zhu, J. K. (2003). Overexpression of a plasma membrane Na$^+$/H$^+$ antiporter gene improves salt tolerance in *Arabidopsis thaliana*. *Nature Biotechnology*, 21, 81–85.

260. Sindhu, S. S., Suneja, S., & Dadarwal, K. R. (1997). Plant growth promoting rhizobacteria and their role in crop productivity. In: *Biotechnological Approaches in Soil*

Microorganisms for Sustainable Crop Production. Dadarwal, K. R. (Ed.). Scientific, Jodhpur, 149–193.

261. Singh, R. B. (2014). Climate Change and Abiotic Stress Management in India. In: *Climate Change and Plant Abiotic Stress Tolerance.* Tuteja, N., Gill, S. S. (Eds.). Wiley-VCH Verlag GmbH & Co. KGaA, Weinheim, Germany, 57–78.

262. Smil, V. (2000). *Feeding the World: A Challenge for the Twenty-first Century.* MIT Press Cambridge, MA.

263. Spaepen, S., Vanderleyden, J., & Okon, Y. (2009). Plant growth-promoting actions of rhizobacteria. *Advances in Botanical Research*, 51, 283–320.

264. Spaepen, S., Vanderleyden, J., & Remans, R. (2007). Indole-3-acetic acid in microbial and microorganism-plant signaling. *FEMS Microbiology Reviews*, 31, 425–448.

265. Sreenivasulu, N., Sopory, S. K., & Kavi Kishor, P. B. (2007). Deciphering the regulatory mechanisms of abiotic stress tolerance in plants by genomic approaches. *Gene*, 388, 1–13.

266. Staswick, P. E., & Tiryaki, I. (2004). The oxylipin signal jasmonic acid is activated by an enzyme that conjugates it to isoleucine in *Arabidopsis. Plant Cell*, 16, 2117–2127.

267. Suarez, R., Wong, A., Ramirez, M., Barraza, A., OrozcoMdel, C., Cevallos, M. A., Lara, M., Hernandez, G., & Iturriaga, G. (2008). Improvement of drought tolerance and grain yield in common bean by overexpressing trehalose-6-phosphate synthase in rhizobia. *Molecular Plant-Microbe Interactions*, 21, 958–966.

268. Sunarpi Horie, T., Motoda, J., Kubo, M., Yang, H., Yoda, K., Horie, R., Chan, W. Y., Leung, H. Y., Hattori, K., Konomi, M., Osumi, M., Yamagami, M., Schroeder, J. I., & Uozumi, N. (2005). Enhanced salt tolerance mediated by *AtHKT1* transporter-induced Na^+ unloading from xylem vessels to xylem parenchyma cells. *Plant Journal*, 44, 928–938.

269. Suzuki, W., Sugawara, M., Miwa, K., & Morikawa, M. (2014). Plant growth-promoting bacterium *Acinetobacter calcoaceticus* P23 increases the chlorophyll content of the monocot *Lemna minor* (duckweed) and the dicot *Lactuca sativa* (lettuce). *Journal of Bioscience and Bioengineering*, 118, 41–44.

270. Taiz, L., & Zeiger, E. (2006). Ethylene: The Gaseous Hormone. In: *Plant Physiology*, 4th edn. Hopkins, W. G., & Hüner, N. P. A. (Eds.). Sinauer Associates, Sunderland, MA, USA, 571–591.

271. Teale, W., Paponov, I., & Palme, K. (2006). Auxin in action: signaling, transport and the control of plant growth and development. *Nature Reviews Molecular Cell Biology*, 7, 847–859.

272. Tegeder, M., & Rentsch, D. (2010). Uptake and partitioning of amino acids and peptides. *Molecular Plant*, 3, 997–1011.

273. Tester, M., & Davenport, R. (2003). Na^+ tolerance and Na^+ transport in higher plants. *Annals of Botany*, 91, 503–527.

274. Thomashow, L. S., Bonsall, R. F., & Weller, D. M. (1997). Antibiotic production by soil and rhizosphere microbes in situ. In: *Manual of Environmental Microbiology*. Hurst C. J., Knudsen G. R., McInerney M. J., Stetzenbach L. D., & Walter, M. V. (Eds.). ASM Press, Washington DC, 493–499.

275. Thomine, S., & Lanquar, V. (2011). Iron transport and signaling in plants. In: *Transporters and Pumps in Plant Signaling*. Geisler M., & Venema, K. (Eds.). Springer, Berlin, Heidelberg, 99–131.

276. Thrane, C., Olsson, S., Nielsen, T. H., & S☐rensen, J. (1999). Vital fluorescent stains for detection of stress in *Pythium ultimum* and *Rhizoctonia solani* challenged with viscosinamide from *Pseudomonas fluorescens* DR54. *FEMS Microbiology Ecology*, 30, 11–23.

277. Toklikishvili, N., Dandurishvili, N., Tediashvili, M., Giorgobiani, N., Lurie, S., Szegedi, E., Glick, B. R., & Chernin, L. (2010). Inhibitory effect of ACC deaminasepro-ducing bacteria on crown gall formation in tomato plants infected by *Agrobacterium tumefaciens* or *A. vitis*. *Plant Pathology*, 59, 1023–1030.

278. Tortora, M. L., Díaz-Ricci, J. C., & Pedraza, R. O. (2011). *Azospirillum brasilense* siderophores with antifungal activity against *Colletotrichum acutatum*. *Archives of Microbiology*, 193, 275–286.

279. Trivedi, P., & Sa, T. (2008). *Pseudomonas corrugata* (NRRL B-30409) mutants increased phosphate solubilization, organic acid production, and plant growth at lower temperatures. *Current Microbiology*, 56, 140–144.

280. Turan, M., Esitken, A., & Sahin, F. (2012). Plant growth promoting rhizobacteria as alleviators for soil degradation. In: *Bacteria in Agrobiology: Stress Management*. Maheshwari, D. K. (Ed.). Springer, Berlin, Heidelberg, 41–63.

281. Uozumi, N., Kim, E. J., Rubio, F., Yamaguchi, T., Muto, S., Tsuboi, A., Bakker, E. P., Nakamura, T., & Schroeder, J. I. (2000). The *ArabidopsisHKT1*gene homolog mediates inward NaR currents in *Xenopus laevis* oocytes and NaR uptake in *Saccharomyces cerevisiae*. *Plant Physiology*, 122, 1249–1259.

282. United Nations. (2013). *World Population Prospects: The 2012 Revision*. United Nations Population Division. New York.

283. USDA. *National Nutrient Databases* (http://www.usda.gov/foodcomp/search/) http://nutritiondata.self.com/facts/custom/2697353/2.

284. Valcheva, V., Trendafilov, K., & Todorova, S. (2012). Influence of mineral fertilization on the harmful soil acidity and chemical composition of wine grape varieties. *Agricultural Science and Technology*, 4, 260–264.

285. Van de Poel, B., & Van Der Straeten, D. (2014). 1-aminocyclopropane-1- carboxylic acid (ACC) in plants: more than just the precursor of ethylene! *Frontiers in Plant Science*, 5, 640.

286. Van der Ent, S., Van Wees, S. C. M., & Pieterse, C. M. J. (2009). Jasmonate signaling in plant interactions with resistance-inducing beneficial microbes. *Phytochemistry*, 70, 1581–1588.

287. Van Loon, L. C., Bakker, P. A. H. M., & Pieterse, C. M. J. (1998). Systemic resistance induced by rhizosphere bacteria. *Annual Review of Phytopathology*, 36, 453–483.

288. Van Wees, S. C. M., Van der Ent, S., & Pieterse, C. M. J. (2008). Plant immune responses triggered by beneficial microbes. *Current Opinion in Plant Biology*, 11, 443–448.

289. Vu, B., Chen, M., Crawford, R. J., & Ivanova, E. P. (2009). Bacterial extracellular polysaccharides involved in biofilm formation. *Molecules*, 14, 2535–2554.

290. Vyrides, I., & Stuckey, D. C. (2009). Adaptation of anaerobic biomass to saline conditions: Role of compatible solutes and extracellular polysaccharides. *Enzyme and Microbial Technology*, 44, 46–51.

291. Wang, C., Knill, E., & Glick, B. R. (2000). Effect of transferring 1-aminocyclo-propane-1-carboxylic acid (ACC) deaminase genes into *Pseudomonas fluorescens*

strain CHA0 and its *gac*A derivative CHA96 on their growth-promoting and disease-suppressive capacities. *Canadian Journal of Microbiology*, 46, 898–907.

292. Wang, F. Y., Lin, X. G., Yin, R., & Wu, L. H. (2006). Effects of arbuscular mycorrhizal inoculation on the growth of *Elsholtzia splendens* and *Zea mays* and the activities of phosphatase and urease in a multi-metal-contaminated soil under unsterilized conditions. *Applied Soil Ecology*, 31, 110–119.

293. Wang, W., Vinocur, B., Shoseyov, O., & Altman, A. (2004). Role of plant heat-shock proteins and molecular chaperones in the abiotic stress response. *Trends in Plant Science*, 9, 244–252.

294. Wang, Y., Li, L., Ye, T., Zhao, S., Liu, Z., Feng, Y. Q., & Wu, Y. (2011). Cytokinin antagonizes ABA suppression to seed germination of *Arabidopsis* by downregulating ABI5 expression. *Plant Journal*, 68, 249–261.

295. Waters, S., Gilliham, M., & Hrmova, M. (2013). Plant high-affinity potassium (HKT) transporters involved in salinity tolerance: structural insights to probe differences in ion selectivity. *International Journal of Molecular Sciences*, 14, 7660–7680.

296. Weyens, N., Lelie, D. V. D., Taghavi, S., Newman, L., & Vangronsveld, J. (2009). Exploiting plant-microbe partnerships to improve biomass production and remediation. *Trends in Biotechnology*, 27, 591–598.

297. Whipps, J. M. (1997). Developments in the biological control of soil-borne plant pathogens. *Advances in Botanical Research*, 26, 1–134.

298. Willems, A. (2006). The taxonomy of rhizobia: an overview. *Plant and Soil*, 287, 3–14.

299. Wolters, H., & Jurgens, G. (2009). Survival of the flexible: hormonal growth control and adaptation in plant development. *Nature Reviews Genetics*, 10, 305–317.

300. Wrzaczek, M., Vainonen, J. P., Gauthier, A., Overmyer, K., & Kangasjärvi, J. (2011). Reactive Oxygen in Abiotic Stress Perception-From Genes to Proteins. In: *Abiotic Stress Response in Plants: Physiological, Biochemical and Genetic Perspectives*. Shanker, A. K., & Venkateswarlu, B. (Eds.). InTech, New York, 27–54.

301. Wu, G., Fanzo, J., Miller, D. D., Pingali, P., Post, M., Steiner, J. L., & Thalacker-Mercer, A. E. (2014). Production and supply of high-quality food protein for human consumption: sustainability, challenges, and innovations. *Annals of the New York Academy of Sciences*, 1321, 1–19.

302. Wu, S. C., Cao, Z. H., Li, Z. G., Cheung, K. C., & Wong, M. H. (2005). Effects of biofertilizer containing N-fixer, P and K solubilizers and AM fungi on maize growth: a greenhouse trial. *Geoderma*, 125, 155–166.

303. Wu, S. J., Ding, L., & Zhu, J. K. (1996). SOS1, a genetic locus essential for salt tolerance and potassium acquisition. *Plant Cell*, 8, 617–627.

304. Xiao, J. H., Chen, D. X., Liu, J. W., Liu, Z. L., Wan, W. H., Fang, N., Xiao, Y., Qi, Y., & Liang, Z. Q. (2004). Optimization of submerged culture requirements for the production of mycelial growth and exopolysaccharide by *Cordyceps jiangxiensis* JXPJ 0109. *Journal of Applied Microbiology*, 96, 1105–1116.

305. Yakes, F. M., & Van Houten, B. (1997). Mitochondrial DNA damage is more extensive and persists longer than nuclear DNA damage in human cells following oxidative stress. *Proceedings of the National Academy of Sciences USA*, 94, 514–549.

306. Yamaguchi-Shinozaki, K., & Shinozaki, K. (2006). Transcriptional regulatory networks in cellular responses and tolerance to dehydration and cold stresses. *Annual Review of Plant Biology*, 57, 781–803.

307. Yang, J., Kloepper, J. W., & Ryu, C. M. (2009). Rhizosphere bacteria help plants tolerate abiotic stress. *Trends in Plant Science*, 14, 1–4.

308. Yang, J., Xie, X., Wang, X., Dixon, R., & Wang, Y. P. (2014). Reconstruction and minimal gene requirements for the alternative iron-only nitrogenase in *Escherichia coli*. *Proceedings of the National Academy of Sciences USA*, 111, E3718–E3725.

309. Yao, L., Wu, Z., Zheng, Y., Kaleem, I., & Li, C. (2010). Growth promotion and protection against salt stress by *Pseudomonas putida* Rs-198 on cotton. *European Journal of Soil Biology*, 46, 49–54.

310. Yim, W., Seshadri, S., Kim, K., Lee, G., & Sa, T. (2013). Ethylene emission and PR protein synthesis in ACC deaminase producing *Methylobacterium* spp. inoculated tomato plants (*Lycopersicon esculentum* Mill.) challenged with *Ralstonia solanacearum* under greenhouse conditions. *Plant Physiology and Biochemistry*, 67, 95–104.

311. Yoon, G. M., & Kieber, J. J. (2013). ACC synthase and its cognate E3 ligase are inversely regulated by light. *Plant Signaling and Behavior*, 8, e26478.

312. Yu, X., Ai, C., Xin, L., & Zhou, G. (2011). The siderophore-producing bacterium, *Bacillus subtilis* CAS15, has a biocontrol effect on *Fusarium* wilt and promotes the growth of pepper. *European Journal of Soil Biology*, 47, 138–145.

313. Yu, X., Liu, X., Zhu, T. H., Liu, G. H., & Mao, C. (2012). Co-inoculation with phosphate-solubilzing and nitrogen-fixing bacteria on solubilization of rock phosphate and their effect on growth promotion and nutrient uptake by walnut. *European Journal of Soil Biology*, 50, 112–117.

314. Yue, H., Mo, W., Li, C., Zheng, Y., & Li, H. (2007). The salt stress relief and growth promotion effect of Rs-5 on cotton. *Plant and Soil*, 297, 139–145.

315. Zamioudis, C., & Pieterse, C. M. (2012). Modulation of host immunity by beneficial microbes. *Molecular Plant-Microbe Interaction*, 25, 139–150.

316. Zhang, H., Kim, M. S., Krishnamachari, V., Payton, P., Sun, Y., Grimson, M., Farag, M. A., Ryu, C. M., Allen, R., Melo, I. S., & Paré, P. W. (2007). Rhizobacterial volatile emissions regulate auxin homeostasis and cell expansion in *Arabidopsis. Planta*, 226, 839–851.

317. Zhang, H., Kim, M. S., Sun, Y., Dowd, S. E., Shi, H., & Paré, P. W. (2008a). Soil bacteria confer plant salt tolerance by tissue-specific regulation of the sodium transporter *HKT1*. *Molecular Plant-Microbe Interaction*, 21, 737–744.

318. Zhang, H., Murzello, C., Sun, Y., Kim, M. S., Xie, X., Jeter, R. M., Zak, J. C., Dowd, S. E., & Paré, P. W. (2010). Choline and osmotic-stress tolerance induced in *Arabidopsis* by the soil microbe *Bacillus subtilis* (GB03). *Molecular Plant-Microbe Interaction*, 23, 1097–1104.

319. Zhang, H., Sun, Y., Xie, X., Kim, M. S., Dowd, S. E., & Paré, P. W. (2009a). A soil bacteria regulates plant acquisition of iron via deficiency-inducible mechanisms. *Plant Journal*, 58, 568–577.

320. Zhang, H., Xie, X., Kim, M. S., Kornyeyev, D. A., Holaday, S., & Paré, P. W. (2008b). Soil bacteria augment *Arabidopsis* photosynthesis by decreasing glucose sensing and abscisic acid levels in planta. *Plant Journal*, 65, 264–273.

321. Zhang, S., Lu, S., Xu, X., Korpelainen, H., & Li, C. (2009b). Changes in antioxidant enzyme activities and isozyme profiles in leaves of male and female *Populus cathayana* infected with *Melampsora larici-populina*. *Tree Physiology*, 30, 116–128.

322. Zhao. Y., Bian, S. M., Zhou, H. N., & Huang, J. F. (2006). Diversity of nitrogenase systems in diazotrophs. *Journal of Integrative Plant Biology*, 48, 745–755.
323. Zhu, J. K. (2000). Genetic analysis of plant salt tolerance using *Arabidopsis*. *Plant Physiology*, 124, 941–948.
324. Zhu, J. K. (2002). Salt and drought stress signal transduction in plants. *Annual Review of Plant Biology*, 53, 247–273.

CHAPTER 9

SUSTAINABLE RECLAMATION AND MANAGEMENT OF SODIC SOILS: FARMERS' PARTICIPATORY APPROACHES

Y. P. SINGH

CONTENTS

9.1 INTRODUCTION

Water logging and soil salinity has been a serious concern for global agriculture throughout the human history [13]. Both are persistent ecological issues and major stress factors [2, 3, 18], which affects plant growth and productivity especially in the arid and the semiarid regions [30]. Apart from affecting the plants, salinity affects economic welfare, environmental health and agricultural production [17] endangering food security and sustainability. Although worldwide figures of actual salt affected areas vary widely but as per Food and Agricultural Organization (FAO) Land and Nutrition Management Service [7], about 800 million ha (M ha) area suffers from salinity and is spread in 100 countries covering over 6% of the world's land. Ghassemi et al. [9] reported that soil salinization causes huge losses in annual income assessed at about US$12 billion globally. Unfortunately, the salinity problem is becoming more prevalent all over the world because of the extensive utilization of lands and expansion of irrigation [6, 14]. Expansion of irrigation coupled with poor drainage has exacerbated the rate of soil salinization in most of the irrigated soils [27]. Although millions of hectares of such salt affected soils are suited to agricultural production yet remain unexploited because of excessive salts or sodicity in the soil profile along with other soil and water related problems [1]. According to FAO, salinization of arable land is likely to result in 30% to 50% land loss by the year 2050, if no remedial actions are initiated in right earnest.

In India, salt affected soils occupy about 6.73 M ha. Indo-Gangetic plains comprising of the states of Punjab, Haryana, Uttar Pradesh and part of Bihar (North), West Bengal (south) and Rajasthan (north) alone having about 2.7 M ha [15,16]. Salt affected soils in Indo-Gangetic plains has excessive exchangeable Na^+ associated with high pH (>8.5) that impairs the physical condition of the soils and adversely affects water and air movement, nutritional and hydrological properties of the soils [8, 10, 11, 24, 25]. So far, a sizeable area in the Indo-Gangetic plains has been reclaimed through physical, chemical and biological amelioration techniques and is now supporting intensive rice-wheat cropping system with a productivity of about 10 t ha^{-1}. The people are now quite sensitized and wish to ameliorate such soils to improve livelihood yet high cost of reclamation of about 50,000 Rs./ha (1 US$ = Rs. 60.00) often hinders the implementation of the technological package. Stakeholders living below poverty line or even marginally above cannot afford the reclamation cost because the money required to reclaim a unit ha area and to undertake initial crop production on reclaimed lands may be equal to the total income of one to two years of such households. Salt affected land owners belong to most disadvantaged population group having small land holdings. Their quality of life and social profile are amongst the lowest because of low per capita income, lesser opportunity for employment and economic activities resulting in their migration in search of livelihood.

Top-down decision making with poor dialog with the beneficiary participants has been the major weakness of the current programs of reclamation and management of salt affected soils. Farmer's participatory management involving stakeholders in the decision making process from the beginning itself (planning stage) to the end of the project can prove to be a sustainable model for enhancing farm productivity and livelihood security of the resource poor farmers.

The overview in this chapter highlights participatory approaches developed and used to manage resources (soil, water, and crop, nutrient) and to establish diverse and profitable agricultural systems on salt affected soils. The overview especially deals with farmers participation in reclamation of alkali or sodic soils (with and without water logging) as these are synonymously referred (but hereinafter only referred as sodic soils) and their matching management practices, post reclamation management, developing salt tolerant varieties and integrated farming system for waterlogged sodic soils. It is emerging that such kind of involvement empowers the stakeholders in decision making while researchers and extension workers providing support to the program as and when required.

9.2   PARTICIPATORY APPROACHES FOR RECLAMATION

After the first concerted effort to reclaim sodic lands in Uttar Pradesh (India) under Phase I and Phase II, most leftovers lands are now in the possession of resource poor farmers belonging to most economically backward communities or the farmers who have been allotted salt affected lands under a government policy to allot lands to landless people. They have become first time land owners. To take care of their problems, UP *Bhumi Sudhar Nigam* (UPBSN) initiated an integrated reclamation program following farmer's participatory approaches. The protocol for sustainable reclamation of salt affected soils developed by Central Soil Salinity Research Institute (CSSRI), Karnal has been adopted. At the village level, a site implementation committee (SIC) is formed, each SIC having several water users groups (WUG) controlling 4 ha command around a tube well. In each village, one *Mitra Kisan (friend farmer)* and one *Mahila Mitra Kisan (Lady friend farmer)* and animators in other fields like animal husbandry, agronomy, fish farming and boring mechanics, etc., are identified to help farmers to reduce their dependency on the external agencies. All the decisions on group formation and implementation are taken by the farmers participating in the SIC meetings. The success of the on-going reclamation program is attributed to the participatory methodology in the project implementation which results in transparency, decentralized decision making, capacity building at village level and accountability of all actors at every level. Community led micro planning, participatory execution, community led monitoring and evaluation, community led maintenance of drainage network and farmers to farmers technology dissemination are the focal points of the approach followed in the project. Since success and sustainability of any reclamation program greatly depends upon post reclamation management activities, involvement of farmers continues from pre-reclamation to post reclamation activities. List of farmers participatory activities are given in Figure 9.1.

The whole implementation strategy of the project includes:
- Effective technology dissemination;
- Human resource development (beneficiaries and staff);
- Monitoring and evaluation (internal and external).

Under the exit protocol, it is ensured that the whole village is self-reliant and has demand driven farmers to farmer's extension system by and within the community itself and other organizations engaged in rural development. It is believed that it should lead to sustainable reclamation and crop production wherein the targets are not based on quantity (say area reclaimed) but on quality.

9.3   PROJECT ACTIVITIES

9.3.1   AREA SELECTION

As per this approach, selection of area is made on the basis of wasteland maps of salt infested villages prepared by Remote Sensing Application

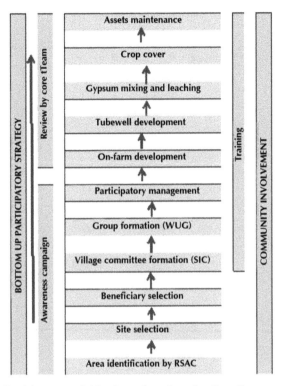

FIGURE 9.1   Participatory activities for reclamation of sodic soils.

Centre (RSAC). Cadastral maps collected from revenue department are sent to RSAC for classification. The classification of land is done on the basis of land use through the satellite imagery using IRS-KISS-III satellite. The soils of the selected sites are categorized into three groups namely B⁺, B and C categories according to the land use. The salt affected lands being utilized for double cropping but with poor yields are categorized as B⁺ category. The lands being utilized for single crop with poor yield are categorized as B whereas completely barren lands unable to support any crop are categorized as C category land. The details of sodic land categorized under phase I to phase III (1994–2015) of Uttar Pradesh Sodic Land Reclamation Project are given in Table 9.1. Apparently, 80% of the categorized lands fell under the C category (Table 9.1).

TABLE 9.1 Classification of sodic land under UP land reclamation projects

Year	No. of Districts	Villages	Sodic area (ha) B+	B	C	Total
Phase I						
1994–2001	10	785	6469.7	12558.3	50001.5	69029.5
Phase II						
2001–2007	21	3369	28868.1	63685.6	245874.6	338428.3
Phase III						
2010–11	22	463	5092.96	9353.36	29427.54	43873.87
2011–12	28	440	1659.12	5315.14	32250.57	39224.83
2012–13	26	389	1140.88	2873.17	19978.32	23992.37
2013–14	30	435	2069.06	2541.02	37837.36	42447.44
2014–15	25	361	2711.62	2358.00	21017.24	26086.86
Total		2,088	12,673.64	22,440.69	1,40,511.03	1,75,625.37

9.3.2 SELECTION OF BENEFICIARIES

Following criteria was used to include the beneficiaries in the project:
- He/She must have salt affected land having pH more than 8.5
- He/She should not be a share cropper
- He/She must be residing of the selected village
- He/She should be prepared to work in a group for common interest and follow the technical advice given by UPBSN

9.3.3 FORMATION OF SITE IMPLEMENTATION COMMITTEE (SIC)

A SIC is constituted at village level comprising of all the beneficiaries, *Gram Pradhan* being the Ex-officio chairman of the committee. SIC serves as a forum for awareness generation, exchange of information, program implementation and monitoring through implementing bodies like WUGs and Core team that also comprises of beneficiaries. Each SIC has a bank account wherein funds collected from the beneficiaries for maintenance of drainage infrastructure of the village are deposited.

9.3.4   WATER USERS GROUP (WUG)

The beneficiaries themselves with the help of reclamation agency form WUGs. Each WUG is formed to cover 4 ha area in geographical proximity. Numbers of beneficiaries in each WUG vary according to the size of holdings. Project's extension strategy revolved around WUG instead of individual contact for planning, execution, monitoring and evaluation. The basic purpose of formation of WUG was irrigation and drainage management within the WUG's lands. For capacity building and empowering women, special efforts are made to address women's need through constitution of women self-help groups. These groups are linked with commercial banks for credit facility.

9.3.5   SITE IMPLEMENTATION PLAN (SIP)

A Site Implementation Plan is prepared for each village using Participatory Rural Appraisal Technique. The Main purpose of preparation of SIP was to make planners, land owners and users aware of the status of natural resources along with their spatial distribution and also to know the problems that affected productivity. On the basis of SIP, an action plan is prepared for development activities and accordingly memorandum of understanding (MOU) is developed and signed between the *Mitra Kisan* on behalf of all the beneficiaries and agency officials. All activities to be carried out by beneficiaries and reclaiming agency are clearly spelt out in the MOU.

9.3.6   SOIL AND TOPOGRAPHICAL SURVEYS

Detailed survey including reconnaissance and topographical surveys are conducted involving all the beneficiaries for the purpose of resource mapping, recording the socio-economic status of beneficiaries and for planning drainage and irrigation networks. Before initiating the reclamation work ground water levels and quality of ground water is also ascertained through selective sampling.

9.3.7 ON-FARM DEVELOPMENT WORKS

After the sites are selected, the main project activities comprises of the design and rehabilitation of the drainage network, provision of shallow tube well with pump sets for irrigation to have an assured water resource to leach the salts and support crop growth; on farm development consisting of land leveling, *bunding* (formation of field dykes) and construction of field drains and irrigation channels and provision of chemical and organic amendments to make salt affected soil free from exchangeable sodium salts. These activities are followed by crop cultivation that includes selection of crops and cropping patterns, crop varieties and appropriate cultural practices. Project funds towards land development, leaching, irrigation infrastructure and drainage are channelized through bank account of the WUGs.

9.3.8 DRAINAGE: THE KEY ELEMENT

Since the water intake rate of sodic soils is very low initially and continues to be slow even several years after reclamation (semi-reclaimed lands), water is likely to accumulate on the land surface following a heavy irrigation or a rainstorm, which is not uncommon in the regions having sodic soils. The stagnation of water reduces the yield even in semi-reclaimed lands as revealed by the data given in Table 9.2 showing 2–13% decrease in yield if water stagnates for 1 day which accentuates to 20–48% if the water stagnates for 6 days [12].

When the soil surface dries up, a crust is likely to be formed that also affects soil aeration. It is because of these reasons; irrigation and drainage activities are of paramount importance. The works related to the construction of link drains are distributed to WUGs in SIC meetings and constructed under the supervision of technical staff. Each WUG is paid for the work executed on the link drain on the basis of work measured and the rate prescribed for. The fields are *bunded* and leveled to ensure: a) uniform distribution of amendment, b) prevent run off from the fields after the amendment application and c) distribute water uniformly for leaching and moisture storage in the soil profile. The cost of field *bunds* and irrigation channels are borne by the farmers, although little incentive of Rs. 100.00

TABLE 9.2   Yield in Drained Plots and Yield Reduction Due to Water Stagnation for Tested Crops

Crop	Yield (t ha^{-1}) Drained	Yield reduction[#] over drained (%) for water stagnation (days)			
		1	2	4	6
Barley	3.65	4	7	13	25
Berseem (seed)	0.48	2	21	35	48
Mustard	1.43	8	16	22	29
Pearl millet	2.22	6	15	22	27
Pigeon pea	1.52	4	14	18	21
Sorghum	4.13	3	11	16	20
Sunflower	1.86	13	19	26	30
Wheat	4.20	8	17	27	39

[#] Yield reduction due to water stagnation is relative to drained plots.

(1 US\$ = Rs. 60.00) and 125 Rs./ha for field drains construction and 400 Rs./ha for land leveling is given to the project farmers. These activities are carried out under the supervision of reclamation agency.

9.3.9   AMENDMENT APPLICATION AND LEACHING

Sodic soils have poor physical structure because of high exchangeable sodium and that makes these lands almost impervious to water. The high sodium on the exchange complex must be replaced with calcium to leach the soluble and exchanged salts out of root zone. Calcium sulfate (gypsum, $CaSO_4 2H_2O$) is often used to replace sodium being the cheapest source in India for this purpose. This input is supplied to the farmers by the state reclamation agency on the basis of category of land holding. Each WUG than accomplishes leaching using tube well or canal water as may be the case for about 7–10 days.

9.3.10   SELECTION OF CROPS AND CROPPING SEQUENCES

Involvement of the farmers in selection of proper crops and varieties at the initial stages is very crucial because crop differs widely in their tolerance to soil salinity/ sodicity. Some crops are sensitive, whereas others are semi tolerant or tolerant to a given level of sodicity (Table 9.3). Field tests on some of the oil seed crops like sunflower, mustard, safflower, linseed, groundnut, soybean and sesame revealed that mustard, rapeseed and sunflower were moderately tolerant, linseed and groundnut semi tolerant and rests of the crops were sensitive to sodicity.

The selected crops should not only be tolerant but should also initiate self-reclaiming process in the soil. As such, rice in monsoon (*Kharif,* June–October) season is most ideal as the first crop because it can tolerate standing water, is tolerant to salinity/sodicity (transplanted rice), has extensive shallow root system and is able to accelerate availability of native Ca for replacement of exchangeable Na through root activities. In the winter (*Rabi*, November–April) season only shallow rooted crops like wheat, barley and Egyptian clover (*berseem*) should be grown in the initial years [31]. In fact, earlier studies in Uttar Pradesh have established that rice-*berseem*, rice-wheat or rice-barley proved to be better sequence than others. Recent studies with various rice based cropping sequences have shown that rice-wheat-*dhaincha* (*dhaincha* for green manuring) and rice-*berseem* cropping sequences are more remunerative in sodic soils. One of these rotations should continue for at least first three years without leaving

TABLE 9.3   Relative Crop Tolerance to Sodicity

Exchangeable sodium percentage (ESP) range		
30–50, Moderately tolerant	**20–30, Semi tolerant**	**<20, Sensitive**
Barley	Linseed	Bengal gram
Mustard	Garlic	Soybean
Rapeseed	Sugarcane	Maize
Wheat	Cotton	Safflower
Sunflower	Cluster bean (*Guar*)	Peas
Sorghum	Groundnut	Lentil
Persian clover (*Shaftal*)	Onion	Pigeon pea
Egyptian clover (*Berseem*)	Pearl millet	Black gram (*Urd* bean)
	Holy basil (*Tulsi*)	
	Matricaria	
	Fababean (*Bakla*)	

the fields fallow to ensure continuity of the reclamation process as well as to avoid reversion of sodic conditions. Since, growing of rice is known to promote a more favorable physical condition in sodic soils, studies on relative merits of rice and sorghum (fodder based cropping sequences) based cropping sequences revealed that rice based cropping sequences were better than sorghum based cropping sequences in terms of yield as well as degree of reclamation of sodic soils.

Within the crops, varieties also differ in their tolerance to salts/ESP as indicated by pH (Table 9.4). To sensitize the farmers on this issue, farmer's managed participatory adaptive trials on evaluation of the performance of salt tolerant and traditional varieties of rice and wheat were conducted on 20 farmer's field in Santaraha village of Hardoi district. Seed of salt tolerant varieties along with technical folder containing package of practices for cultivation of rice and wheat were given to the farmers and all the cultural practices for both salt tolerant as well as traditional varieties were managed by the farmers themselves. The yield enhancement of wheat with salt tolerant varieties over traditional varieties ranged from 9.7 to 62.4% (Table 9.5).

TABLE 9.4 Promising Tolerant Crops and Their Varieties/Identified/Developed by CSSRI, Karnal

Crops	Varieties	Tolerance level
Barley	CSB 1, CSB2, CSB 3, Ratna	pH 9.3
Gram	Karnal *chana* (chick pea) No. 1	pH 8.8–9.0
Mustard	CS52, CS54, CS56	pH 9.0–9.2
	Pusa bold, Varuna, Kranti	pH 8.8–9.0
Rice	CSR10	pH 9.8–10.2
	CSR 13, CSR 23, CSR 27 CSR 36, CSR 43	pH 9.4–9.8
	CSR30 (Basmati)	pH 9.4
Sugarbeet	Ramonaskaya-06, Polyrava-E	pH 9.5–10
Sugarcane	CO453, CO1341, CO6801, CO62329, CO1111	pH < 9.0
Wheat	KRL 1–4, KRL 19, KRL 210, KRL 213, WH157,	pH 9.2–9.3
	Raj 3077	pH 8.7–9.0
	HD2009, HD2285, PBW343, HD2329,	

TABLE 9.5 Grain Yield of Salt Tolerant and Traditional Varieties of Wheat

S.No.	Name of the farmer	Area sown (ha)	Grain yield of salt tolerant variety	Grain yield of traditional variety	Increase over traditional varieties (%)
1	Hetum	0.379	37.2	24.6	51.2
2	Ramdas	0.935	30.4	21.5	41.4
3	Jwala Prasad	0.506	38.0	23.4	62.4
4	Chhote Lal	0.253	28.2	22.5	25.3
5	Ali Raza	0.253	34.3	27.3	25.6
6	Kuldeep	1.0	35.0	28.4	23.2
7	Rakesh	1.0	32.0	24.6	30.1
8	Shanti Devi	1.0	28.3	25.8	9.7
9	Rajbhadur	0.277	29.2	21.6	35.2
10	Suvedar	0.594	32.5	27.4	18.6
11	Ramutar	0.529	42.2	32.5	29.8
12	Rajendra	0.683	47.0	36.2	29.8
13	Bhure	0.202	30.5	24.4	25.0
14	Rajaram	0.253	36.4	28.8	26.4
15	Ramchandra	0.576	9.8	3.5	46.6
16	Shivratan	0.364	38.2	32.6	17.2
17	Balla	0.288	36.3	32.5	11.7
18	Ramadhar	1.175	13.5	11.7	15.4
19	KusumTandan	0.935	39.3	32.9	19.4
20	Prihalad	0.273	38	32.2	18.0
	Total area/average yield	11.475	32.8	25.7	27.6

After the salt tolerant varieties of rice/wheat are grown in sequence for three years, the surface soil (0–15 cm) pH is reduced to make it possible to grow high valued crops like oil seed (mustard, linseed) and medicinal and aromatic crops (*Tulsi*, Matricaria and garlic) in the subsequent years. Studies conducted to determine the time frame for crop diversification revealed that diversification of rice-wheat cropping system depends on the extent of reclamation although it can be possible after three years of continuous rice-wheat/*berseem*/mustard cropping systems (Table 9.6).

TABLE 9.6   Cropping Sequences Recommended at Different Stages of Reclamation

Likely soil pH$_2$	Reclamation period (years)	Cropping sequences
9.2–9.8	1–3	Rice-Wheat-*Dhaincha*, Rice-*Berseem*, Rice-Mustard, Rice-Barley
9.0–9.2	4 to 5	Sorghum-Wheat or Mustard, Pearl millet-Wheat or Mustard, Cotton-wheat, Groundnut-Mustard or wheat
8.8 to 8.9	6 to 8	Sunflower-Wheat, Maize-Linseed, *Tulsi*-Matricaria, Chilli-Garlic, Sorghum-Mustard, Sugarcane, Sorghum-Gram or Pea
8.6 to 8.7	9 to 10	Pearl millet-Lentil or Gram, Pigeon pea-Wheat, Soybean-Wheat
8.5 to 8.6	After 10	Any cropping sequence including vegetables and flowers

9.3.11   NUTRIENT AND WATER MANAGEMENT APPROACHES

Time and method of irrigation and fertilizer application in sodic soils are quite different than the normal soils and have a bearing on crop yields. Thus, improved cultural practices were propagated through the WUGs as follows:

Rice being the principal crop in sodic soils, it requires submerged moisture regime for optimum grain yield. Application of 7 cm irrigation water after a day of disappearance of ponded water, produced higher grain yield as compared to continuous submergence [19]. The yield of wheat crop significantly improved when first irrigation at crown root initiation (CRI) stage was given 30 days after sowing (DAS) than at 21 DAS in sodic soil. Five irrigation's scheduled at CRI, tillering, jointing, milking and dough stages resulted in higher yield, which was closely followed by treatment in which irrigation at tillering and dough stages were skipped and flowering stage was added. Besides, to avoid water stagnation, light and frequent irrigation are recommended for which emphasis is placed on proper leveling especially through laser land leveling.

Based on field experimental results, around 20–25% higher amount of N-fertilizer than the recommended dose of N for normal soils is recommended because of high N losses through volatilization. The optimum dose of nitrogen for rice and wheat in sodic soils has been found to be 150 kg ha⁻¹ except in case of short duration varieties of rice, where it could be reduced to about 120 kg ha⁻¹. It is also recommended that N should be applied in split applications. In rice and wheat crops, nitrogen should be applied in three splits, half or one third at transplanting/sowing and remaining in two equal splits at 3 and 6 weeks after transplanting/sowing. If available, 10 t ha⁻¹ farm yard manure/compost should be applied at the time of land preparation. No response to phosphorus application has been reported in rice- wheat cropping sequence initially for 3 to 4 years. Sodic soils are generally deficient in zinc and most of the crops respond favorably to application. Application of zinc at the rate of 25 kg $ZnSO_4$ ha⁻¹ on a regular basis to rice- wheat/*berseem* crop sequence was sufficient to produce significantly higher yields.

9.4   IMPACT OF FARMERS PARTICIPATORY APPROACHES

9.4.1   SOIL RECLAMATION

Soil reclamation in terms of reduced pH is an indicator of soil quality and health, lower the pH better the soil health. Studies revealed that with the adoption of this approach the pH value of the reclaimed fields came down (Table 9.7). Before reclamation, about 87% of the B class plots had pH of more than 9.0 but after five years no plot had such high pH. In the case of C class plots about 78% had pH more than 9.0 before treatment but after the participatory reclamation protocol only a small percent (2.33%) of C class plot has pH more than 9.0. The average pH of C class land has come down from 9.5 to 8.1 after five years.

9.4.2   CROP YIELDS

Before reclamation paddy yield under B class soils was about 1.5 t ha⁻¹ whereas under C class soils the farmers were not able to cultivate the fields. After reclamation, all categories of salt affected lands came under paddy cultivation with paddy yield increasing to about 3.0 t ha⁻¹ during first year, which further increased to 4.0 t ha⁻¹ over the years. Farmers,

TABLE 9.7   Percentage Distribution of Reclaimed Plots by pH Range [22]

pH value	B category plots (%)		C category plots (%)	
	Pre-reclamation status	Post-five years status	Pre-reclamation status	Post-five years status
7.0–7.5	0.00	12.73	0.00	7.60
7.5–8.0	0.00	60.00	0.58	54.39
8.0–8.5	9.09	20.00	2.34	29.83
8.5–9.0	3.64	7.27	19.88	5.85
9.0–9.5	47.27	0.00	30.41	2.33
9.5–10.0	32.73	0.00	36.26	0.00
>10.0	7.27	0.00	10.53	0.00
Average pH	9.5	8.0	9.5	8.1

who adopted salt tolerant varieties like CSR 36 and CSR 43, could even achieve higher yields of 5.0 t ha^{-1} with recommended dose of fertilizers and matching management practices. Wheat yield with salt tolerant varieties KRL 210 and KRL 213 was up to 3.0–3.2 t ha^{-1} during first year of reclamation, which increased to 4.0 t ha^{-1} over the year.

9.4.3   CROPPING PATTERNS AND CROPPING INTENSITY

While the C category lands could be greened where nothing grew before (Figure 9.2), there was significant improvement in crops yield of paddy and wheat or other crops like oil seeds, pulses, vegetables and fodder crops grown in a very limited area in B and B+ category lands. After five years of reclamation, the proportion of gross sown area under paddy and pulses declined while under wheat/barley, oilseed, vegetables, spices and fodder crops increased because of improved soil quality and increasing area under cultivation. Reclamation provided opportunity to the farmers to diversify the traditional rice-wheat cropping system with high value crops like medicinal and aromatic crops and sugarcane. It also increased the cropping intensity, being 200% in most cases as revealed by the data from Santaraha village (Table 9.8). Some farmers used short duration rice variety and cultivated a short duration crop like *toria*, spinach, etc., in between rice and wheat to have a cropping intensity of 300%.

TABLE 9.8   Cropping Intensity (%) by Soil Sodicity Class [26]

Soil sodicity class	Pre-reclamation (2009–2012)	Post reclamation (2012–2014)
Normal	198.5	198.5
Slight (B+ class)	191.5	199.7
Moderate (B class)	100	199.9
Severe (C class)	0.0	200.0
Annual average	122.5	199.5

9.4.4   EMPLOYMENT GENERATION

One of the important benefits that did stem from reclamation was employment generation, which provided work to the land less laborers and marginal and small farmers right in the village, eliminating their migration outside for livelihood. The employment generation in the first year of reclamation was about 165 person-days ha^{-1} (B and B+ lands). It ranged between 207–237 person-days ha^{-1} in areas with low degree of mechanization of agricultural operations.

9.4.5   LAND VALUE

The value of land, besides a social asset, decides the credit worthiness of the farmers. The increased value of land therefore is an important although intangible benefit of land reclamation. The higher land value is attributed to improved health and crop productivity. It was observed that after five years of reclamation, the value of B category salt affected soils increased by 48% whereas value of C category land rose by 317%. C category land being totally unproductive, its value zoomed after reclamation. Earlier, Tripathi [28] also reported that land value increased from 115,000 to 170,000, 55000 to 155,000, and 33,000 to 125,000 for B+, B and C class lands in Uttar Pradesh.

The overall impact on the landscape as brought out by land reclamation is shown in Figure 9.2, which reflects upon the discussions made in previous sections.

Before Reclamation   **After Reclamation**

FIGURE 9.2   Overall impact of participatory reclamation and management approaches on landscape after reclamation.

9.5   PARTICIPATORY APPROACHES TO VARIETAL DEVELOPMENT

Farmer's Participatory approaches to varietal selection have been reviewed by Weltzein et al. [29] and described for many crops and proven to be effective in increasing genetic diversity and harnessing productivity potential of crops [4]. In view of these advantages and to take advantage of the rapport developed between the stakeholders, reclamation agency and the research organization, a Participatory Varietal Development Program was initiated at ICAR-central Soil Salinity Research Institute, Regional Research Station, Lucknow to develop high yielding, salt tolerant, adaptable and acceptable rice variety for salt affected especially sodic areas through participatory varietal selection (PVS) approach [21, 22]. The complete process of farmer's participatory varietal development is given in Table 9.9.

In conventional methods of varietal development, plant breeders develop rice varieties through on station trials and often consider yield, ability to withstand salt stress, flowering duration and height as important traits. However, farmers may have some other important considerations as is often reflected in the mismatch between the breeding lines selected by the farmers and lines offered by the scientists. Farmers also emphasize on traits such as straw yield, suitable plant height for easy threshing, hardiness in threshing and disease resistance etc., which many times

TABLE 9.9   PVS Methodology Adopted for Varietal Development

2001 screening of genotypes	126 genotypes (18 selected)	Multi-location testing (2 location)
2002 PVS of selected genotypes	18 genotypes (12 selected)	Multi-location testing (5 location)
2003 PVS of selected genotypes	12 genotypes (9 selected)	Multi-location testing (6 location)
2004 PVS of selected genotypes	9 genotypes (6 selected)	Multi-location testing (6 location)
2005 Final selection	6 genotypes (2 selected)	Multi-location testing (6 location)
2006-2007 Baby trials	2 genotypes planted and compared with traditional varieties	60 locations in two districts
1 selected (CSR-89IR-8)		
Released as 'CSR 43'		

are ignored by the breeders [5]. Moreover, varieties which perform well in controlled well-managed on-station experiments might yield lower or even fail in farmers' fields, where situation is more challenging as a result of wide differences in the level of management and resources use.

A large number of genotypes, 126 in all, including local types, genotypes collected from different parts of the country and the advanced materials developed by the CSSRI were screened through on-station and on-farm trials at pH ranging from 9.6–9.8. On the basis of grain yield, eighteen genotypes/varieties were selected and evaluated next year through farmers' participatory trials under two on-station (Shivri and *Krishi Vigayan Kendra* (KVK), Ainthu) and two on-farm trials (Dhora and Itaara) where pH_2 ranged from 9.4 to 9.7. Preference analysis was conducted at all the four locations following the participatory varietal selection (PVS) protocol. Farmers were asked to vote for two best and two worst verities according to their choice of the desired trait. They were also asked to briefly state the reasons for selecting the best and worst genotypes. After the voting, preference scores were computed.

On the basis of preference ranking /voting of the farmers, twelve geno-types/ varieties were again planted under three on station (Shivri, KVK Dhora and KVK Ainthu) and two on-farm (Dhora and Itaara) trials having pH_2 9.5–9.7. The data for plant height, tillers/plant, days to 50% flowering, panicle length and grain yields were collected and voting for preference analysis was arranged at two locations following the PVS protocol (Figure 9.3). Based on grain yield and preference score, nine top yielders were again planted at 6 locations (three on-station and three on-farm) at pH_2 9.2–10.3 in Unnao, Pratapgarh and Raebareli districts of Uttar Pradesh.

At the time of maturity, preference analysis was done by the farmers. Thirty farmers including 16 male (52%) and 15 female (48%) from the neighboring villages having similar kind of sodicity problems participated in the preference analysis. On the basis of preference ranking, the list was pruned to six genotypes, which were again evaluated at six locations, three on-station and three on-farm. Forty-three farmers including 32 male (74%)

FIGURE 9.3 A view of the participatory varietal selection exercise at site.

and 11 female (26%) took part in preference ranking exercise organized at one on-farm trial. To monitor the post-harvest evaluation for cooking and eating qualities of the genotypes identified through PVS, sensory analysis was also conducted. Forty farmers including 25 women and 15 men participated in this exercise.

On the basis of this PVS exercise, one most preferred salt tolerant genotype (CSR-89IR-8) was selected by the farmers. Later it was released as salt tolerant variety "CSR 43." This variety is highly preferred by the farmers and was adopted by a large number of farmers even before its release as a variety. The yield advantage over locally cultivated varieties at 20 locations shows the superiority of the variety (Figure 9.4).

9.6 PARTICIPATORY APPROACH FOR MANAGEMENT OF WATERLOGGED SODIC LANDS

Traditional approaches to reclaim waterlogged sodic soils pose several insurmountable problems because of the combination of shallow water table and water stagnating on the land surface (Figure 9.5). Although the pH can be brought down with the application of gypsum and even

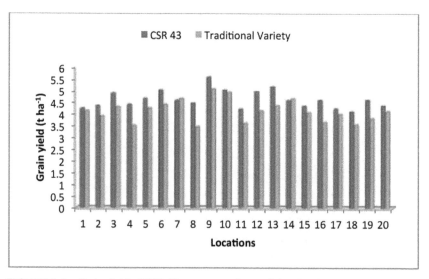

FIGURE 9.4   Grain yield of CSR 43 as compared to traditional varieties on farmer's field.

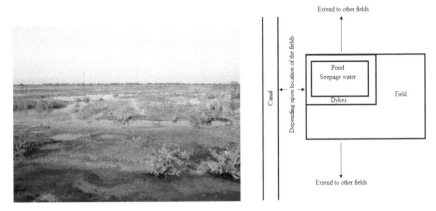

FIGURE 9.5   View of waterlogged salt affected soils in the command.

crops can be grown for a year or two but thereafter soil again becomes sodic. Since such a situation is spread over large areas especially in Uttar Pradesh (India), a need was felt to develop an alternate approach for the sustainable reclamation and management of such lands. Land modification was hypothesized as one of the effective methods to control water logging and reclaim sodic soils of canal commands [23]. A farmer's participatory pond based integrated farming system approach was adopted and found effective.

9.6.1   POND BASED INTEGRATED FARMING SYSTEM APPROACH

A farmers' participatory pond based integrated farming model of 1.0 ha comprising of: (a) 0.40 ha on-farm pond; (b) 0.25 ha under field crops; (c) 0.25 ha under fruit crops; (d) 0.10 ha under forage; and (e) 0.10 ha under vegetable crops, was developed at the Sharda Shahayak canal command at a site situated at 26°23' N latitude to 81°14'E longitude and 10 m above mean sea level. The pond (80 m x 50 m) was excavated to a depth of 1.75 m depth, at about 200m distance from the canal (Figure 9.5). The pond was so designed that it could harvest maximum seepage water and sustains the fish farming without the need of water from outside source. The excavated soil was used to construct dykes and spread over the adjoining field

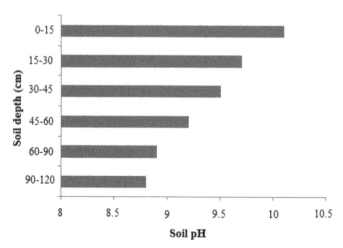

FIGURE 9.6   Variation in soil pH with depth at the study site.

putting the top soils at the bottom and the bottom soil at top. It was in fact a physical reclamation as the bottom soil had lesser pH than the top soil (Figure 9.6). As a result, the pH of the surface soil (0–15 cm) after land modification reduced to 9.1.

Aquaculture was practiced in the pond; the species composition for the first year being Rohu 30%, Catla 25%, Mrigal 25% and Silver carp 20%. The slopes of the embankment and the elevated field beds were utilized for the eucalyptus plantation, which served as bio-shield and bio-drainage purpose in the system. The slope of the pond was also used for cultivating fodder grasses. Farmer's participation in the project was assured from the initiation of the project to the completion of the project with the conditions that he will undertake all the field activities from sowing to harvesting including management of fish production and its timely disposal in con-sultation with the project scientists.

9.6.2   PERFORMANCE OF THE SYSTEM

Rice, wheat, sorghum, mustard, garlic and onion were grown successfully on elevated field beds and pond embankments. In the first year, 4.0 and 2.7tha^{-1} yield of rice and wheat were obtained, respectively. Sorghum and *berseem* crops were grown to meet the fodder requirement of dairy animals.

In the first year forage yield of 15.0 t ha^{-1} and 15.4 t ha^{-1} of sorghum and *berseem* was recorded. All these activities could be successfully carried out using canal seepage water alone. Benefit-cost ratios of different cropping systems are given in Table 9.10. Maximum benefit: cost ratio of 15.60 was obtained for garlic followed by 4.42 for onion. Apparently, wheat had the lowest B: C ratio followed by rice. Overall benefit cost ratio of the system was quite impressive, being 3.64.

9.7 SUMMARY

Disadvantaged and underprivileged sections of the rural society require special attention in any welfare program of the government. As such a participatory reclamation and management program was initiated by UPBSN, CSSRI regional Research Station, Lucknow with the stakeholders owning salt affected sodic soils. The participatory mode was used from the beginning to the end of the project before exiting under a well-defined exit code. Site implementation committees, Water Users Groups and self-help groups were created in each village to undertake the projects. The site implementation plan as approved by the SIC is implemented by all the stakeholders as per MoU. In the reclamation process, while on-farm development was entrusted to WUGs, amendment was supplied by the reclamation agency at subsidized rates. The technological inputs on water

TABLE 9.10 Economic Analysis of Various Components of Farmer's Participatory Integrated Farming System Model

Crops	Area (m^2)	Yield (kg)	Gross income (Rs./ha)	Cost of pro- duction (Rs.)	Net returns (Rs.)	B:C ratio
Fish	4000	849.0	40970.0	8500.0	32470	4.82
Garlic	92	110.0	8800.0	264.0	8536.0	15.60
Mustard	531	65.0	1430.0	594.0	836.0	2.43
Onion	266	200.0	3000.0	678.0	2322.0	4.42
Rice	2473	950.0	8075.0	4278.0	3797.0	1.88
Sorghum	531	1063.0	1063.0	392.0	671.0	2.71
Wheat	2473	372.0	4389.0	3857.0	532.0	1.13
Grand total	-	-	67727.0	18563.0	49164.0	3.64

management and nutrient management including varietal selection were facilitated by the CSSRI, Regional Research Station. The complete package in participatory mode led to reduced pH (reclamation), higher crop yield, employment generation and increase in land value.

All the economic parameters were found encouraging to claim that land reclamation is economically viable and socially acceptable. Farmers participatory approach changed the life of the stakeholder's altogether and enhanced their socio-economic status. The special venture of varietal development through PVS has been a success with release of the variety CSR 43, but more so because it inculcated the scientific approach in the community and enhanced the chances of technology adoption in a sustainable manner. For waterlogged sodic lands, an on-farm pond technology has been developed and hopefully it will get upscaled soon as the work in this direction is progressing well.

KEYWORDS

- alkali soils
- amendment
- aquaculture
- cadastral maps
- canal seepage
- crop salt tolerance
- crop selection
- cropping intensity
- cropping pattern
- drainage
- economic analysis
- employment generation
- ESP
- farm pond
- farmers' participation

- **gypsum**
- **integrated farming system**
- **nutrient management**
- **on-farm development**
- **participatory varietal selection**
- **pH**
- **salt affected soils**
- **site implementation committee**
- **sodic soils**
- **soil physical properties**
- **surface stagnation**
- **water logging**
- **water management**
- **water users' groups**

REFERENCES

1. Abrol, I. P., Yadav, J. S. P., & Massoud, F. I. (1988). *Salt affected Soils and Their Management*. FAO Soils Bulletin, Rome, 39, 131 p.
2. Ali, Y., Aslam, Z., Ashraf, M. Y., & Tahir, G. R. (2004). Effects of salinity on chlorophyll concentration, leaf area, yield and yield components of rice genotypes grown under saline environment. *International Journal of Environmental Science & Technology*, 1, 221–225.
3. Balal, R. M., Ashraf, M. Y., Khan, M. M., Jaskani, M. J., & Ashfaq, M. (2011). Influence of salt stress on growth and biochemical parameters of citrus rootstocks. *Pakistan Journal of Botany*, 43, 2135–2141.
4. Bellon, M. R., & Reeves, J. (Eds.). (2002). *Quantitative Analysis of Data from Participatory Methods in Plant Breeding*. DF: CIMMYT, Mexico.
5. Ceccarelli, S., Grando, S., Baily, E., Amri, A., Et-Felah, M., Nassif, F., Rezgui, S., & Yahyaoui, A. (2001). Farmer participation in barley breeding in Syria, Morocco and Tunisia. *Euphytica*, 122, 521–536.
6. Egamberdieva, G. M. E., Albacete, A., Martý' nez-Andujar, C., Acosta, M., Romero-Aranda, R., Dodd, I. C., Lutts, S., & Peçrez-Alfocea, F. (2008). Hormonal changes during salinity-induced leaf senescence in tomato (*Solanum lycopersicum* L.). *Journal of Experimental Botany*, 59, 3039–3050.

7. FAO, 2008. Land and Plant Nutrition Management Service. http://www.fao.org/ag/agl/agll/spush.

8. Garg, V. K. (1998). Interaction of tree crops with a sodic soils environment: Potential for rehabilitation of degraded environments. *Land degradation and Development*, 9, 81–93.

9. Ghassemi, F. A. J., & Nix, H. A. (1995). *Salinization of Land and Water Resources. Human Causes, Extent, Management and Case Studies*. The Australian National University Pub. CAB. International, Wallingford, Oxon.

10. Gupta, R. K., & Abrol, I. P. (1990). Salt affected soils: Their reclamation and management for crop production. *Advance in Soil Science*, 11, 223–228.

11. Gupta, S. K., & Gupta, I. C. (2014). *Salt Affected Soils: Reclamation and Management*. Scientific Publishers, Jodhpur. 321 p.

12. Gupta, S. K., Sharma, D. P., & Swarup, A. (2004). Relative tolerance of crops to surface stagnation. Journal of Agricultural Engineering, 41, 44–48.

13. Lobell, D. B., Ortiz-Monsterio, J, I., Gurrola, F. C., & Valenzuuela, L. (2007). Identification of saline soils with multiyear remote sensing of crop yields. *Soil Science Society of America Journal*, 71, 777–783.

14. Meloni, D. A., Oliva, M. A., Martinez, C. A., & Cambraia. J. (2003). Photosynthesis and activity of superoxide dismutase, peroxidase and glutathione reductase in cotton under salt stress. *Environmental and Experimental Botany*, 49, 69–76.

15. Mondal, A. K., Sharma, R.C and Singh, G. (2009). Assessment of salt affected soils in India using GIS. *Geocarto International*, 24, 437–456.

16. NRSA and Associates (1996). *Mapping Salt Affected Soils of India, 1:250,000 Map Sheets, Legend*. NRSA, Hyderabad.

17. Rengasamy, P. (2006). World salinization with emphasis on Australia. *Journal of experimental Botany*, 57, 1017–1023.

18. Saadia, M., Jamil, A., Ashraf, M., & Akram, N. A. (2013). Comparative study of SOS2 and a novel PMP3–1 gene expression in two sunflower (*Helianthus annuus* L.) lines differing in salt tolerance. *Applied Biochemistry and Biotechnology*, 170, 980–987.

19. Singandhupe, R. B., & Rajput R. K. (1989). Ammonia volatilization from rice fields in alkali soil as influence by soil moisture and nitrogen. *Journal of Agriculture Science* (Camb.), 112, 185–90.

20. Singh, Y. P., Nayak, A. K., Sharma, D. K., Gautam, R. K., Singh, R. K., Singh, R., Mishra, V. K., Paris, T., & Ismail, A. M. (2014). Farmers participatory varietal selection: a sustainable approach for the 21st century. *Agro ecology and Sustainable Food System*, 38, 427–444.

21. Singh, Y. P., Nayak, A. K., Sharma, D. K., Gautam, R. K., Singh, R. K., Singh, R., Mishra, V. K., Paris, T., & Ismail, A. M. (2013). Varietal selection in sodic soils of Indo-Gangetic plains through farmer's participatory approach. *African Journal of Agricultural Research*, 8, 2849–2860.

22. Singh, R. P., & Ojha, R. K. (2004). *Socio-economic Impact of Reclamation of Sodic Land: A Case Study of Uttar Pradesh, India*. Report to UP *Bhumi Sudhar Nigam*, Lucknow.

23. Sharma, D. K., Rathore, R. S., Nayak, A. K., & Mishra, V. K. (2011). *Sustainable Management of Sodic Lands*. Central Soil salinity Research Institute, Regional Research station, Lucknow, 416 p.

24. Suarez, D. L., Rhoades, J. D., Savado, R., & Grieve, C. M. (1984). Effect of pH on saturated hydraulic conductivity and soil dispersion. *Soil Science Society of American Journal*, 48, 50–55.

25. Sumner, M. E. (1993). Sodic soils: New perspectives. *Australian Journal of Soil Research*, 3, 683–750.

26. Thimmappa, K., Singh, Y. P., Raju, R., Kumar, S., Tripathi, R. S., Pal, G., & Reddy, A. A. (2015). Reducing farm income losses through land reclamation: a case study from Indo-Gangetic plains. *Journal of Soil Salinity and Water Quality*, 7, 68–76.

27. Town, M. H., Chandrasekhar, T., Mahamed, H., Zafar, S., Brhan, K. S., & Rama Gopal G. (2008). Recent advances in salt stress biology – a review. *Biotechnology and Molecular Biology Review*, 3, 8–13.

28. Tripathi, R. S. (2011). Socio-economic impact of reclaiming salt affected lands in India. *Journal of Soil Salinity and Water Quality*, 3, 110–126.

29. Weltzein, E., Smith, M., Meitzner, L. S., & Sperling, L. (2000). *Technical Issues in Participatory Plant Breeding from the Perspective of Formal Plant Breeding. A Global Analysis of Issues, Results and Current Experiences.* Working Document No. 3. CGIAR, System wide Program on Participatory Research and Gender Analysis for Technology Development and Institutional Innovation.

30. Win, K. T. (2011). Genetic analysis of Myanmar *Vigna* species in responses to salt stress at the seedling stage. *African Journal of Biotechnology*, 10, 1615–1624.

31. Yadav, J. S. P., & Agarwal, R. R. (1959). Dynamics of soil changes in the reclamation of saline alkaline soils of the Indo-Gangetic alluvium. *Journal of Indian Society of Soil science*, 7, 214–222.

CHAPTER 10

POKKALI RICE CULTIVATION IN INDIA: A TECHNIQUE FOR MULTI-STRESS MANAGEMENT

A. K. SREELATHA and K. S. SHYLARAJ

CONTENTS

10.1   INTRODUCTION

Man has been confronted with saline water and saline soils since time immemorial. The encounter may be as long as the history of Agriculture. While in earlier times, management was to earn livelihood, currently management requirement has taken wide overtones considering the requirement of food, fiber and other requirements of the burgeoning population. While our knowledge is expanding on reclamation and management of salt affected soils and water, so are the problems, which are further being compounded by climate change.

This chapter discusses one of the popular technologies enshrined in *Indian Traditional Knowledge* (ITK) to manage rainwater and seawater in a holistic manner for rainfed rice-saline fish/prawn culture. The technology in common parlance in Kerala (India) is known as *Pokkali* Cultivation. *Pokkali* is also unique saline tolerant rice variety that is cultivated in an organic way in the water logged coastal regions of Alappuzha, Thrissur and Ernakulam districts of Kerala in Southern India. In view of the uniqueness of the system, *Pokkali* rice was awarded the status of registered Geographical Indication (GI) by the Geographical Indications Registry Office, Chennai, Tamil Nadu in 2007.

The GI registration permits the exclusive global right on the concerned farmers to cultivate *Pokkali* paddy and sell the finished product in the brand name of '*Pokkali*' the world over. To ensure full protection of the community's rights, addresses of 12,000 *Pokkali* cultivating farmers have been included in the application for Geographical Indication. The registered *Pokkali* farmer group (Varappuzha-Kadamakudy Jaiva *Pokkali* Society) received the Plant Genome Saviour Community award (2011–12) for protecting this valuable genotype through cultivation by the Protection of Plant Variety and Farmer's Right Authority of Government of India. There is a great need to value the location specific technologies developed in India to manage the salts and integrate them with scientific principles and developments to overcome major problems and/or to make them more effective. Some of the scientific interventions such as breeding rice varieties to manage acid saline soils also form the part of this chapter.

10.2 LOCATION OF THE STUDY AREA

The state of Kerala, India lying between 8°18′ to 12° 48′ Latitude and 74°25′ to 77°22′ Longitude is a small but beautiful narrow strip in the southwestern corner on the Indian Peninsula having a geographical spread of 1.18% of the total geographical area of India. The long coastline of about 580 km in the state has sixteen backwaters linked to the sea covering an area of 650 km². Most of the coastal lands, deltaic areas at river mouths and reclaimed backwaters are either at sea level or 1.0 to 1.5 m below Mean Sea Level (MSL). As a result sea water intrudes up to a distance of 10 to 20 km upstream during high tides. The special zone for problem areas lies on the coastal line covering an area of 428,540 ha.

The saline soils of Kerala occur in these periodically saline water inundated lands and are divided into four different physiographic zones (Table 10.1). *Pokkali* lands are known after the *Pokkali* type of cultivation and are located between the Thannermukkam and Enamakkal bunds and besides extend to Ernakulam, Thrissur and Alappuzha districts [6]. Saline hydromorphic soils are usually seen within the coastal tracts of the districts of Ernakulam, Alappuzha, Thrissur and Kannur covering an area of 63,974 ha (Table 10.2). The origin, genesis and development of these soils have been under peculiar physiographic conditions. They are, therefore, not comparable with the saline soils occurring in the other parts of the country or the globe. The *Pokkali* lands come under the special agro ecological unit-5 delineated for the lowlands, often below mean sea level (Figure 10.1).

TABLE 10.1   Categorization of Special Zone [4, 6]

Physiographic tract	*Taluks* (Blocks)	Districts
Kuttanad	Kuttanad, Ambalappuzha, Cherthala, Vaikom, Changancherry, Kottayam	Alappuzha, Kottayam, Pathanamthitta
Pokkali	Paravoor, Kochi and Kannayannur	Ernakulam
Kole	Thalappalli, Mukundapuram, Thirssur, Chavakkadu and Ponnani	Thrissur, Malappuram
Kaipad	Thaliparamba, Kannur, Thalassery	Kannur, Kozhikode and Kasargod

TABLE 10.2   The Total Area Under Saline Soils of Kerala

Soil type	District	Area (ha)
Pokkali	Ernakulam, Thrissur, Alappuzha	39,765
Kaipad	Kannur, Kozhikode, Kasargod	24,209
Total		63,974

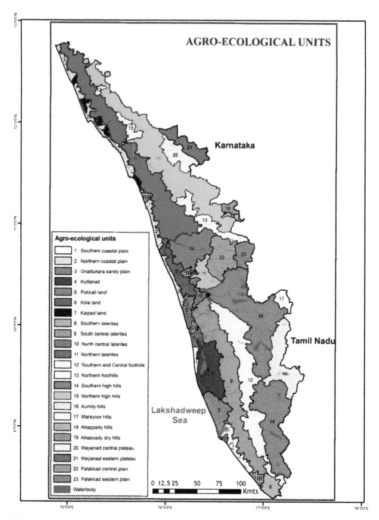

FIGURE 10.1   Agro ecological units of Kerala [2].

10.3   CHARACTERISTICS OF *POKKALI* LANDS/SOILS

Pokkali lands are low lying marshes and swamps situated near the estuaries of streams and rivers. *Pokkali* fields remain inundated with water because of their topography and being naturally connected to the Arabian Sea through backwaters and canals [1]. The fields are poorly drained and subjected to tidal action throughout the year [1]. *Pokkali* soils have developed from river borne alluvium. They belong to fine, mixed, iso-hyperthermic typic sulfaquents as per Soil Taxonomy [2]. A detailed study of the physical and chemical characteristics of *pokkali* soil was carried out by Varghese et al. [12], who revealed that the soils are light gray on surface, intensity of which increases with depth. The soil texture becomes finer with depth. These soils have high water holding capacity. The soil is stiff impervious clay but rich in organic matter. It is bluish black in color and is hard and creates deep fissures when dry and sticky when wet [12].

Periodical inundation by sea water and backwater make these soils saline. The soluble salts comprise mainly chlorides and sulfates of Na, Mg and Ca. Most of saline soils of Kerala are acidic with pH ranging from 3.0 to 6.8 (Table 10.3). The electrical conductivity of soils during the high saline phase (November–May) varies from 12–24 dS m^{-1} and average salt content reaches up to 20 ppt [3]. During low saline phase (June–October) water becomes almost fresh, salt content reduces to traces, EC

TABLE 10.3   Physico-Chemical Properties of *Pokkali* Soils

Parameters	Rice Low saline phase	Prawn High saline phase
pH	2.62 to 5.97	3.31 to 6.46
EC (dS m^{-1})	0.01 to 7.80	0.10 to 9.80
Organic C (%)	0.45 to 2.90	0.48 to 2.79
Al (mg kg^{-1})	0 to 866	4 to 289
B (mg kg^{-1})	0.13 to 0.75	Trace to 1.20
Ca (mg kg^{-1})	76 to 256	60 to 218
Cu (mg kg^{-1})	2 to 13	1 to 10
Fe (mg kg^{-1})	171 to 232	172 to 2028
K (kg ha^{-1})	13 to 1777	52 to 1086
Mg (mg kg^{-1})	9 to 20	6 to 8
Mn (mg kg^{-1})	2 to 26	1 to 34
P (kg ha^{-1})	0.24 to 88.24	12.32 to 119.24
S (mg kg^{-1})	8 to 6846	3 to 3625
Zn (mg kg^{-1})	2 to 173	2 to 125

ranging from 1 to 2 dS m^{-1}, although inherent acidity dominates. The SAR values are very high. Innumerable varieties of microorganisms have been detected in these soils. They are deficient in P and Ca, high in K and Mg and thus require application of phosphatic fertilizers with lime [9].

The high tide and low tides occurring twice a day regulate the fertility and productivity of the *Pokkali* soils. While the tide brings nutrients to the *Pokkali* fields, low tide removes toxic concentrations. The tidal influx is also helpful to the growth of a broad spectrum of beneficial microbes. Clearly, the lands under this scenario face multiple stresses of water logging, soil salinity and acidity compounding the situation to the detriment of the farming communities.

10.4   THE TECHNOLOGY OF POKKALI RICE CULTIVATION

The traditional *Pokkali* rice cultivation and shrimp farming (rainfed rice-saline fish/prawn system) practiced in the wet lands of Ernakulam, Alappuzha and Thrissur districts [10] is one of the most eco friendly farming practices in the world (Figure 10.2). This practice has passed down from one generation to another generation from more than 3000 years. It relies on symbiotic nature of rice and shrimp, all processes based on natural conditions following the lunar cycle. The distinguishing feature of *Pokkali* cultivation is that the *Pokkali* varieties grow in low to medium saline condition.

Pokkali cultivation begins at the onset of the southwest monsoon in May and goes until October to November. This is completely a natural

FIGURE 10.2   Rice (left) and fish/prawn culture (right) in *Pokkali* fields.

process which doesn't use fertilizers, insecticides or any other chemicals [8]. The *Pokkali* fields, after the harvest of rice are used for fish/prawn farming. At this time the backwaters and canals become saline and juvenile prawns and fingerlings of other fishes come in large number in the outer canals. These fingerlings are guided to the fields through trap sluices and the sluice gates prevent them from going out. The waste materials of *Pokkali* rice forms the food for the fish/prawn. In this system, no selective stocking or supplementary feedings is done [11]. This practice has been effectively pursued for centuries as it maintains ecological balance and ensures a reasonable profit to the farmer.

10.4.1   RICE CULTIVATION

The soil management practices and varietal selection are meticulously followed to get optimum benefits. Rice cultivation begins from the month of April by strengthening the outer bunds and setting up of sluices to control the level of water. The fields are then drained during low tide and the sluices are closed to avoid water entry during high tide. After the fields get dried, soil is heaped to form mounds with base width of 1 m to a height of 0.5 m (Figure 10.3). The mounds are also allowed to dry. With the onset of monsoon, the salt is washed off from the soil and water with dissolved salt is drained from the field so that salinity levels are brought to levels that can be tolerated by the rice varieties. These mounds later act an elevated *in-situ* nursery and provide protection to the seedling from flash floods. The

FIGURE 10.3   Bunding and drying the fields (left) and fields with soil heaped to form mounds (right).

most popular traditional variety used is '*Pokkali.*' Other varieties include, Chettivirippu, Cheruvirippu, Kuruka, Ponkuruka, Karutha, Anakandan, Eravapandy, Bali, Orpandi, etc. High yielding varieties released from Rice Research Station, Kerala Agricultural University, Vyttila are Vyttila 1 to Vyttila 9.

In *Pokkali* cultivation, a high seed rate of 100 kg ha^{-1} is used to compensate the damages caused to seedling by the rats, pigeons and parrots. A special method is adopted by the farmers for sprouting the seeds. The seeds are tightly packed in baskets made of plaited coconut leaves lined with leaves of banana and teak. These baskets are then immersed in fresh water ponds for 12 to 15 h, after which they are stored in shade. The radicle just sprouts and remains quiescent in this condition for more than 30 days. The seed are sown in the mounds when the soil and weather conditions are favorable. Before sowing, the basket containing the seeds is re-soaked for 3 to 6 h. The mounds are then raked and the tops are leveled. The sprouted seeds are sown on the mounds (Figure 10.4).

The *Pokkali* varieties have early seedling vigor and attain a height of 40–45 cm in 30–35 days. When the field conditions become favorable the mounds are cut into pieces using a spade and spread uniformly in the field with a few seedlings intact with a clod of earth. The tall seedlings survive in the flooded conditions. The clods of earth attached to the clumps give anchorage to the seedling. Since the *Pokkali* fields are naturally fertilized, chemical fertilizers and pesticides are not used. The decaying paddy stubbles remaining after the harvest of paddy and the fauna and flora particularly the profuse floating vegetation die out and decompose and add on to

FIGURE 10.4   Sprouted seeds at the top of the mounds (left) and seedlings (right).

soil fertility. After transplanting no intercultural operations are adopted in *Pokkali* farming. Since water management is very important to minimize the loss due to flash floods, water level is raised as per the growth of the plant through the sluice gate.

Aquatic and semi aquatic weeds (such as *Echinochloa crusgalli, Eliocaris fistula, Fimbristylis milliaceae, Monochoria vaginalis, Valisneria spiralis, Nymphea* sp., *Marsilea quadrifolia, Astiracantha longifolia, Lymnophylla heterophylla, Sphenoclea zeylanica, Cyperus difformis, Ludwigia octovalois*)are encountered in the *Pokkali* fields. Floating migratory weeds (like *Salvinia auriculata, Eichornia crassippes* and recently *Lemna minor*) also create problems in rice cultivation. The weed management is usually done by raising the water level and allowing the free flow of water into and out of the field. During high saline phase, these weeds and the flora and fauna decompose and add to soil fertility.

Generally pest and disease attack in *Pokkali* is below threshold level. The prevalence of natural enemies, predators, varietal resistance, single cropped system, etc., can be attributed to the low incidence of pests and diseases. The non-insect pests peculiar to *Pokkali* crop include fishes, tortoises, rats and parrots. Fishes like *Ambass* sp., *Anabas* sp., *Channa* sp. Etc. feed on the radicle and plumule of the sprouting seeds, thus damaging the seedlings. Tortoises cause serious damage by burrowing the mounds and sprawling over them. The damage by rats is encountered right from the seedling stage to harvest. Sparrows, doves and pigeons pick seeds at the time of sowing. At harvest stage, while sparrows feed on single grains, the parrots carry away the entire panicle. A new bird pest, Purple moor hen (*Porphyrio porphyrio*), a migratory bird swarms in large numbers and cause extensive damage to the crop by making nest inside the field.

The crop is ready for harvest in about 110–120 days, usually by October-November coinciding with northeast monsoon. At this stage, water level is high in the field. So only the panicles are harvested leaving behind the plants. Yield of traditional *Pokkali* varieties is low to the tune of 1.5–2.0 t ha^{-1}. Farmers use small country boats to carry the harvested panicles to the field bunds. Threshing is done using power operated thresher. Cost of rice production under the *Pokkali* system is generally lower than that of other rice ecosystems in view of the saving in fertilizer, pesticide, and herbicide costs [10]. The average cost of cultivation of *pokkali* is Rs. 30,000 ha^{-1} (1.00 US\$ = 60.00 Rs.) [11].

10.4.2   SCIENTIFIC BASIS OF SOME PRACTICES IN "POKKALI" CULTIVATION

In the practical application of this ITK, some of the processes are based on sound scientific principles. Since the soils after the saline fresh/prawn culture are highly saline with inadequate drainage, formation of mounds helps in leaching of salts faster and effectively due to the following reasons.

- The annual rainfall in Kerala is more than 3000 mm, the major fraction of which is received from June to August. Water is available in plenty and washing with this best quality water is the easiest and cheapest method to reclaim highly saline soil. The process of forming the fields into mounds expedites removal of salts as the mounds have larger surface area and soil is quite loose in nature. The leaching as such is easy and perfect.

- Due to impeded drainage under coastal condition, fields become flooded as soon as heavy rains set in. The mounds with about half a meter height serve as '*in-situ*' nursery and save the young seedlings from total submergence.

- Once the monsoon commences, the fields remain flooded with 45–50 cm of standing water. Under such a condition, normal system of transplanting is impossible. When mounds are leveled, the clumps of seedlings get fixed due to weight of soil attached to these clumps.

Similarly, sprouting of seeds in plaited coconut baskets has perfect scientific reasoning as follows:

- The sprouted seeds remain viable and quiescent for 30–40 days, so that these can be used at any time when the field conditions are favorable. It is achieved due to the high pressure and anaerobic condition developed within the tightly packed baskets.

- When seeds in tightly packed baskets are soaked for 12–15 h, seeds expand as a result of imbibition of water. It creates pressure and raises the temperature inside the basket. When the excess water in the basket is drained off, atmospheric air (oxygen) enters in the interspaces. The seeds start sprouting utilizing available moisture and oxygen inside the basket. Sprouting of seeds and respiration of seeds and sprouts (radicle) exhaust the available oxygen inside the basket. Carbon-dioxide exhaled by the sprouts creates anaerobic condition inside the basket and prevents further growth of the radicle. As a result, the sprouts (radicle) remain quiescent.

- On the periphery of the basket, the sprouts having direct contact with atmospheric oxygen continue to grow to form matting on the baskets. This condition results in preventing the contact of the seeds inside the basket with atmospheric oxygen, thus contributing for their quiescent condition.
- When re-soaking is done, the carbon-dioxide inside the basket is displaced by water. On draining the basket again atmospheric air (oxygen) gets in, which expedites further growth of sprouts.

10.4.3 TRADITIONAL PRAWN INFILTRATION/CULTIVATION

The integrated farming system involving rice, fish or prawn is the most viable and eco-friendly practice complementing each other by the organic recycling and generation of live feed. After the harvest of rice, the land preparation for prawn infiltration begins in November and actual harvest by the middle of January coinciding with the lunar phase. The traditional technique of prawn infiltration has been developed by the innovative *Pokkali* farmers. Prawn infiltration in fields starts with the strengthening of outer bunds and installation of sluice gates when there is medium flow of water from the canal or backwaters. Removal of floating weeds and desilting of canals, wherever necessary is carried out for holding more water in the field [7]. Water is let in through the sluice during high tide at night. A hurricane lamp is hung for luring the prawn seedlings. During the low tide, the water is let out through a closely packed bamboo screen or a large mesh conical bag net to prevent the trapped fishes and prawns to escape (Figure 10.5). This process is repeated cautiously during each tide so as to build up the population of shrimp and fish seeds.

Trapping of prawns and the harvest continues simultaneously till the end of March. Since the panicles of the rice plant alone are harvested, the stubbles left in the field get decomposed and provide shelter and feed for the growth of microorganism which forms the ideal feed for the fishes and prawns.

FIGURE 10.5   Conical net placed in sluice.

The production cost of shrimp is substantially low since the rice crop provides adequate dietary supplement for fishes. At this time fields are finally drained for taking up paddy cultivation. The yield during five months is about 800 to 1000 kg ha^{-1}. About 80% of the catch constitutes prawns. The main advantage of prawn filtration is that it is a source of subsidiary income to the farmers without much investment.

10.5   SCIENTIFIC INTERVENTIONS

The discussions made in the foregoing section reveal that the ITK of *Pokkali* cultivation is low input-low output technology although in the past it has played its role to ensure livelihood to provide some means to the poor farming community facing the multiple stress threat. Realizing the limitations of labor-intensive nature of the technology, drudgery to the farmers and low productivity, scientific interventions are being introduced.

10.5.1   *MECHANIZATION OF RICE CULTIVATION*

Mechanization in *Pokkali* farming has so far remained confined to threshing since the unfavorable soil physical conditions restrict the use of normal agricultural machinery in these lands. Experimental evidences have been generated to show that making ridges instead of the traditional mounds for sowing the seeds not only saves labor cost but give almost equal yield. It has been brought out that 30 cm high-ridge constructed 40 cm apart are equally good as the traditional mounds. If implemented, this system will allow the use of light weight ridge former with power tillers.

10.5.2 RICE AND FISH CULTURE

Since low yields in rice are weaning away the farmers from this traditional practice, simultaneous culture of fish species *Etroplus suratensis, Oeochemis mossamticus, Cyprinus caspio, Labio rohita, Chanos chanos and Mugil cephalus* with *Pokkali* rice genotypes have been conducted with some success [7]. Although such systems have been successfully tried in other parts of India as well, the simultaneous rice fish culture in this set-up has not become popular as the sequential rice-prawn culture.

10.6 CROP IMPROVEMENT FOR HIGHER RICE PRODUCTIVITY

To improve rice productivity, any varietal development program must look at the requirements of *Pokkali* ecosystem, which are:
- plant height not less than 125 cm;
- crop duration not more than 125 days;
- tolerance to salinity, acidity and submergence;
- should possess dormancy for about 2 weeks to one month.

Accordingly Rice Research Station, Vyttila of Kerala Agricultural University, took up concerted efforts and developed 9 varieties for this ecosystem. As a first step, land races from different *Pokkali* ecosystem like *Pokkali*, Cheruvirippu, Chettivirippu, Kuruka, KaruthaKuruka, Ponkuruka, Anakomban, Mundakan, etc., were collected. Besides, saline tolerant accessions from other places were also introduced, evaluated and utilized for the breeding program. For example; the national accessions CSR-10, SR-26-B, etc., and international accessions IR-5, Nona Bokra, etc., were part of the program.

10.6.1 PURE LINE SELECTION

'Mass' and 'pureline' selection was attempted in the two widely cultivated local varieties viz. Choottupokkali and Cheruvirippu. The first high yielding salinity tolerant rice variety VTL-1 was evolved by pureline selection from Choottupokkali and released for cultivation. This variety has 10–15% higher yield than the local variety. The second variety VTL-2 was evolved by pureline selection from Cheruvirippuis also having 10–15 % higher yield advantage over the local parent variety (Table 10.4).

10.6.2 HYBRIDIZATION AND SELECTION

The rice hybridization program worked on released high yielding varieties and abiotic stress tolerant *Pokkali* land races. The first high yielding variety VTL-3 was developed by crossing VTL-1 and Taichung Native-1. VTL-4 was developed by crossing the land race Chettivirippu with IR 4630-22-2-17.

To take care of peoples' choice for red rice and taking note of the importance of export of *Pokkali* rice, being organic, a mutant high yielding variety VTL-5 was developed from the widely accepted variety Mahsuri. Considering the increasing labor shortage and enhanced labor wages, a semi tall high yielding rice variety having tolerance to the multiple abiotic stresses of the ecosystem suitable for mechanical harvesting was developed as VTL-6, a cross of Cheruvirippu/IR 5/Jaya. For a similar situation but white kernelled rice variety VTL-7 suitable for export purpose was developed with its parentage as IR-8 x Patnaik 23. The variety VTL-8 evolved from the cross IR 47310-94-4-1 X CSR-10 suited to slightly deep waterlogged areas has been developed. Now, the farmers have many alternatives to suit various conditions with good high yield potential of many varieties. A brief description of these varieties is given in Table 10.4.

Although, Pokkali rice is known for its salt-resistant genes and the University of Arizona in the United States has developed a DNA library of *Pokkali* rice. But the center at Vyttila is now concentrating on molecular breeding by the introgression of abiotic stress tolerant QTL/genes through Marker Assisted Backcrossing. The *Saltol* QTL and *Sub* 1genes are being

TABLE 10.4   Characteristics of Various Multi-Stresses Tolerant Rice Varieties for *Pokkali* Ecosystem

Variety	Characters
VTL-1	Pureline selection from Choottupokkali, photo-insensitive tall lodging variety, medium duration (115 days), good cooking quality, tolerant to salinity, acidity and submergence. Grains are long bold and red with special taste. Donor of *Saltol* QTL, realizable yield 2000–2500 kg ha^{-1} with an average of 1500 kg ha^{-1}.
VTL-2	Pureline selection from Cheruvirippu, Photo-insensitive tall lodging variety, medium duration (115 days), good cooking quality, tolerant to salinity, acidity and submergence, Tip awned grains are long bold and red with special taste. Realizable yield 2500–3000 kg ha^{-1} with an average of 1750 kg ha^{-1}
VTL-3	Hybridization and selection VTL-1 X TN 1, Photo-insensitive tall lodging variety, medium duration (115 days), good cooking quality, tolerant to salinity, acidity and submergence, tip-awned grains are long bold and red with special taste. Realizable yield 3000–3500 kg ha^{-1} with an average of 2000–2500 kg ha^{-1}
VL-4	Combination breeding Chettivirippu X IR 4630-22-2-17, Photo-insensitive, medium duration (120–125 days) variety with good cooking quality. Tolerant to salinity and soil acidity, escapes flash flood and water stagnation. Tolerant to major pests and diseases. Tall plants (160–165 cm) having lodging habit and kneeing ability. Grains are straw colored, medium bold with red kernel. Realizable yield 3000–4000 kg ha^{-1} with an average of 3250 kg ha^{-1}.
VTL-5	Mutation breeding – Mahsuri, Photo-insensitive tall lodging variety, medium duration (115 days), good cooking quality, tolerant to salinity, acidity and submergence. Tall lodging plants (140–150 cm) with kneeing ability. Grains are having golden seed coat, medium bold with white kernels. The rice has a special taste. Realizable yield 3500–4000 kg ha^{-1} with an average of 3250–3500 kg ha^{-1}.
VTL-6	Hybridization and selection- Cheruvirippu/IR 5/Jaya, Medium duration (105–110 days), good cooking quality, non photosensitive, tolerant to salinity, acidity and submergence, good hulling percent (80%), generally semi tall (115–120 cm) non lodging plants with light green leaves. Auricles absent Grains with golden seed coat and red kernels. Realizable yield 5000–6000 kg ha^{-1} with an average of 3500–4000 kg ha^{-1}.
VTL-7	Hybridization and selection- IR-8 x Patnaik 23, medium duration (115–120 days) variety with long slender white kernelled grains. It is a non-photosensitive variety tolerant to salinity, acidity and submergence. Generally semi tall (115–120 cm) non-lodging plants with erect green leaves. Long slender grains with golden seed coat and white kernels. Realizable yield 5000–6000 kg ha^{-1} with an average of 4000–4500 kg ha^{-1}.

TABLE 10.4 (Continued)

Variety	Characters
VTL-8	Hybridization and selection- IR 47310–34–4-1 X CSR 10. Generally semi tall (130–135 cm) non-lodging plants with drooping green leaves. The grains are medium bold and awnless. Realizable yield 4500–6000 kg ha^{-1} with an average of 4100–4500 kg ha^{-1}.
VTL-9	Mutant of the land race Chettivirippu. It is a semi-tall non lodging medium duration variety having high level of salinity tolerance (up to 12 dSm^{-1} at seedling stage) with erect plant canopy. The grains are deep straw colored with brownish tint in the grooves. The cooking quality and taste of the rice is very good. The realizable yield is 5000–6000 kg ha^{-1} with an average yield of 4250 kg ha^{-1}.

introgressed into three major high yielding varieties of rice viz. Jyothi, Jaya and Uma. The introgression of *Saltol* Jyothi is complete and the introgressed lines are under field evaluation. Some of the introgressed lines are better than the parent with respect to yield also. The introgression of *Saltol* QTL and Sub 1 genes into other varieties are almost complete and ready for field evaluation. At present the combined introgression of both abiotic stresses is being done. The varieties thus evolved will be a boon to get good yield under salinity and submergence throughout the coastal areas of Kerala.

10.7   THE NUTRACEUTICAL PROPERTIES OF *POKKALI* RICE

Pokkali rice is known for its taste and quality and is different from the conventional varieties of rice. The medicinal properties of *Pokkali* rice have always been in the news as claimed by the farmers. But the actual components contributing to the medicinal values were still unknown.

A scientific investigation conducted at the Rice Research Station, Vyttila revealed that the *Pokkali* bran was nutritionally rich in bran oil and antioxidants like oryzanol (202.9 mg 100 g^{-1} bran), tocopherol (2.26 mg 100 g^{-1} bran), tocotrienols (5.13 mg 100 g^{-1} bran) etc. as compared to the most popular varieties like Jyothi and Uma in the region. Oryzanol has antioxidant, anti-cancerous and anti-lipidaemic properties. Tocopherols and tocotrienols are vitamin E analogs which have antioxidant and anti-cancerous properties. These contents were also higher than the

TABLE 10.5   Bran Oil and Micronutrient Content of *Pokkali* Rice

Fat content (%)	Fe (mg kg⁻¹)	Zn (mg kg⁻¹)	Cu (mg kg⁻¹)	Mn (mg kg⁻¹)	B (mg kg⁻¹)
9.83	25.0	22.6	1.4	20.4	76.7

concentration reported in the medicinal rice Njavara. It has also been reported that *Pokkali* rice is rich in micronutrients like Fe, Zn, Mn and B as shown in Table 10.5. Therefore, recommendations have emerged to have only mild polishing of *Pokkali* rice so that medicinal properties of rice bran remain intact [11].

10.8   CURRENT STATUS

Like several ITKs, which are numbering their days, *Pokkali* cultivation is also facing a host of problems such as: unavailability of farm workers, especially for harvesting, high labor wages, low income from rice and intensive aquaculture activities [5]. These are some of the main causes for the peoples' increasing disinterest in this practice. As per the *Pokkali* Land Development Agency (PLDA), *Pokkali* cultivation has shrunk from 25,000 ha a few decades back, to a mere 8,500 ha today. Out of this, only rice is cultivated on 5500 ha. Although in the short run, intensive aquaculture activities may give higher returns, in the long run this may prove to be environmentally and socially unviable. Besides what is stated, serious anthropogenic threats are emerging and these lands are being converted for other purposes like roads, bridges, residential or commercial activities. Invasion of weed and overexploitation of fish and prawn are some of the major reasons for the decline of the lands.

10.9   SUMMARY

The lowlands or coastal areas, in the state of Kerala, are essentially the lands of coconut and rice. About 25% of total paddy lands in Kerala are water logged especially in *Pokkali*, Kole, Kuttanadand Kaipad areas. The *Pokkali* lands are low-lying marshes and swamps situated near the estuaries of streams and rivers not far from the sea. Integrated rice-shrimp farming is traditionally practiced by the farmers, providing them the necessary livelihood. The system has proved to be quite useful for economic utilization of land, family labor along with a host of ecological benefits. The rice cultivation in the *Pokkali* ecosystem is becoming difficult and non profitable as a result of non-availability of farm workers, high labor wages, climate change, non availability of quality seeds of high yielding or potential multi-stress tolerant varieties etc. To rejuvenate the rice cultivation in this potential organic farming area with the antioxidant rich *Pokkali* varieties, crop improvement works have led to the release of several improved varieties for cultivation in this ecosystem. Policy initiatives are needed for incentives so that these lands are saved from encroachment and peoples' apathy. For example, an organic certification for the produce may allow the farmers to sell the produce at a premium. A dedicated processing center may also help the farmers to get good remuneration.

KEYWORDS

- acidity
- anti-oxidants
- aquaculture
- backwaters
- crop improvement
- fish
- geographical indication
- hybridization
- integrated farming system
- inundation
- kerala
- lowlands
- micro-nutrients
- nutraceutical properties

- organic farming
- *Pokkali*
- prawn culture
- quantitative trait locus (QTL)
- rice
- saline soils
- salinity
- *Saltol*
- sluice gate
- VTL
- water logging

REFERENCES

1. Jayan, P. R., & Sathyanathan, N. (2010). Overview of farming practices in the water-logged areas of Kerala, India. *Int. J. Agric. Biol. Eng.*, 3, 18–33.
2. KSPB (Kerala State Planning Board) (2013). *Soil Fertility Assessment and Information Management for Enhancing Crop Productivity in Kerala.* Kerala State Planning Board Thiruvananthapuram. 514 p.
3. Kuruvila, V. O. (1974). *Chemistry of Low Productive Acid Sulphate Soils of Kerala and Amelioration for Growing Rice.* PhD thesis, Utkal University, Bhuvaneswar, 156p.
4. Mathew, E. K., Nair, M., Raju, T. D., & Jaikumaram, U. (2004). *Drainage Digest.* Kerala Agricultural University, Thrissur and ICAR, New Delhi. 200p.
5. Nair, K. N., Vineetha, M., & Mahesh, R. (2002). *The Lure of Prawn Culture and the Waning Culture of Rice-Fish Farming: A Case Study from North Kerala Wetlands.* Kerala Research Program on Local Level Development, Centre for Developmental studies, Thiruvananthapurram. Discussion Paper No. 53.
6. Padmaja, P., Geethakumari, V. L., Harikrishnan, N. K., Chinnamma, N. P., Sasidharan, N. K., & Rajan, K. C. (1994). *A Glimpse to Problem Soils of Kerala.* Kerala Agricultural University, Thrissur. pp. 76–89.
7. Rajendran, C. G., George, T. U., Mohan, M. V., & George, K. M. (1993). Problems and prospects of integrated agriculture in *Pokkali* fields. In: *Rice in Wetland Ecosystems. Nair, R. R., Vasudevan, K. P., & Joseph, C. A. (Eds.).* Kerala Agricultural University, Vellanikkara, Thrissur. pp. 276–279.
8. Ranga M. R. (2006). *Transformation of Coastal Wetland Agriculture and Livelihoods in Kerala, India.* MSc Thesis. Canada: Faculty of Graduate Studies, University of Manitoba.

9. Samikutty, V. (1977). *Investigations on the Salinity Problems of Pokkali and Kaipad Areas of Kerala state.* MSc (Ag) Thesis, Kerala Agricultural University, Thrissur.

10. Sasidharan, N. K. (2004). *Enhancing the Productivity of the Rice – Fish/Prawn Farming System in Pokkali Lands.* PhD Thesis, Kerala Agricultural University, Thrissur, 190p.

11. Shylaraj, K. S., Sreekumaran, V., & Annie, K. (2013). *Five Decades of Glorious Research at the Rice Research Station, Vyttila (1958–2013).* Directorate of Extension, Kerala Agricultural University, Thrissur, 80p.

12. Varghese, T., Thampi, P. S., & Money, N. S. (1970). Some preliminary studies on the *Pokkali* saline soils of Kerala. *J. Indian Soc. Soil Sci.*, 18, 65–69.

CHAPTER 11

PREVENTION, RECLAMATION AND MANAGEMENT OF ACID AND ACID SULPHATE SOILS

A. KRISHNA CHAITANYA, SHRIKANT BADOLE,
ARBIND KUMAR GUPTA, and BIPLAB PAL

CONTENTS

11.1   INTRODUCTION

During the last four decades in India, agriculture sector has witnessed several revolutions, i.e., green, white and blue yet the country is still aspiring to have a second or an evergreen revolution to ensure a sustainable food security regime. Although, India is largest or second largest producer of many agricultural commodities in the world [28], yet the productivity levels in many cases continue to be low and worrisome. In order to feed the ever-growing population, efforts are needed to have a profound increase in agricultural productivity without overexploitation of natural resources to minimize ecologically vulnerable resources from further degradation and pollution. Anon-sustainable land management system is likely to further trigger land degradation [35] vastly experienced during the first green revolution achieved through unsustainable exploitation of land and water resources.

According to ICAR-NAAS [10] document, a preliminary assessment puts the total area under degraded and wastelands in the country at 114.01 million hectare (M ha). The extent of area under major degradation processes is water erosion 23.62 M ha, wind erosion 8.89 M ha, salinization/alkalization 6.73 M ha and acidification (<5.5 pH) 16.03 M ha. Other degradation processes such as mining and industrial waste, barren rocky/stony waste and snow-covered/ice caps occupy 12.17 M ha. The low agricultural productivity can be largely attributed to loss in soil fertility due to these degradation processes, excess of toxic elements coupled with intensive cropping and poor management practices [6, 12].

This chapter deals with the reclamation and management of acid soils in India while briefly discussing acid sulfate soils. It may be noted that permanent reclamation of acid soils in the real sense may not be achievable and only course is to manage these soils for higher productivity.

11.2   EXTENT AND DISTRIBUTION

One of the important limiting factors for optimal use of land resources for higher crop productivity especially in sub-humid and humid regions is soil acidity. Globally, acid soils occupy about 3.95 billion ha that accounts for 30% of ice-free land area [34] in the world. Tropic and sub-tropic regions alone account for about 60% of the affected area. In the Indian context, 89.94 M ha (Table 11.1) is affected by some degree of soil acidity out of which 48 M ha is under cultivation [1]. Of it, 25 M ha has pH below 5.5 (moderately to strongly acidic), a slightly higher figure than the one given by ICAR-NAAS [10], while about 23.0 M ha is slightly acidic with pH ranging between 5.6 and 6.5 [29]. In India, acid soils occur in the

TABLE 11.1   State Wise Distribution Area (M ha) Under Acid Soils in India [18]

State	Strongly acidic (pH< 4.5)	Moderately acidic (pH 4.5–5.5)	Slightly acidic (pH 5.5–6.5)	Total
Arunachal Pradesh	4.78	1.74	0.27	6.79
Assam	0.02	2.31	2.33	4.66
Bihar	–	0.04	2.32	2.36
Chhattisgarh	0.15	6.3	4.39	10.84
Goa	–	0.11	0.19	0.30
Himachal Pradesh	–	0.16	1.62	1.78
Jammu & Kashmir	–	0.09	1.48	1.57
Jharkhand	–	1.00	5.77	6.77
Karnataka	–	0.06	3.25	3.31
Kerala	0.14	2.87	0.75	3.76
Madhya Pradesh	–	1.12	10.6	11.72
Maharashtra	–	0.21	4.33	0.54
Manipur	0.42	1.44	0.32	2.19
Meghalaya	–	1.19	1.05	2.24
Mizoram	–	1.27	0.78	2.05
Nagaland	0.12	1.48	0.05	1.65
Odisha	–	0.26	8.41	8.67
Sikkim	0.28	0.32	–	0.60
Tamil Nadu	0.21	0.35	4.29	4.85
Tripura	0.06	0.75	0.24	1.05
Uttarakhand	–	1.18	2.30	3.48
West Bengal	–	0.56	4.20	4.76
Total	6.19	24.81	58.94	89.94

Himalayan region, the Eastern and north-eastern plains, peninsular India and coastal plains under varying topography, geology, climate and vegetation. Most of these soils belong to the soil order: Ultisols, Alfisols, Mollisols, Spodosols, Entisols and Inceptisol [11]. The acid soils are mostly distributed in Assam, Manipur, Tripura, Meghalaya, Mizoram, Nagaland, Sikkim, Arunachal Pradesh, West Bengal, Jharkhand, Odisha, Madhya Pradesh, Himachal Pradesh, Jammu & Kashmir, Andhra Pradesh, Karnataka, Kerala, Maharashtra and Tamil Nadu (Table 11.1). Considering the magnitude of the problem, soil acidity management to increase crop productivity is an important issue to ensure food for the burgeoning population especially for the eastern and north-eastern regions of India, having the largest potential to contribute to India's food basket.

11.3   CAUSES OF LOW PRODUCTIVITY

Low productivity of acid soils is a syndrome of problems associated with low pH. This complex of problems in acid soils mostly arises as a result of direct and indirect influences.

11.3.1   DIRECT EFFECTS

- Toxic effects of low pH due to H+ ion concentrations on root tissues.
- Influence of soil acidity on the permeability of the plasma membrane for cations.
- Disturbance in the balance between basic and acid constituents through roots.
- Effect on enzymatic processes since enzymes is particularly sensitive to pH changes.

11.3.2   INDIRECT EFFECTS

- Availability of various nutrients: Phosphorous (caused by the high fixation capacity of soil) and copper, boron, molybdenum, and zinc, etc.
- High solubility and availability of elements: aluminum, manganese and iron in toxic amount due to high acidity in the soil.
- Deficiency of bases: Calcium, magnesium and potassium.
- Prevalence of plant diseases.
- Adverse effects on soil microbial activities resulting in impairment of biological activities including nitrogen fixation by legumes.
- Low organic carbon.

A number of reports have investigated these factors to identify the most important ones. For example, crop productivity on these soils is mostly constrained by aluminum (Al), manganese (Mn) and iron (Fe) toxicity, phosphorus (P) and Mo deficiencies, low base saturation, impaired biological activity and other acidity-induced soil fertility and plant nutritional problems [14, 19]. Toxic concentrations of aluminum have been reported in sedentary soil with pH below 5.0 and alluvial soil with pH below 4.5. Aluminum toxicity in soil reduces length and branching of roots [27]. A high concentration of Al^{3+} also has an indirect effect on legume growth by inhibiting the symbiotic relationship between the host plant and rhizobia. Fertility status of acid soils is very poor; and the plant growth and development is affected to a great extent under strongly to moderately acidic condition. For example, except for the brown forest soils, the other group of acid soils of Odisha is low in organic matter and total nitrogen status. Most of the red and lateritic acid soils of Odisha have low available Phosphorous, although the total content is adequate. Soil acidity has been found to create unfavorable micro-environment for soil micro flora and fauna, which reduces the rate of organic matter breakdown that makes primary, secondary and micronutrients to remain in fixed or in insoluble form.

11.4   GENESIS OF SOIL ACIDITY

Acid soils develop mainly due to the influence of relief, acidic parent material such as granite and hot and humid climate. Weathering and dissolution of parent materials result in the hydrolysis of CO_2 under the hot and humid climate with heavy precipitation. Sloppy places with good drainage conditions are supposed to be places where acid soils develop. In acid soils, the concentration of H^+ ions exceeds that of OH^- ions. For a long time, the soil acidity was considered to be related to exchangeable H^+ ions only. The dominance of Al in the soil acidity was reported only in only early thirties. Chernov [4] promulgated the Al theory and claimed that H-clay must get converted to Al-clays due to its stronger adsorption on clay than H+ ions. Coleman and Harward [5], based on their work on titration curves of H-and Al-resins, concluded that clay prepared by concentrated HCl or H-resin treatment is a strong acid soil, which gets changed on storage to an Al-clay [22]. Heavy precipitation resulting in leaching of appreciable amounts of exchangeable basic cations like Ca^{2+}, Mg^{2+}, Na^+ and K^+ from the surface soil leaves relatively insoluble compounds of Al and Fe to remain in the soil. The nature of these compounds is acidic and its oxides and hydroxides react with water to release hydrogen ions in the soil solution causing colloidal exchange sites to be dominated by H^+ and Al^{3+} ions [23, 30]. Under the leaching environment, the dissolution of clay occurs as follows [22]:

$$[X_a - Al_b - Si_c]_{clay} + 8H^+ \longleftrightarrow bAl^{3+} + cH_4SiO_4 + aX^{n+} \tag{1}$$

where, $[X_a - Al_b - Si_c]_{clay}$ represents a generalized clay; and X represents non-Al cations.

As weathering continues, soils become more and more acidic. Soil acidity also results from anthropogenic activities such as improper crop management practices like high level use of nitrogenous fertilizers (ammonium based like ammonium sulfate, urea) and the associated removal of lime-like elements. Ammonium applied through fertilizers is oxidized by microbes and produces strong inorganic acids contributing to soil acidity as follows:

$$(NH_4)_2SO_4^- \longleftrightarrow 2NH_4^+ + SO_4^{2-} \tag{2}$$

$$2NH_4^+ + SO_4^{2-} + Ca\,[clay] \longrightarrow CaSO_4 + [clay]_{NH_4} \tag{3}$$

$$\text{Leached out} \downarrow$$

$$[clay]^{NH_4}_{NH_4} + 3O_2 \longrightarrow [clay]^{H}_{H} + 2\,HNO_3 \tag{4}$$

$$\text{Acid soil}$$

Besides, basic portion of ammonium sulfate is NH_4, which undergoes biological transformation in the soil and forms acid forming nitrate ions. Similarly, Sulphur also produces acid forming sulfate ions through oxidation. Divalent cations of soluble salts usually have a greater effect on lowering soil pH than monovalent metal cations.

Other sources of soil acidity are: humus and organic acids, alumina-silicate minerals, carbon dioxide produced within the soils, and presence of hydrous oxides. Humus in soils occur as a result of microbiological decomposition and has different functional groups like carboxylic (−COOH), phenolic (−OH) etc., which can protonate (H^+) to cause soil acidity. Carbon dioxide is the end product of soil respiration and decomposition of organic matter and produces carbonic acid (H_2CO_3). Acid rains in some regions may also contribute to soil acidity.

It may be noted that soil acidity does not significantly develop in one or two years of intensified use of N fertilizers as shown by a classical long-term field experiment started in 1956 at Kanke, Ranchi, India on a Paleust-alf (acid red loam) of pH 5.8. With application of different combinations of N, NP, NPK, farmyard manure (FYM) and lime, the continuous application of N alone or in combination with P and K decreased the soil pH from 5.8 to 4.0–3.8 in 28 years while application of FYM and lime helped to maintain the initial soil pH [15]. In this set-up, lime was applied at the rate of 2.5 t ha^{-1}, once in four years, while FYM was applied on equivalent N basis. Excessive application of acid forming organic materials like sewage sludge, animal manures to agricultural fields can also decrease soil pH, both by oxidation of ammonium nitrogen released and by organic and inorganic acids formed during decomposition.

11.4.1 TYPES OF ACIDITY AND ASSOCIATED PROBLEMS

Acidity is categorized as active, potential or exchangeable and residual acidity, total being the sum of the three acidities. Active acidity is the amount of hydrogen and aluminum ions in the soil solution or free H^+ ions. Its magnitude is limited. In spite of smaller concentration, active acidity is important since the plant roots and the microbes around the rhizosphere are influenced by it. Potential acidity, on the other hand, represents the amount of hydrogen and aluminum ion adsorbed on cation exchange capacity (CEC). Since the two forms of acidity are in equilibrium with each other, it is necessary to neutralize not only the active but also the potential acidity. Aluminum hydroxyl ions and H^+ and Al ions present in non-exchangeable form with organic matter and clay account for the residual acidity. Most widely used method to measure active acidity is pH of the saturated paste or pre-decided diluted soil water mixture. Potential acidity on the other hand, requires titration of the soil with a base.

11.5   MANAGEMENT OF ACID SOILS

The slightly acidic soils pose little problem with respect to crop production. The problem of phosphorus fixation can be safely managed by applying relatively large quantities of organic matter and/ or mixture of rock phosphate and super phosphate in combination. Application of organic matter, tank slit, conservation tillage, contour and strip cropping, intercropping of cereals with legumes or oilseeds and *in-situ* rainwater harvesting are some of the suitable methods for improving crop production in such upland soils. The real challenges lie in management of acidic soil having pH < 5.5. The management of these soils should aim to correct the acidity by applying amendments to: Reduce the toxicity of aluminum, manganese and iron; improve the nutrient supplying capacity; and to manipulate the agricultural practices for obtaining optimum crop yields. One of the most widely accepted practices for reclamation of soil acidity is liming. Cultivation of acid tolerant crops/varieties is another option, where liming is un-economical or unavailable.

11.5.1   LIMING

Liming is widely adopted practice for amelioration of soil acidity and alleviation of acidity-induced soil fertility and plant nutritional problems [7, 19]. Liming is the application of an alkaline compound of calcium (Ca) or magnesium (Mg) or both to soil to raise its pH to an optimum value. Application of lime eliminates active acidity, minimizes residual acidity and raises the calcium content in the soil [33]. Liming improves base saturation and availability of Ca and Mg. Fixation of P and Mo is reduced by inactivating the reactive constituents. Toxicity arising from excess soluble Al, Fe and Mn is corrected and thereby root growth is promoted and uptake of nutrients improved. Liming also stimulates microbial activity and encourages N_2 fixation and nitrogen mineralization, and therefore, legumes are highly benefited from liming.

11.5.2   LIMING MATERIALS

Liming materials are used to neutralize soil solution hydrogen (H^+) ions to decrease soil acidity (raise the pH). The soil is usually amended with alkaline materials that provide conjugate bases of weak acids. Such conjugate base includes carbonate, hydroxide, oxides and silicates. Because of their common use in agriculture, carbonates, oxides and hydroxides of calcium and magnesium are referred to as agricultural lime. Liming materials (Agricultural lime) include limestone, burned lime, marl, dolomite, stormatolitic and oyster shells. Since calcite and dolomite have industrial use, their application in agriculture may not be economical yet these are the most widely used amendments. Deposits of stormatolitic lime stone, a poor grade lime containing 28–32% CaO, 12% MgO and 0.5% P_2O_5 found in Odisha needs detailed study to reclaim acid soils [11]. Industrials wastes like steel mill slag, blast furnace slag, lime sludge from paper mills, press mud from sugar mills that use carbonation process in manufacturing sugar, wood ashes, cement wastes, precipitated calcium carbonate, etc. are equally effective as ground limestone.

All Ca supplying materials are not liming materials. For example: gypsum, though containing calcium produces sulfate ions, which reacts with soil moisture to produce sulfuric acid as follows:

$$CaSO_4.2H_2O \longrightarrow Ca^{2+} + SO_4^{2-} + 2H_2O \qquad (5)$$

$$SO_4^{2-} + H_2O \longrightarrow H_2SO_4 \qquad (6)$$

Moreover, Ca^{2+} replaces adsorbed Al^{3+}, which results in further lowering of soil pH. Although gypsum $(CaSO_4)$ or calcium chloride $(CaCl_2)$ can be applied to supply calcium ions to the soils but it is worth considering what will be the effects of these salts on soil acidity? Researchers in southeastern United States, Brazil and South Africa have found that gypsum does not increase soil pH, but can ameliorate aluminum toxicity and improve root growth and crop yield. Addition of phospho-gypsum (PG) @ 60 kg of S ha^{-1} to acid soil of Bhubaneswar reduced the exchange acidity, exchangeable Al^{3+} and exchangeable H^+. Researchers need to look at these contradicting issues in more details.

Capacity of liming materials is compared with standard material, i.e., pure calcium carbonate taking its neutralizing value as 100%. Calcium carbonate equivalent (CCE) or neutralizing value (NV) is acid neutralizing capacity of an agricultural liming material expressed as a weight percentage of calcium carbonate.

CCE of a liming material = (Molecular weight of calcium carbonate)/
                                              (Molecular weight of liming material)   (7)

For example, CaO will neutralize 1.78 times more acidity as the same weight of $CaCO_3$. CCE values of different liming materials are calculated using Eq. (7), and these values are given in Table 11.2.

Efficiencies of the liming materials also depend on purity and degree of fineness of liming materials. The finer the material, more rapidly it dissolves into the solution and increases the surface contact with the soil resulting in a higher rate of reaction. The efficiency rating of the materials on the basis of size is given in Table 11.3.

Field studies have revealed that best results are obtained with the liming materials 90% of which can pass through 10 mesh sieve and 50% through 60 mesh sieve. Further at least 20% should pass through a 100 mesh sieve [22]. A good liming material is one that is locally available, can be properly and easily grounded and has high neutralizing value and is of low cost so that it can be used by small and marginal farmers.

TABLE 11.2   CCE Values of Different Liming Materials

Common name	Chemical formula	% $CaCO_3$ equivalent
Basic slag	$CaSiO_3$	86
Burned lime (Quick lime)	CaO	178
Calcite	$CaCO_3$	100
Dolomite	$Ca(CO_3)_2$; $Mg(CO_3)_2$	108.7
Hydrated lime (Slaked lime)	$Ca(OH)_2$	136
Marl	$CaCO_3$	40–70
Wood ash	CaO, MgO, K_2O, etc.	40

TABLE 11.3   Efficiency Rating of Different Fineness of Liming Materials

Size of the sieve	Efficiency rating (%)
Material passing through a 60 mesh sieve	100
Material passing through a 20 mesh but not a 60 mesh sieve	60
Material passing through an 8 mesh but not a 20 mesh sieve	20

11.5.3 CHEMICAL REACTION OF LIMING MATERIALS

Most liming materials whether oxides, hydroxide or carbonates react with carbon dioxide and water to produce bicarbonate when applied to an acid soil. Calcitic or dolomitic limestone reacts slowly with soil colloids, whereas burnt lime and hydrated lime react faster and bring about changesin soil pH within a few days. The reaction of calcite ($CaCO_3$) in acid soil is shown below:

$$CaCO_3 + CO_2 + H_2O \longrightarrow Ca^{2+} + OH^- + 2HCO_3^- \qquad (8)$$

The Ca^{2+} (Ca^{2+} and Mg^{2+} in dolomite) helps in displacing the H^+ and Al^{3+} ions from the exchange sites, and OH^- ions liberated in solution to neutralize the acidic components as follows:

$$Al^{3+} + 3OH^- \longrightarrow Al(OH)_3 \qquad (9)$$

$$H^+ + OH^- \longrightarrow H_2O \qquad (10)$$

$$[Colloid]_{Al^{3+}}^{H^+} + Ca^{2+} + 2HCO_3^- \longrightarrow [Colloid] + Al(OH)_3 + H_2O + CO_2 \quad (11)$$

The insolubility of $Al(OH)_3$, the weak dissociation of water and the release of CO_2 gas to the atmosphere, all pulls these reactions to the right. In addition, the adsorption of the calcium and magnesium ions reduces the percentage acid saturation of the colloidal complex and the pH of the soil solution increases correspondingly.

11.5.4 LIME REQUIREMENT (LR)

Liming involves a considerable cost in crop production programs and therefore it is quite important to assess the required amount of lime to be applied to raise the soil pH to an optimum value. This value is usually in the range of 6.0–7.0. In addition to soil pH, the amount of lime to be applied is affected by a number of other factors such as surface and sub-surface texture and structure, clay content, CEC, base saturation, amount of organic matter, crops to be grown, fineness of liming materials, and economic returns likely to be achieved through the lime application. A fine textured acid soil requires much larger quantity of lime than a sandy soil or a loamy soil with the same pH value. Mainly the lime requirement is governed by the following factors:

- required change in soil pH;
- the buffer capacity of soil;
- depth of soil to be ameliorated; and
- efficiency of the liming material.

Several principles have been developed to estimate lime requirement, mainly based on soil-lime incubation, soil based titration and buffer solutions (specific known pH). Some of the commonly used methods are based on buffer solution principle. When buffer solution (usually $\geq$ 7.3 pH) is added to the soil, the buffer pH is depressed in proportion to the original soil pH and its buffering capacity, the magnitude in the change in pH representing the lime requirement. For example, a large drop in buffer pH would indicate a low pH soil with a great potential acidity, and a high lime requirement. The Shoemaker et al. [31] indicated that single buffer method is widely adopted and commonly used for soils with pH<5.8, OM <10%, and appreciable quantities of soluble Al^{3+}. For soils with low lime requirement, this method often overestimates the lime requirement. Mehlich [16] buffer is another method that is predominantly used in highly weathered soils. Modified Woodruffs buffer method [3] has been successfully used to assess the lime requirement of acid soils in Odisha (2.02 to 6.08 tha^{-1}). A preliminary research findings on "Comparison of Modified Woodruff Buffer in Missouri Soils" revealed a good correlation ($R^2 = 0.952$) between the modified Woodruff and Mehlich buffers [17]. Exchangeable Al as lime requirement index is also used for highly weathered soils with pH < 5.0, low CEC where exchangeable Al is equal to or exceeds the total exchangeable bases. Field trials on the other hand proved that the yields were optimized by applying the limestone 1 to 2.5 equivalents of exchangeable Al^{3+} and soil pH around 5.5 depending on soil characteristics [24]. To ease the process of calculating lime requirement, a number of ready reckoners for location specific situation have been prepared. One of such ready reckoner to raise the soil pH to 6.5 from different degradation is reported as Table 11.4.

TABLE 11.4   Requirement of Lime for Reclamation of Acid Soils in Odisha (kg ha^{-1})

pH range	Soil type		
	Sandy soil	Loam	Clay and loamy clay
4.5–5.0	1750*	2650	4100
	2300#	3500	5450
5.1–5.5	1100	1600	2550
	1450	2100	3400
5.6–6.0	600	900	1350
	800	1200	1800

Soils having pH < 4.5 are not considered.

* Pure CaCO$_3$.

Equivalent quantity of paper mill sludge.

Source: https://www.nabard.org/english/ld_acidsoils.aspx [9].

However, results of the field trials reveal that to get the optimum yields of most crops one half of the lime requirement (LR) dose determined by Woodruffs buffer (pH 7.0) is good enough for maize and cotton grown in lateritic soils. Application of full LR dose suppressed P availability and caused B deficiency. Scientists also recorded good economic response to lime sludge applied at 0.5 LR dose in a variety of crops in red loam soils of Semiliguda having pH around 5.0.

11.5.5   METHOD AND TIME OF LIME APPLICATION

Lime requirement is dictated by the root zone depth of the plants as this depth must be reclaimed. Liming for general arable cropping is most effective if lime is broadcasted uniformly over the field and mixed thoroughly into the soil (10–15 cm), since this depth represents most of the active root zone from where the plants get water and nutrients. For horticultural crops, the depth should be increased suitably. Since uniform spreading without drifting of the applied lime is increasingly difficult with finely ground limestone products, a moisture content of 7 to 10% in fine limestone is required to minimize drift and uniform spreading pattern. Liming agents should be applied to soils at least one month before sowing the crops or they should be applied and thoroughly mixed with soil just after harvesting the crops.

Furrow application is another method to avoid dusting and save on the amount of lime application. Yields of crops were found to be higher when 250–500 kg limestone ha^{-1} was applied in furrows every year as compared to broadcasting at higher rates [20, 22]. In hilly terrains of India, where inputs are carried to distant agricultural fields by head load, furrow application at lower rates suits the requirement as it is easier to carry the amendments to distant fields.

11.5.6   FREQUENCY OF APPLICATION OF LIME

Frequency of liming mostly depends on soil type (texture) and amount and intensity of precipitation of the region. For coarse textured soils in heavy rainfall regions, lime is applied more frequently in split applications because, liming material get little time to react with soil due to leaching. It has been shown that in northeast India where rainfall is about 3,000 mm yr^{-1}, the lime applied once at optimum dose is sufficient for two years only, because the precipitated exchangeable Al^{3+} starts to reappear with cropping seasons. After two years, limestone application is needed to sustain productivity although its rate can be about half the previous rate [23]. The areas where rainfall is around 2,000 mm yr^{-1}, the frequency of lime application can be increased to 3 to 4 years depending on the crops grown. A small dose of lime @10–20 % of LR to each crop instead of one time application of full dose gave better results. For row crops, placement is the best way to apply the amendment. How to apply the amendment uniformly in fields where sowing is done through broadcasting method is still an issue to be resolved?

11.5.7   IMPACT OF LIMING ON CROP YIELD

Kumar et al. [11] studied the effect of furrow application of lime (control, 0.2, 0.4 and 0.6 t ha^{-1}) on rice bean (*Vigna umbellata*) and observed that liming not only increased the growth and yield but also enhanced the crop quality. Maximum gross returns (INR 39,098 ha^{-1}; 1 US$ = Rs. 64), net returns (INR 27,281 ha^{-1}), benefit-cost (B: C) ratio (2.29) and production efficiency at 0.6 t ha^{-1} were observed with the application of lime. Indian Council of Agricultural Research (ICAR) formulated Network Project on

Acid Soils (NPAS) for optimized liming dose for different crops in acid soil areas throughout the country. The general guideline emerging from the studies revealed that application of lime @ 0.2–0.4 t ha^{-1} in combination with 50% recommended dose of NPK results in equal or a slightly higher yield than application of 100% NPK alone (Table 11.5).

11.5.8 ORGANIC MANURES AS AMENDMENT

Application of lime along with the organic manures is more efficient having synergic relation. Organic manures applied to acid soils counters Al^{3+} toxicity by forming organo-Al-complexes which are non phyto-toxic compounds. It has been observed that fertilizers respond well when farmyard manure (FYM) is applied to each crop in Al-toxic acid soils [21]. The periodic analysis of soil samples following FYM application revealed that

TABLE 11.5   Yield (t ha^{-1}) of Crops in Acid Soils with 100% NPK and 50% NPK+ Lime

State	Crop	100% NPK	50% NPK+ lime	Change in yield (%)
Assam	Rapeseed	0.97	1.01	+4.1
	Summer green gram	0.44	0.52	+18.2
Himachal Pradesh	Maize	3.40	3.31	−2.6
	Wheat	2.79	2.37	−15.0
Jharkhand	Maize + Pigeon pea	6.90	6.50	−5.8
	Pea	3.84	5.08	+32.3
Kerala	Cowpea	0.86	1.06	+23.2
	Black gram	0.64	0.81	+26.6
Meghalaya	Maize	3.05	3.03	−0.7
	Groundnut	1.42	2.13	+50.0
Orissa	Groundnut	2.25	2.36	+4.9
	Pigeon pea	1.20	1.22	+1.7
West Bengal	Mustard	0.82	0.84	+2.4
	Wheat	1.67	1.71	+2.4
All States		**2.16**	**2.28**	**5.5**

RDF is recommended dose of fertilizers;

* Equivalent yield of maize.

application of FYM initially decreased the soluble Al without affecting soil pH and favored growth of plant roots. However, the difference was eliminated with time as a result of release of Al from unstable organo-Al-complexes and subsequent adsorption on soil colloid. Apparently, application of FYM is required to each crop to avoid Al toxicity in acidic soils.

Bhat et al. [2] studied the effect of various low cost liming materials alone and in combination with organics. The results revealed that 1/5th of LR through basic slag in combination with organics (poultry manure (PM) and FYM) gave better response through higher yield (Table 11.6) as compared to liming materials alone as a result of better nutrient availability and plant uptake (Table 11.7) and resulted in higher B:C ratios (Table 11.8).

TABLE 11.6   Effect of Liming on Nutrient Availability and Uptake in Mustard and Rice Crops

Treatment	Availability (kg ha^{-1})			Uptake (g kg^{-1})		
	N	P	K	N	P	K
Mustard						
No lime	145.99	10.28	154.4	7.40	2.04	5.83
BS 1/5th LR	262.59	26.39	268.31	9.00	3.54	6.86
BS 1/10th LR	195.95	17.89	211.07	8.20	2.89	6.58
Ca 1/5th LR	241.75	2.54	236.11	8.70	3.13	6.33
Ca 1/10th LR	182.20	16.85	192.87	7.90	2.82	6.23
BS 1/5th LR+ PM	313.44	32.63	321.38	9.40	3.83	7.18
BS 1/5th LR+FYM	299.10	30.16	303.74	9.20	3.58	6.96
Rice						
No lime	175.54	12.53	155.68	6.22	0.79	11.33
BS 1/5th LR	229.4	18.62	243.29	7.17	1.24	15.16
BS 1/10th LR	229.40	19.55	226.75	6.80	0.92	13.98
Ca 1/5th LR	230.13	22.83	206.05	6.99	1.04	13.03
Ca 1/10th LR	212.40	16.36	213.54	6.96	0.94	11.93
BS 1/5th LR+ PM	305.38	32.58	344.08	7.48	1.42	16.98
BS 1/5th LR+FYM	287.24	29.65	322.35	7.28	1.32	16.46

BS – basic slag; LR – Lime requirement; PM – poultry manure; FYM – farmyard manure.

TABLE 11.7   Economics of Lime Application in Wheat Crop

Treatment	Price of lime (Rs.)	Yield (t ha⁻¹)*	Yield increased over check	Percent response	Price of increased yield (Rs.)	Profit over check (Rs. ha⁻¹)	B:C ratio
No lime	–	1.71	–	–	–	–	–
BS 1/5th LR	1093	2.48	0.77	45.6	6223	5130	4.7
BS 1/10th LR	546	2.00	0.29	17.6	2366	1820	3.3
Ca 1/5th LR	2352	2.32	0.61	35.7	4874	2522	1.1
Ca 1/10th LR	1176	1.87	0.16	9.4	1309	133	0.1
BS 1/5th LR+ PM	1093	2.74	1.03	60.2	8240	7147	6.5
BS1/5th LR+FYM	1093	2.68	0.97	56.7	7760	6667	6.1

*Mean of 14 experiments, price of wheat @ Rs 8 per kg, price of basic slag Rs. 800 per ton; price of calcite @ Rs 2000 per ton; 1 US$ = Rs. 64).

BS – basic slag; LR – lime requirement; PM – poultry manure; FYM – farmyard manure (1 US$ = Rs. 64).

TABLE 11.8 Optimum pH Value of Various Crops

Crops	Optimum pH
Cotton	5.0–6.5
Egyptian clover(*Berseem*)	6.0–7.5
Field beans, soybean, pea, lentil	5.5–7.5
Groundnut	5.3–6.6
Maize, sorghum, wheat, barley	6.0–7.5
Millets	5.0–6.5
Oats	5.0–7.7
Potato	5.0–5.5
Rice	4.0–6.0
Sugarcane	6.0–7.5
Tea	4.0–6.0

The results of a permanent manurial experiment at Ranchi on an acid red loam Alfisol also revealed that fertilizers alone increased the yields of maize and wheat up to 15th year but sharp reduction occurred in subsequent two years under N, NP and NPK treatments [15]. Maize failed to grow at the end of 27th year while wheat still continued to yield 0.8 t ha^{-1} under NPK treatment. The soil pH decreased from 5.7 to 4.5 as a result of continuous fertilizer application. Application of NPK with lime sustained higher yield of crops and maintained soil fertility. Organic manure alone or in combination with fertilizer also maintained yield and soil fertility at a moderate level. The results of a study in Meghalaya suggest that liming along with integrated nutrient management practices, if adopted properly, can lead to more than three-fold increase in maize productivity on acidic soils of north-eastern states of India [14].

Although liming is beneficial, over liming, i.e., application of lime in such quantities that the resultant soil pH is too high for optimal plant growth, besides cost may also decrease the yields. Over liming causes imbalance of nutrients in the soil system. It increases the deficiency of phosphorus, potassium, iron, copper and zinc in the soil and alters the ratio of Ca and Mg. The over liming is found to increase the disease infestation especially in root crops like Scab in potatoes.

11.5.9 INDUSTRIAL WASTES AS AMENDMENTS

Industrial wastes like basic slag, blast furnace slag, lime sludge, cement kiln wastes have been used as liming materials. Since the industries producing these by-products are located in acid soil regions of Odisha, West Bengal, Karnataka, Chhattisgarh and Jharkhand, there is a great opportunity to use these products as soil amendments. While Basic slag, blast furnace slag are the by-products of steel industries, lime sludge is a by-product of the paper mills. Indian slag contains 1–7% P_2O_5, 24–50% CaO and 2–10% MgO. Lime sludge contains 65–85% $CaCO_3$, 2% R_2O_3 (sesquioxides), 1% free CaO and 1.5% free alkali.

Basic slag, a byproduct of the basic open-hearth method of making steel contains 32% total calcium and 3.5–8% total phosphorus out of which 62–94% is citrate soluble. Its performance compares well with superphosphate in acid soils with leguminous crops. Several studies in India indicated that the application of basic slag at 1–1½ times the lime requirement of acid soils resulted in higher yield of paddy. In a field experiment with clay-loam soil of pH 5.5 in CRRI, Cuttack, application of basic slag resulted in higher pH and higher total and extractable P. The rice yield increased by 0.3 t ha^{-1}. In a field trial at Palampur with soil pH 5.2, the yield of rice-wheat cropping system was significantly higher with basic slag from Bhilai steel plant. The beneficial effects of basic slag on maize, wheat, gram and groundnut in Bihar have been documented. Basic slag may contain some heavy metals (Cd, Cr, Ni). However, high amount of Fe and Al present in red and laterite soils bind these toxic heavy metals tightly and thus prevent them to remain in available form in soil and plant. Thus, there is hardly any possibility of contamination of underground aquifers through percolation. The main disadvantage of basic slag is its hardness and poses difficulties in grounding to optimum particle size of 60 mesh. The grinding using ball mill is quite expensive.

Paper mill sludge is capable to ameliorate acidic soils resulting in higher yields of various crops by 34–68%. The Odisha Government adopted this technology and ameliorated more than 0.24 M ha acidic land by supplying Paper Mill Sludge at subsidized rate (Rs. 10.00 for 50 kg bag) resulting in unit cost of Rs. 12,000 to 15,000 ha^{-1} depending on the soil pH [8].

11.6   CROP SELECTION: ACID TOLERANT CROPS

Crops differ in their tolerance to soil acidity, there being inter and intra-genic differences amongst various crops. Thus, where liming materials are either unavailable or farmers are unable to afford it because of cost involved, selection of crops in favor of acid tolerant crops is the best management option to get sustainable yields. Different crop plants have their specific optimum pH requirement. Rice, oat and linseed can endure a fairly acidic reaction (pH = 5.0) while barley, sugar beet, lucerne, etc., can tolerate a fairly alkaline reaction (pH = 8.0). Optimum pH values for various crops given in Table 11.8 show that rice and tea can be grown at a low pH of 4.0. Rice has certain amount of tolerance to soil acidity; and flooding of the field also creates favorable conditions (increase in pH and availability of P, Si, and K) for growth of rice. Minor millet (*Panicum miliare*) and finger millet (*Eleusine coracana*) are also quite tolerant to acidity. While Bengal gram, lentil, groundnut, sorghum, maize and field peas are medium tolerant, pigeon pea, soybean and cotton are sensitive. Even in limed lands, it is preferable to cultivate high response crops like cotton, soybean, pigeon pea etc. in the first year of liming, followed by medium response crops like maize and wheat in the subsequent seasons. The low responsive crops may be grown when the effect of liming is further reduced. Some varieties/cultivars of various crops show acid tolerance (Table 11.9). Since this option is most easy to adopt and is cost effective, there is a need to broaden research on breeding of acid tolerant crops especially pulses, oilseeds and cereals.

In the Western Ghats of India, commercial cultivation of horticultural and plantation crops is practiced as these are best suited to the acid soil conditions and are remunerative. In the northeast, farmers are accustomed to cultivate ginger, turmeric, sweet potatoes, colossal, alocasia, *taro*, and potatoes in acidic soils. These crops are also known to survive well at low P as compared to the recently developed high yield yielding agricultural crops. Pineapple (*Ananas cosmosus*), coffee (*Coffea* sp.), tea (*Camellia* sp.), rubber (*Achra zapata*), cassava (*Manihot* sp.), sweet potato (*Ipomea batata*), rice (*Oryza sativa*), finger millet (*Eleusine coracona*), buckwheat (*Fagopyrum esculentum*) as well pasture species such as guinea grass

TABLE 11.9   Acid Tolerant Cultivars/Verities of Different Crops [28]

Acid soil region/State	Crop	Varieties
Assam	Rapeseed	Varuna, Sonmukhi
	Summer green gram	K851, Sonmugu
Himachal Pradesh	Soybean	Bragg, Ph1, Harasoya
	Gobhi sarson	ONK 1, Hoyala, HPN-3
Jharkhand	Black gram	KU-301
Kerala	Vegetable cowpea	Bhagyalakshmi
	Cowpea	V-16
Meghalaya	French bean	HUR-15
Orissa	Groundnut	Smruti
	Pigeon pea	UPAs-120
West Bengal	Mustard	Sanjukta, Pusa bold
	Wheat	K9107, PBW 343

(*Panicum maximum*), jajagua (*Hypaahanea rufa*), molasses grass (*Melinis minutiflora*) and *Brachiaria decumbens* are known to tolerate aluminum toxicity to a great extent.

In Odisha, about 0.1 M ha medium-low land rice suffers due to iron toxicity resulting in yield loss to the extent of 60 to 80%. Several rice genotypes were evaluated in iron toxic laterite soils of Bhubaneswar for three consecutive years. Rice genotypes like *Kalinga III, Udayagiri, Panidhan* and *Tulasi* are tolerant to iron toxicity while, *IR 36, Konark, Birupa, Gajapati, Samalai* and *Indrabati* are moderately tolerant to iron toxicity. Genotypic difference in the degree of susceptibility to excess Fe was confirmed by changes in the content of metabolically active Fe^{2+}, chlorophyll and enzymatic activity in plant parts. Metabolically-active Fe^{2+} contents in the leaves, stems and roots of tolerant genotype (ASD-16) were 212, 392 and 4674 ppm as compared to highly susceptible rice genotype (*ADT-36*) having 281, 442 and 5933 ppm, respectively. Genotypic differences in degree of susceptibility to excess Fe were attributed to leaf tissue tolerance of high level of Fe, reduced translocation from root to shoot and ability of roots to resist its entry inside the plant.

11.7   ACID SULPHATE SOILS

Acid sulfate soils are emerging as a global environmental problem, particularly in coastal regions where sulphidic materials (FeS_2 and others) dominate/accumulate in the soil. In India, acid sulfate soils are mostly found in Kerala, Odisha, Andhra Pradesh, Tamil Nadu, and West Bengal. Usually, these materials remain submerged by sea water which maintains them in reduced conditions. Soils containing sulfidic minerals that have not yet been oxidized are referred as potential acid sulfate soils. When these soils are drained or altered enough for cultivation, sulfides begin to oxidize. It may also happen as a result of tectonic uplift or oceanic regression. Under humid or moist aerobic conditions, sedimentary sulfide minerals can oxidize chemically, but this is a slow process, probably due to particular rate-limiting reactions. Various micro-organisms are adapted to oxidize sulfides either directly through sulfur transformations or by facilitating (catalyzing) such rate-limiting reactions. Oxidation of these sulphidic materials leads to formation of sulfuric acid, reducing the soil pH below 4.0 as follows:

$$2S + 3O_2 + 2H_2O \longrightarrow 2H_2SO_4 \tag{12}$$

$$H_2S + 2O_2 \longrightarrow H_2SO_4 \tag{13}$$

Accelerated by bacteria:

$$4FeSO_4 + O_2 + 2 H_2SO_4 \longrightarrow 2Fe_2(SO_4)_3 + 2H_2O \tag{14}$$

Such strong acidity in acid sulfate soils causes toxicities of aluminum, iron, manganese and hydrogen sulfides (H_2S) gas. Hydrogen sulfides (H_2S) formed in lowland rice soils causes a kiochi disease that prevents rice plant roots from absorbing nutrients. Recently, elevated levels of dissolved iron from acid sulfate soils drainage have been implicated in bloom of the cynobacterium (*Lyngbya majascula*) in coastal water. Toxins from these blooms have major impacts on most aquatic organisms but also pose a significant health risk to humans.

11.7.1   MANAGEMENT OF ACID SULPHATE SOILS

The best management of acid sulfate soil requires knowledge of their distribution, depth of sulfide layer from the soil surface, the acidity stored in the sulfide layer and drainage system. A primary preventative consideration with acid sulfate soils should be an avoidance strategy that includes the decision not to drain a potential acid sulfate soil. This solution almost limits the use of the area to rice growing.

Acid sulfate soil may require up to 224 Mt ha^{-1} of lime within 10 year period or less. It is not a cost effective method. Neutralization of acid sulfate soil by the alkalinity of seawater is the ultimate natural process of management. Under lowland rice growing areas, application of MnO_2 along with lime may reduce the sulfide formation that may ultimately prevent the chances the formation of H_2SO_4.

11.8   SUMMARY

Acid soils occupy approximately 60% of land area of the world. In India, around one-third of the cultivated land is affected by soil acidity resulting in low productivity. Such low productivity is not only due to acidity *per se* of the soils, but also to deficiency of some micronutrients like B, Mo, and Zn and toxicity of others like Fe and Al. Although, soil acidification is a natural process but it is greatly accelerated by intensive cultivation with faulty management practices. The good option to reclaim acid soils is liming. Low cost liming materials and industrial wastes like basic slag, lime sludge and press mud are effective in neutralizing acidity.

The best results are achieved when liming materials are applied along with organics as the organic matter reduces the iron and aluminum toxicity in acid soils, hence supporting sustainable crop production. Liming also helps to some extent in correcting the deficiencies/toxicity of the micronutrients. Other soil/crop management practices include selection of acid tolerant crops/crop cultivars that may not only reduce the lime requirement and cost of reclamation but also may help to partially alleviate the problem.

There is a great need to expand research on breeding of acid tolerant crops especially pulses, oilseeds and cereals. Occurrence of acid sulfate (cat clay) soils is pronounced in areas dominated by soils having sulfidic material. Sulfurization leads to reduction in soil pH below 4.0 in active acid sulfate soils that limits the crop growth. Prevention of oxidation is the only way to prevent active acid sulfate soil formation. Neutralization of such soils with liming material is uneconomical, but it can provide bases for better crop growth. Avoidance is often the most environmentally responsible and cheapest option. To make these areas more productive keep them in flooded by growing rice and use basic fertilizers.

KEYWORDS

- acid soils
- acid sulfate soils
- acidity
- active acidity
- alumino-silicates
- amelioration
- basic slag
- buffering capacity
- carbonic acid
- CCE
- degradation
- dolomite
- fineness of liming materials
- fixation capacity
- FYM
- hydrolysis
- hydroxides
- industrial waste
- leaching
- lime requirement
- liming

- **management of acid soils**
- **neutralizing value**
- **nutrient management**
- **organic manure**
- **oxidation**
- **parent materials**
- **pH**
- **potential acidity**
- **poultry manure**
- **protonation**
- **reclamation**
- **soil fertility**
- **tolerant crops**
- **toxicity**

REFERENCES

1. Badole, S., Datta, A., Basak, N., Seth, A., Padhan, D., & Mandal, B. (2015). Liming influences forms of acidity in soils belonging to different orders under sub-tropical India. *Communications in Soil Science and Plant Analysis,* 46, 2079–2094.
2. Bhat, J. A., Kundu, M. C., Hazra, G. C., Santra, G. H., & Mandal, B. (2010). Rehabilitating acid soils for increasing crop productivity through low-cost liming material. *Science of the Total Environment,* 408, 4346–4353.
3. Brown, J. R., & Cisco, J. R. (1984). An improved Woodruff buffer for estimation of lime requirement. *Soil Science Society of America Journal,* 48, 587–591.
4. Chernov, V. A. (1947). *The Nature of Soil Acidity.* (English Translation Sciences, USSR, Moscow). *Soil Science Society of America* (1964).
5. Coleman, N. T., & Howard, M. E. (1953). The heats of neutralization of acid clays and cation exchange resins. *Journal of American Society of Chemistry,* 75, 6045–6046.
6. Crawford, T. W. (2008). *Solving Agricultural Problems Related to Soil Acidity in Central Africa's Great Lakes Region.* An International Center Soil Fertility and Agriculture Development. Alabama.
7. Haynes, R. J. (1984). Lime and phosphate in the soil–plant system. *Advances Agronomy,* 37, 249–315.
8. http://icar.org.in/en/node/1566 (Opened on 30.09.2015).

9. https://www.nabard.org/english/ld_acidsoils.aspx (Opened on 30.09.2015).

10. ICAR and NAAS. (2010). *Degraded and Wastelands of India – Status and Spatial Distribution.* Indian Council of Agricultural Research and National Academy of Agricultural Sciences, New Delhi, 158 p.

11. Jena, D. (2013). Potential of industrial by-products in ameliorating acid soils for sustainable crop production. *The Fourth Dr. G. S. Sekhon Memorial Lecture delivered on 2nd September 2013 at Birsa Agricultural University, Kanke, Ranchi.* Ranchi Chapter of the Indian Society of Soil Science, Ranchi.

12. Kiiya, W. W., Mwwoga, S. W., Obura, R. K., & Musandu, A. O. (2006). Soil acidity amelioration as a method of sheep sorrel (*Rumex acetosella*) weed management in Potato (*Solanum tuberosum* L.) in cool highlands of the north Rift, Kenya. *KARI Biannual Scientific Conference.* KARI. Nairobi.

13. Kumar R., Chatterjee, D., Kumawat, N., Pandey, A., Roy, A., & Kumar M. (2014). Productivity, quality and soil health as influenced by lime in rice bean cultivars in foothills of north-eastern India. *The Crop Journal, 2,* 338–344.

14. Kumar, M., Khan, M. H., Singh, P., Ngachan, S. V., Rajkhowa, D. J., Kumar, A., & Devi, M. H. (2012). Variable lime requirement based on differences in organic matter content of iso-acidic soils. *Indian Journal of Hill Farm, 25,* 26–30.

15. Lal, S., & Mathur, B. S. (1989). Effect of long-term manuring, fertilization and liming on crop yield and some physicochemical properties of acid soil. *Journal of the Indian Society of Soil Science, 36,* 113–119.

16. Mehlich, A. (1976). New buffer method for rapid estimation of exchangeable acidity and lime requirement. *Communications in Soil Science and Plant Analysis, 7,* 637–52.

17. Nathan, M., Scharf, P., & Sun, Y. (2005). *Comparison of Woodruff Buffer and Modified Mehlich Buffer Tests for Determining Lime Requirement in Missouri Soils.* Report for First Year. http://aes.missouri.edu/pfcs/research/prop205a.pdf.

18. Panda, N., Sarkar, A. K., & Chamuah G. C. (2009). Soil acidity. In: *Fundamentals of Soil Science*, Indian Society of Soil Science. 2nd ed., New Delhi, India, 317–328.

19. Patiram, P. (1991). Liming of acid soils and crop production in Sikkim. *Journal of Hill Research, 4,* 6–12.

20. Patiram, P. (1996a). Efficacy of furrow application of limestone on soybean and wheat grown in sequence on an acid soil. *Indian Journal of Agricultural Sciences, 66,* 81–85.

21. Patiram, P. (1996b). Effect of limestone and farmyard manure on crop yields and soil acidity on an acid Inceptisol of Sikkim, India. *Journal of Tropical Agriculture (Trinidad), 73,* 238–241.

22. Patiram, P. (2007). Management and future research strategies for enhancing productivity of crops on the acid soils. The 14th Dr. D.P. Motiramani Memorial Lecture. *Journal of the Indian Society of Soil Science, 55,* 411–420.

23. Patiram, Rai, R. N., & Prasad, R. N. (1990). Frequency of lime application to wheat-maize crop sequence on an acid soil of Sikkim. *Journal of the Indian Society Soil Science, 38,* 723–727.

24. Patiram, P., Singh, K. P., & Rai, R. N. (1989). Liming for maize production on acid soils of Sikkim. *Journal of the Indian Society of Soil Science, 37,* 121–125.

25. Ritchey, K. D., & Snuffer, J. D. (2002). Limestone, gypsum and magnesium oxide influence the restoration of an abandoned Appalachian pasture. *Agronomy Journal,* 94, 830–839.

26. Sarkar, A. K. (2005). *Annual Report, ICAR Network Project on Soil Characterization and Resource Management of Acid Soil Regions for Increasing Productivity.* Birsa Agricultural University, Ranchi.

27. Sarkar, P. K., & Debnath, N. C. (1989). Effect of aluminum concentration in rooting medium on the growth and nutrition of two cultivars of wheat. *Journal of the Indian Society of Soil Science,* 37, 506–512.

28. Sen, H. S. (2003). Problem soils in India and their management; Prospect and Retrospect. *Journal of the Indian Society of Soil Science,* 51, 388–408.

29. Sharma, P. D., & Sarkar, A. K. (2005). *Managing Acid Soils for Enhancing Productivity.* Indian Council of Agricultural Research NRM Division, *Krishi Anusandhan Bhavan*-11 New Delhi, p. 22.

30. Sharma, U. C., & Singh, R. P. (2002). Acid soils of India: their distribution, management and future strategies for higher productivity. *Fertilizer News,* 47, 45–52.

31. Shoemaker, H. E., McLean, E. O., & Pratt, P. F. (1961). Buffer methods for determining the lime requirement of soils with appreciable amounts of extractable aluminum. *Soil Science Society of America Proceedings,* 25, 274–277.

32. Singh, K., Mishra, A. K., Singh, B., Singh, R. P., & Patra, D. D. (2014). Tillage effects on crop yield and physicochemical properties of sodic soils. *Land Degradation & Development. Published online in Wiley Online Library (wiley-on-line library.com) doi: 10.1002/ldr.2266.*

33. Somani, L. L., Totawat, K. L., & Sharma, R. A. (1996). *Liming Technology for Acid Soils.* Agrotech Publishing Academy, Udaipur, 240 p.

34. Von Uexkull, H. R., & Mutert, E. (1995). Global extent, development and economic impact of acid soils. In: *Plant-Soil Interaction at Low pH: Principles and Management.* Date, R. A. (Ed.). Klüwer Academic Publishers, Dordrecht, pp. 5–19.

35. Zhao, G., Mu, X., Wen, Z., Wang, F., & Gao, P. (2013). Soil erosion, conservation, and eco-environment changes in the loess plateau of China. *Land Degradation & Development,* 24, 499–510.

CHAPTER 12

NUTRIENT MANAGEMENT FOR SUSTAINED CROP PRODUCTIVITY IN SODIC SOILS: A REVIEW

N. P. S. YADUVANSHI

CONTENTS

12.1 INTRODUCTION

The world population will soar to 9.7 billion by the end of 2050. India, with a higher rate of population growth than China, may soon become the most populous country in the world. The projections reveal that to ensure food to the teeming billions, globally as well as in India around 60% more food grains need to be produced besides additional production of fodder and fuel. Since most productive lands have already been committed to various sectors, there is little scope to bring additional area under crops. Rather, due to diversion of good fertile lands for non-agricultural purposes like urbanization, roads and industry, the area under food crops may shrink. Therefore, one way to increase production is to increase productivity (yield per unit area) and the other to restore the productivity of marginal soils like salt affected lands and expanding irrigation utilizing saline water irrigation technologies in arid and semiarid regions having abundance of poor quality ground water. Out of an estimated 952.2 million ha (M ha) of salt affected soils in the world, India occupies about 6.73 M ha, that are either lying barren or produce very low and uneconomical yields of various crops [42].

Based on the soil pH, exchangeable sodium percentage (ESP), concentration and nature of soluble salts and the reclamation procedure to be adopted, salt affected soils from management point of view in India have been classified in two major categories namely: Saline soils and sodic soils, 2.96 M ha being saline and 3.77 M ha sodic/alkali (henceforth in this chapter described as sodic) in nature. The affected area is expected to increase further with expansion of canal irrigation and intensive exploitation of poor quality ground waters for agriculture in non-canal command areas. Amongst the two kinds of soils, sodic soils in general have low fertility status due to high pH, excess soluble and exchangeable Na, high amounts of $CaCO_3$, negligible to low organic matter content and adverse soil physical conditions. Therefore, any attempt to sustainable reclamation of sodic lands, nutrient management is likely to play a very significant role.

This chapter addresses this issue in the light of extensive work carried out at Central Soil Salinity Research Institute (CSSRI), Karnal, India.

12.2 SODIC SOILS

Sodic soils are high in exchangeable sodium compared to calcium and magnesium. Electrical conductivity of the saturation paste extract (EC_e) is variable and could be less/more than 4 dS m^{-1}, pH of the saturation paste greater than 8.2 and exchangeable sodium percentage greater than 15. These soils thus, have been either categorized as saline-sodic or sodic although management principles being adopted remain similar. High exchangeable sodium, high pH, and low calcium and magnesium combine to cause the soil to disperse, meaning that individual soil particles act independently. The dispersion of soil particles destroys soil structure and prevents water movement into and through the soil by clogging pore spaces. Sodic soils often have a black color due to dispersion of organic matter and a greasy or oily-looking surface with little or no vegetative growth. These soils are also referred as "black alkali" or "slick spots' and upon drying develop small cracks (Figure 12.1).

Growth of most crop plants on sodic soils is adversely affected because of impairment of soil physical conditions, disorder in nutrient availability and transformations and suppression of biological activity due to high pH, exceeding even 10 in severe cases. Salt solution in such soils has preponderance of sodium carbonates and bicarbonates capable of alkaline hydrolysis, thereby saturating the absorbing complex with sodium, exchangeable sodium percentage going up to 90% or more [2, 28]. The sodic soils of the Indo-Gangetic plain are generally devoid of gypsum $(CaSO_4 \cdot 2H_2O)$ but are calcareous, with $CaCO_3$ content increasing with depth, which is present in amorphous or concretion form or even as an indurate bed at about 1 m depth (Table 12.1). The accumulation of $CaCO_3$ generally occurs within the zone of fluctuating water table. The dominant clay mineral is illite. The processes that target the dissolution of $CaCO_3$ have significant role in reclamation of sodic soils. These soils are deficient in organic matter, available N, Ca, and Zn. Certain micro-nutrients present problems of either deficiency or toxicity. Recent studies have revealed

FIGURE 12.1 A view of sodic soil in Haryana, India (right) and small cracks that develop upon drying (right).

TABLE 12.1   Some Important Properties of a Sodic Soil of the Indo-Gangetic Plain

Depth (cm), Horizon	EC$_e$ (dS m^{-1})	pH$_s$	CaCO$_3$ < 2mm (%)	Sand (%)	Silt (%)	Clay (%)	CEC (cmol-ckg^{-1} soil)	ESP
0–10 Ap	22.34	10.6	5.1	67.5	17.6	12.2	5.4	90.7
10–48 B21	6.28	10.2	8.9	55.8	23.4	18.5	9.3	87.1
48–76 B22	4.19	9.8	9.4	46.0	29.5	22.2	9.4	88.3
76–104 B2 ca	2.34	9.5	12.6	36.2	28.4	29.3	12.6	84.9
104–163 C ca	1.31	9.6	13.8	27.4	38.4	30.7	13.8	68.8

CEC = Cation exchange capacity

that toxicities of Al, Mn, and Fe may sometimes pose problems for wheat when water stagnates due to heavy rainfall or over- irrigation resulting in yellowing of the crop.

12.2.1 RECLAMATION/MANAGEMENT OF SODIC SOILS

The reclamation technology involves the use of amendments like calcium salts such as gypsum, acid or acidifying materials like S or H_2SO_4 or industrial wastes such as pyrites, press mud and others to lower pH and ESP. It is followed by leaching to remove the reaction products and soluble salts in the soil solution. Soils with a sodicity problem must have adequate drainage to facilitate sodium removal from the root zone. When a shallow water table is part of the problem, it must be lowered before reclamation can proceed. It can also be improved by altering the topography or by installing pipe drains. If canal seepage is the cause of shallow water table, the seepage water must be intercepted before it enters the field. Based on these generic solutions, following package of practices has been recommended by CSSRI to reclaim the sodic lands:

- Land leveling and bunding of fields with 35–40 cm high bunds to check outflow and inflow of water from adjoining unreclaimed fields. Strong bunds are also needed to preserve and utilize rainwater for leaching salts and growing rice crop.
- Since the infiltration rate of these soils is quite low, surface drains must be strengthened and optimum sized drains constructed to regulate excessive surface runoff during heavy rainfall.

- Construction of tube well(s) to ensure timely low depth high frequency irrigation regime. Such a practice when adopted over large area also helps to lower the water table to attain permanent reclamation as these irrigation wells also act as vertical drains.
- Soil sampling and testing to determine gypsum requirement (GR) although depending upon soil pH and texture of the soil, it varies from 10 to 15 t ha^{-1} for barren soils where nothing can be grown. Gypsum needs to be applied @ 50% of the GR only once in the 1st year of soil reclamation.
- Apply gypsum powder in the well-plowed and leveled fields in the month of June or early July. It should be mixed in the upper 8–10 cm soil depth. After gypsum application, water is kept standing in the field for 10–15 days before transplanting rice as the first crop.
- CSSRI recommends that rice-wheat crop rotation should be adopted during the first three years of reclamation period. Salt tolerant rice and wheat varieties should be grown on newly reclaimed sodic soils, which later on can be replaced with high yielding varieties for better economic returns.
- For rice crop, 35–40 days old seedlings grown on the normal soil should be transplanted at a distance of 15 cm keeping 3–4 plants per hill. Seed rate needs to be kept high (40–50 kg ha^{-1}) as compared to the normal lands. Similarly, seed rate of wheat may be increased by 20–25% over the recommended rate for normal soils.
- Apply 25% more nitrogen to rice crop as compared to the normal soil. Nitrogenous fertilizers and zinc sulfate should be applied @ 150 kg and 25 kg ha^{-1}, respectively.
- Phosphorus and potash need not be applied during initial years of the reclamation (5–6 years) although this recommendation needs to be supported by soil tests.
- Wheat crop should be grown during winters. The salt tolerant wheat cultivars should be grown during initial stages of the reclamation. Barseem (Trifolium alexandrinum) or shaftal (Trifolium resupinatum) can also be grown for fodder for the animals after 2–3 years of reclamation.
- Five to six light irrigations are recommended for the wheat crop.
- Sesbania should be grown for green manuring during summers.

- The field should not be allowed to remain fallow during the recla-mation period. After few years of continuous cropping, other crops may be introduced to diversify the cropping system.

Sodic soils are usually the most expensive to reclaim and, in many situ-ations, reclamation is uneconomical. The reclamation procedure discussed herein can improve sodic soils and a good rice crop can be harvested right in the first year of reclamation, yet many years or decades of good soil and crop management are required to fully remediate a sodic soil [74]. In this intermediary period the soils are usually designated as semi-reclaimed.

12.3   NUTRIENT MANAGEMENT

In sodic soils reclamation programs, the main problem arise as a result of adverse effects on plant growth because of limited nutrient availabil-ity resulting from host of unfavorable physico-chemical conditions. Thus, appropriate prescription of macro and micro-nutrients in right quantities at the right time and place, from the right source, and in the right combina-tion would play a major role in sustaining the reclamation benefits. These and similar other aspects of efficient fertilizer management are discussed in this section.

12.3.1   NITROGEN

In one or the other from, nitrogen (N) accounts for about 80% of the total nutrients absorbed by the plants [34]. Salt affected soils in general and sodic soils in particular are deficient in available nitrogen (N) due to low organic matter content, less biological N_2-fixation, slow transformation of NH_4^+-N to NO_3 – N, and high volatilization losses. A host of chemical and biochemical processes are involved in the turnover of N in the soil through various N transformation processes. High soil pH coupled with poor physical conditions adversely affects the transformations and avail-ability of applied nitrogenous fertilizers in sodic soils. Martin et al. [36] reported that threshold pH value for nitrification of ammonia is 7.1 ± 0.1. They observed that nitrification did occur at higher pH values but was accompanied by considerable accumulation of nitrites. Soil inorganic N, particularly NO_3 is readily soluble and is lost in drainage or through leaching depending on soil condition and crop uptake.

NH_3 volatilization is another pathway through which N is lost from the soil-plant system. The ESP has a marked effect on the amount of added N volatilized, increasing with increasing ESP although quantum may vary with soil type. Ammonia volatilization losses are quite high in the initial 24 hours of urea application and sharply decline thereafter during the next 48 hours. Ammonium fertilizers are particularly liable to volatilization loss if:

- these remain on the surface of a damp but drying calcareous soil; and
- the fertilizer anion forms an insoluble calcium salt.

High volatilization losses of NH_3 were noticed as a result of decrease in solubility of reaction products of NH_4^+-N sources with Ca compounds. Jewitt [23] observed that 87% N is lost when ammonium sulfate was applied to a Barber soil having pH 10.5 in northern Sudan. Similarly, Bhardwaj and Abrol [8] observed that 32–52% of the applied nitrogen was lost through volatilization in a sodic soil. The method of N application has been found to influence the ammoniacal N concentration in flood water [31] and therefore, plays an important role in ammonia volatilization. Significant increase in grain yield of rice was obtained when one-third of total urea-N was applied before puddling in a partially reclaimed sodic soil under 1 week pre-submerged conditions. The benefits were attributed to reduced N losses through volatilizations resulting in increased applied N use efficiency [31]. Laboratory and also field studies have shown lesser losses of N from green manuring as compared to urea – N [51, 81]. The loss of N as NH_3 volatilization from green manuring combined with urea was 13.4% as compared to 19.55% when urea was applied alone (Table 12.2).

Symbiotic nitrogen fixation is low in salt affected soils, mainly attributable to the sensitivity of the microbes to high pH [57] and reduced growth of most legumes, the host plants. Bhardwaj [7] reported that though Rhizobia could survive in sodic soils of pH as high as 10.0, the effective contribution of the bacteria to the plants needs of nitrogen is limited because of delayed nodulation and the sensitivity of the host plants to soil sodicity.

The source of nitrogen may have its own influence on growth and yield. Ammonium sulfate proved to be a better source of N for rice and

TABLE 12.2   Ammonia Losses From Integrated Nutrient Management System in Rice
Fields in a Semi-Reclaimed Sodic Soil

Treatment combination	Urea N loss per split			Total urea N lost	% urea N loss	pH
	I^{st}	2nd	3rd			
Control (No fertilizer application)	1.23	-	-	1.23	-	8.56
N_{120}	8.49	8.21	6.76	23.46	19.55	8.49
$N_{120} P_{22}$	8.28	7.35	6.70	22.33	18.61	8.48
$N_{120} P_{22} K_{42}$	8.14	7.24	6.65	21.75	18.13	8.45
$N_{120} P_{22} K_{42}$ + GM	5.82	5.20	5.06	16.08	13.40	8.10
$N_{120} P_{22} K_{42}$ + FYM	6.73	5.74	5.28	17.75	14.79	8.15
$N_{180} P_{39} K_{63}$	12.12	10.60	9.48	32.20	17.89	8.49
Mean	8.26	7.39	6.66			
CD (P = 0.05)						
(i) Treatments	0.51	0.91	1.19			
(ii) Stage of urea application	0.32					

wheat in terms of grain yield, as compared to urea and calcium ammonium nitrate [41]. This was attributed to the beneficial effect of residual acidity of Ammonium sulfate. However, farmers are often attracted towards urea because of the cost differential between the two kinds of fertilizers.

Because of low available N status and slow transformation of organic and inorganic-N in salt affected soils, added N exhibits poor use efficiency and as a result crops grown in these soils generally respond to higher levels of N than those raised in normal but otherwise similar soil and climatic conditions (4, 15, 27). In general, it is recommended that crops grown in salt affected soils be fertilized at 25% higher N over the rates recommended for normal soils. To get the maximum advantage from the applied fertilizer-N, nitrogen application should synchronize with the growth stage at which plants have the maximum requirement for this nutrient. Rice and wheat plants use nitrogen most efficiently when it is applied at the maximum tillering stage. Rice plants use N around the panicle initiation/jointing stage also. Therefore, split application of N for wheat (1/2 at sowing, remaining 1/2 in two splits at tillering (21 days) and 42 days after sowing and for rice (half at transplanting + 1/4 at tillering + 1/4 at panicle initiation) resulted in maximum efficiency [15]. Other field experiments have shown that maximum yields of rice and wheat were obtained when N was applied in 3 equal splits, as basal and at 3 and 6 weeks after transplanting/sowing and foliar application of N (3% solution of urea) [73, 80].

At lower input levels even a small increase in the amount of N alone has a dramatic effect on the crop yields. However, as the amount of the applied fertilizer progressively increases, the additional/proportional response of crop diminishes and eventually levels off at higher doses and may not be economical. In intensive cropping systems, heavy use of chemical fertilizers alone has created economic, environmental and ecological problems adversely affecting the sustainability of agricultural. The crops yields are higher when both chemical and organic sources are used in an integrated manner as compared to either chemical or organic sources alone, as discussed in a forthcoming section of this chapter. It is attributed to the proper nutrient supply as well as creation of better soil physical and biological conditions when the two are used in an integrated manner.

Based on the above, it is recommended that to ensure a sustainable land reclamation program and to achieve high fertilizer-N efficiency in reclaimed sodic soils:

- Apply 25% more N than the recommended doses for normal soils, normally add 120–150 kg N ha^{-1}.
- Split N application in three doses, 50% as basal, 25% as top dressing after 21 days and another 25% after 42 days of transplanting/sowing.
- Apply basal dose at the time of puddling to mix in the soil rather than broadcast on the surface to minimize N-losses.
- Practice green manuring with Sesbania, which will add N equivalent to 60–80 kg N through urea, and minimize N- losses

12.3.2 PHOSPHORUS

Phosphorus, a constituent of a large number of macromolecules like phospholipids, nucleic acids, phosphor-proteins, dinucleotide and adenosine triphosphate etc., is required for several plant processes such as storage and transfer of energy, photosynthesis, the regulation of some of the enzymes and transport of carbohydrates. The plant roots largely absorb it as dihydrogen orthophosphate ion ($H_2PO_4^-$), however, under neutral to alkaline environments; it is also taken up as monohydrogen orthophosphate (HPO_4^{2-}) ion. Sodic soils have been reported to contain high amount of soluble phosphorus. The high amounts of Na_2CO_3 and Na_2HCO_3 present in sodic soils react with native insoluble calcium phosphates to form soluble sodium phosphate to give a positive correlation between the electrical conductivity and the soluble P status.

Research conducted at CSSRI revealed that there is no response to added phosphorus in sodic soils during early years of reclamation. The rice crop responds to phosphorus application only when the Olsen's P in 0–15 cm layer falls below 7.5–8.0 kg P ha^{-1} after five year of cropping [11, 76]. Since wheat being a relatively deep rooting crop, it is able to forage P from the lower layers to meet its P requirement and as such does not respond to applied P for another 3 to 5 years. However, few other studies indicate that sodic soils are not always high in available phosphorus and significant increase in yields of some crops is obtained with application of P fertilizer. It is attributed to decrease in Olsen's extractable P of surface soil due to its movement to lower sub-soil layers, uptake by the crop and increased immobilization when these soils are reclaimed by using amendments and growing rice under submerged conditions [12, 73].

Long-term field studies on a gypsum amended sodic soil (pH$_2$ 9.2, ESP 32) with rice, wheat and pearl millet cropping sequence and NPK fertilizer use for 20 years (1974–75 to 1993–94) revealed that phosphorus applied @ 22 kg P ha^{-1} to either or both rice and wheat crops in rotation significantly enhanced the grain yield of rice after 1978 [76], when Olsen's extractable P in 0–15 cm soil depth had come down to 12.7 kg ha^{-1}, which is very close to widely used critical soil test value of 11.2 kg P ha^{-1}[43]. Wheat responded to applied P when available P came down close to 8.7 kg P ha^{-1} in 0–15 cm soil depth and nearly close to critical level (11.2 kg P ha^{-1}) in the lower depths (15–30 cm). Application of N alone significantly enhanced the grain yield of pearl millet but phosphorus applied either or to both crops had no effect on yield though available P declined to less than the critical soil test value in 0–15 and 15–30 cm soil depths. With regular P application to crops in sequence the available P increased from initial value of 33.6 kg ha^{-1} to 60.0 in NP-NP and to 56.6 kg ha^{-1} in NPK-NPK in the 0–15 cm soil depth. Where N alone was applied P declined progressively to 4.0 kg ha^{-1} from initial 33.6 kg ha^{-1} (Table 12.3). Further studies proved that crop responds to applied P at level, i.e., 11 kg P ha^{-1} in the initial years of cropping and that too only to rice crop in a rice-wheat cropping sequence. At this stage, application of 22 kg P ha^{-1} significantly enhanced the rice and wheat yield. Recent studies on integrated nutrient management showed that continuous use of fertilizer P, green manuring and FYM to crops significantly enhances the yield of rice and wheat and improves available P status of the sodic soils [82].

12.3.3 POTASSIUM

Potassium is not only the most prominent inorganic constituent for osmoticum mediating cell expansion and turgor-driven movements, is also essential for protein synthesis, glycolytic enzymes and photosynthesis as it regulates the opening and closing of the stomata. It plays a major role in transport of water and nutrients throughout the plant. Since potassium is known to influence root growth, it improves the drought tolerance of plants. Being competitor to Na$^+$ uptake by the plants, it also helps to manage salt stress.

Potassium is absorbed as potassium ion (K$^+$), its concentration in healthy plant tissues varying from 1 to 5%, which is more or less equal to that of N. Potassium is a unique element in the sense that plants are able to accumulate it in abundant amounts without exhibiting any toxicity symptoms. Under reduced K supply, translocation of NO_3^-, Ca^{2+}, Mg^{2+} and amino acids is hampered. Plants invariably show a decrease in K concentration under salt stress and therefore, maintenance of adequate levels of K are essential for plant survival in saline habitats [34]. In general, K content of the plants decreases while Na content increases with increasing soil ESP. Numerous studies have shown that K concentration in plant tissue declines with increase in the levels of salt stress or Na/Ca in the root media leading to high Na/K ratio [1, 22, 25, 46, 48, 61, 62, 65]. Several studies have shown reduced uptake of K by plants raised in alkali/sodic soils due to high Na content and deficiency of Ca [35, 62]. Contrary to these observations, an increase in K levels in the cell sap of bean leaves was observed

TABLE 12.3 Effect of Phosphorus Fertilizers on Yield of Crops and Available P on a Long-Term Basis

Treatments		Grain yield (Mg ha⁻¹)						Millet* Mean	Wheat Mean	Avg. P kg ha⁻¹	
		Rice			Wheat					After 1985–86	After 1993–94
Rice/millet	Wheat	1974	1985	Mean	1974–75	1985–86	Mean	1986–94	1986–94	wheat	wheat
Control		3.81	3.17	3.17	0.84	0.78	1.04	0.75	1.10	17.9	9.6
N	N	6.64	4.73	5.33	4.11	4.00	4.53	1.71	3.48	8.7	4.0
NP	NP	6.56	6.92	6.84	3.71	4.73	4.89	1.95	4.78	7.1	60.0
NP	N	6.63	6.97	6.79	4.14	4.62	4.87	1.93	4.49	45.2	14.6
N	NP	7.17	6.47	6.71	3.90	4.89	4.93	1.94	4.66	45.9	15.6
NPK	NPK	7.08	6.97	6.96	4.05	4.58	4.83	2.11	4.90	67.4	56.6
NPK	N	6.45	6.68	6.80	4.02	4.32	4.62	2.07	4.58	4.7	15.0
N	NPK	6.85	6.45	6.59	4.14	4.28	4.70	2.01	4.71	46.5	14.8
LSD (p=0.05)		0.96	0.55	0.52	0.82	0.36	0.41	0.38	0.54	8.9	5.9

* Average of six crops. Pearl millet (*bajra*) crop during 1988 and 1990 was completely damaged due to heavy floods.

with increasing NaCl salinity [37]. Excess of Na^+ in the growth medium has antagonistic effect on K^+ uptake both under sodicity (CO_3^-, HCO_3^- predominant anions) and salinity where Cl^- and SO_4^- are predominant anions [33, 40, 49]. The magnitude of reduction in K^+ uptake might differ under the two conditions. As an example, chickpea and pea shoots grown in the SO_4^- system contained more K than in the Cl^- system [33, 40]. However, absolute K concentration in the plant tissue is nearly always above the lower critical limit. Sodic soils of Indo-Gangetic plains generally contain very high amounts of available K [3, 75]. Application of K fertilizer to either or both the crops had no effect on yields of rice and wheat [76, 77]. Lack of crop responses to applied K in these soils is attributed to:

- High available K status due to presence of K bearing minerals; and
- Large contribution of non-exchangeable K (97%) towards total K uptake by plants.

Kanwar [26] also observed that under the field conditions, crop response to applied K fertilizers has not been observed, possibly because of the presence of micaceous minerals and illite, which are capable of releasing sufficient K to meet the crop needs. Studies conducted so far suggest that application of K fertilizer to rice-wheat system can be avoided without any adverse effect on crop productivity and/or K fertility status. The contribution of the non-exchangeable K towards total potassium removal was about 94.9% in the absence of applied K, which decreased to 69.9% with application of K. The decrease was about 50.7% when application of K was combined with organic manures (Table 12.4) [82, 89]. The K content of the plants is less affected by soil ESP in salt tolerant plants/cultivars as compared to salt-sensitive cultivars resulting in more stable/narrow Na:K ratio in the former [24]. Considering these factors, Chhabra [11] concluded that to correct lower content of K in the plants raised on K rich sodic soils, add recommended doses of amendments (50% of GR) to correct Ca:Na:K balance rather than to apply K fertilizers.

12.3.4   ZINC

The plant availability of micronutrients like zinc (Zn), iron (Fe), manganese (Mn), copper (Cu), and molybdenum (Mo) in saline and sodic soils is limited because solubility of micronutrients is pH and pE (negative logarithm of the

TABLE 12.4 Removal and Addition of Potassium (kg ha–1) in Two Year of Cropping Sequence

Treatments	Available K (kg ha⁻¹)		Fertilizer K added (kg ha⁻¹)	K uptake (kg ha⁻¹)			Contribution of non-exchangeable K (kg ha⁻¹)	Contribution (%)
	Before 1996, rice	After 1997–98, wheat		Rice	Wheat	Total		
T_1	198	187	NIL	106.9	59.2	166.1	155.1	93.4
T_2	195	183	NIL	198.9	165.2	364.1	352.1	96.7
T_3	205	180	NIL	249.7	220.2	469.9	444.9	94.7
T_4	249	261	168	303.3	242.3	545.6	389.6	71.4
T_5	227	230	84	203.4	178.4	381.8	300.8	78.8
T_6	254	270	84+180ᵃ	306.4	184.6	491.1	243.1	49.5
T_7	265	289	168+180ᵃ	339.1	268.1	607.2	283.2	46.6
T_8	250	269	84+140ᵇ	276.1	186.1	462.2	257.2	55.6
T_9	252	275	168+140ᵇ	330.3	248.2	578.5	293.5	50.7
T_{10}	255	272	252	325.5	252.8	578.3	343.3	59.4

T_1, Control; T_2. N_{120} kg ha⁻¹; T_3. N_{120} P_{26} kg ha⁻¹; T_4. N_{120} P_{26} K_{42} kg ha⁻¹; T_5. N_{60} P_{13} K_{21} kg ha⁻¹; T_6. T_5 plus Sesbania green manuring before rice; T_7. T_4 plus Sesbania green manuring before rice; T_8. T_5 plus FYM @ 10 Mg ha⁻¹ before rice; T_9. T_4 plus FYM @ 10 t ha⁻¹ before rice and T_{10}. N_{180} P_{39} K_{63} kg ha⁻¹. A and b denotes K contribution of green manure and FYM.

activity of electrons) dependent. As a result, plants experience deficiencies of one or more of these elements in salt affected soils [44]. High pH of sodic soils is exceptionally notorious to impair the solubility of Zn, Lindsay [32] observing 100-fold decrease in solubility of Zn per unit increase in pH. The solubility of Zn in sodic soils depends upon the solubility of $Zn(OH)_2$ and $ZnCO_3$, which are the immediate reaction products [58]. Dhillon et al. [16] reported that Zn concentration in the soil solution was regulated by both $Zn(OH)_2-Zn^{2+}$ (aq) and $ZnCO_3-Zn^{2+}$(aq) systems during the initial periods (up to 21 days) and thereafter by $Zn(OH)_2-Zn^{2+}$ (aq) system alone because of the buffering effect of soil carbonate equilibria. Thus, availability of Zn to plants in sodic soils is limited in spite of the fact that most sodic soils contain high amount of total Zn (40 to 100 mg Zn kg^{-1} soil) and less than 0.6 mg DTPA – extractable Zn kg^{-1} soil [59]. A negative correlation was observed between extractable Zn and pH and $CaCO_3$ content of the soil [39]. They observed that the value of pZn + 2pOH vary between 16 and 17.5 and that of pZn + pCO_3 between 17.6 and 18.8. The solubility of added Zn also decreases with time but increases with increase in ESP of the soil [59]. The higher extractability at high ESP is attributed to the formation of sodium zincate, which is soluble. Shukla et al. [55] observed that Zn adsorption in soils saturated with various cations was in the order of H < Ca < Mg < K < Na. When an amendments is added to reclaim a sodic soil, the extractability of added Zn decreases as a result of [59]:

- Greater adsorption of Zn by Ca-than by Na saturated soil;
- Retention of added Zn on the surface of freshly precipitated $CaCO_3$, formed as a result of reaction between soluble carbonates and added gypsum; and
- Enhanced competition of added Ca with Zn for the DTPA ligands during extraction.

Rice crop raised in the soils invariably suffers due to Zn deficiency, the deficiency symptoms in the form of brown rusty spots on the third mature leaf appearing after 15 to 21 days of transplanting. The affected plants show stunted growth, poor tillering, delayed maturity and low grain yield. Efficiency of applied Zn depends upon the degree of amelioration brought about in the sodic soils. When the recommended dose of amendments is added to the soil, 10 to 20 kg $ZnSO_4$ ha^{-1} is enough to get optimum yields of crops. At low level of gypsum application, the plants suffer more due to excess of Na and deficiency of Ca and are incapable of utilizing Zn result-ing in poor yields [56, 60]. Since the total requirement of Zn by the plants is small, its continuous application results in build-up of DTPA extractable Zn in reclaimed sodic soils. Based on extensive field studies, it has been observed that once the soil contains more than 1 mg kg^{-1} DTPA extract-able Zn, it is enough to apply Zn only to rice crop in rice-wheat cropping sequence to get the optimum yields [13]. Best results are obtained when $ZnSO_4$ is added along with other fertilizers as a basal dose. It is possible to prevent the occurrence of Zn deficiency in rice grown on sodic soils by the application of FYM and/or *Sesbania* green manure [71,72]. Organic amendments like press mud, poultry manure and farmyard manure could effectively supply zinc from the native and applied sources to rice crop in a saline sodic soil [38].

12.3.5   IRON

Since the solubility of iron decreases with increasing soil pH and $CaCO_3$ in the soil, iron exhibits low levels of solubility in sodic soils, which results in its deficiency in plants leading to poor metabolism and reduced growth. A reported negative relationship between pH and availability of Fe is traced to the following reasons:
*   Decreased solubility with increasing soil pH; and
* CO_3^{2-} and HCO_3^{-} induced iron chlorosis.

Reports reveal that even the application of $FeSO_4$, a soluble salt, fails to overcome the iron deficiency under sodic conditions [66]. It is because of the fact the soluble Fe salts applied to sodic soils are rendered unavail-able because of rapid oxidation and precipitation [67]. Application of Fe (3% solution of $FeSO_4$) can provide limited relief to the suffering crop but it must be supplemented by changes in the oxidation status of the soil brought about by prolonged submergence and addition of easily decom-posable organic matter [52]. Increased availability of Fe is expected in reclaimed sodic soils because cultivation of lowland rice is recommended requiring submerged conditions. Available results on Fe concentration in plants growing under salt stress are quite inconsistent. The concentration of Fe in the shoots of pea [14] and in lowland rice [79] increased, but its concentration decreased in the shoots of barley and corn under salinity [20, 21]. Considerable variations have been reported even amongst the genotypes for Fe concentration in shoots [48]. Soils irrigated with sodic water for 9 years recorded significant decrease in DTPA extractable iron [5], because of increasing pH and ESP.

12.3.6 MANGANESE

The sodic soils are rich in total Mn but are generally poor in water soluble plus exchangeable and reducible forms of Mn [70]. Similar to Fe, solubility and availability of Mn is also governed by pH and oxidation-reduction status of the soils. Pasricha and Ponnamperuma [45] found that the effect of $NaHCO_3$ and NaCl on the kinetics of water – soluble Mn^{2+} were similar to those of Fe^{2+}. Substantial leaching losses of Mnoccur following gypsum application in sodic soils [53]. As a result, Mn deficiency is increasingly observed in wheat grown in rice-wheat cropping system on coarse – textured sodic soils. As a result of oxidation of Mn, it is very difficult to correct Mn deficiency by soil application of $MnSO_4$. It requires repeated sprays of $MnSO_4$ to make up the deficiency of this element in upland crops. Adoption of rice-wheat system for more than two decades on gypsum-amended sodic soils resulted in decline of the DTPA- extractable Mn to a level of 2.7 mg kg^{-1}. At this stage wheat responded to $MnSO_4$ application at a rate of 50 to 100 kg ha^{-1} [64]. In a study with rice genotypes for their tolerance to multiple stresses of sodicity and Zn deficiency, increased Mn concentration was observed under sodic conditions [48].

12.3.7 BORON

Boron is absorbed passively as H_3BO_3. One may expect the toxicity rather than deficiency of boron in sodic soils as its availability increases with increase in soil pH and ESP. Kanwar and Singh [29] observed a positive correlation between water soluble B and pH and EC of soils. Uncultivated sodic soils in Haryana and Punjab are reported to have hot water extractable B up to 25 mg kg^{-1} in 0–15 cm layer [9]. Temporal and spatial variability are also observed in sodic soils with B varying from 85 to 100 mg

kg^{-1} as pH varies from 9.6 to 10.7 respectively [6]. B toxicity in sodic soils is minimized as soon as gypsum is added to reclaim these soils. Gupta and Chandra [19] from a laboratory study observed a marked reduction in water soluble B together with pH and SAR, on addition of gypsum to a highly sodic soil. At high pH/ESP, boron is present as highly soluble sodium metaborate, which upon addition of gypsum is converted into relatively insoluble calcium metaborate. The solubility of calcium metaborate is very low, 0.4% as compared to 26 to 30% of sodium metaborate. Kumar and Abrol [30] reported reduced uptake of B even by grasses grown in gypsum-amended sodic soils.

12.3.8   MOLYBDENUM

Sodic soils often contain high amounts of available Mo as its solubility increases with increasing soil pH. As such, forage crops grown on sodic soils might contain high Mo that may turn to be toxic to the animals causing molybdenosis. Application of gypsum, however subdues the Mo content in plants because of reduced pH. It could also be due to the antagonistic interaction of SO_4 released from the applied gypsum with the Mo. Not much information is available on the effect of salt stress on Mo concentration in plants. Salinity caused increased Mo concentrations in maize when the crop was grown in a saline soil [50], but there was no effect of salinity on Mo uptake from solution culture [22].

12.4   INTEGRATED APPLICATION OF INORGANIC FERTILIZERS AND ORGANIC MANURES

Intensive cropping with heavy doses of chemical fertilizers have created economic, environmental and ecological problems and has adversely affected the sustainability of agricultural productivity. Continuous rice–wheat cropping in several parts of India has led to the emergence of micronutrients deficiencies especially that of Zn and Mn. Organic farming, a catch word in the international scenario, relies only on applying organic and/or green manures alone as a primary source of plant nutrients may also not be able to support the sustainability concept. Therefore, integrated use of organic manures and chemical fertilizers seems to be the kingpin in the long-term sustainability of agricultural production and productivity especially in salt affected soils.

Use of green manure crops like *Sesbania* spp. (*S. aculeata*) proved quite useful in increasing organic carbon and improving physical properties of sodic soils especially in the initial years of reclamation [87]. It can easily fit into the rice-wheat cropping system as it can be cultivated during summer months following wheat harvest and before transplanting paddy. Under favorable climate and proper management, *S. aculeata* in 50–55 days can accumulate as much as 100 kg N ha^{-1}, mostly through biological N fixation. A fifty days old *Sesbania aculeata* was able to supply about 4.2 t ha^{-1} dry matter and accumulate 90 kg N, 11 kg P and 90 kg K. Turning over the crop for green manuring, increased the yield of the following rice crop significantly [63] with a saving of 60 kg N and 11 kg P ha^{-1} [81] (Table 12.5). Ghai et al. [18] reported that a decomposition period of 5-days for *Sesbania* species would suffice in a reclaimed sodic soil as compared to a period of 10–15 days normally practiced in non-sodic soils. However, the advantages of green manuring in sodic soils were more when it was decomposed for one week under submerged conditions prior to transplanting of rice [69]. Long-term field experiment in sodic soils at CSSRI, Karnal showed that the rice and wheat yield could be maintained even when only 50% of the recommended dose of NPK was used in an integrated manner with FYM or *Sesbania* green manuring [83] (Table 12.5).

12.4.1 ORGANIC AMENDMENTS

Use of organic amendments like Sulphitation Press Mud (SPM) and Farmyard Manure (FYM) with inorganic fertilizers has been found to be more effective in improving and maintaining fertility of soil under irrigation with sodic/good quality water [78, 88] (Table 12.6). In sugarcane, gypsum and FYM, either alone or together, decreased the adverse effect of high RSC irrigation water and improved yield contributing parameters like cane number, cane height and cane thickness [10]. Sharma and Minhas [54] reported that yields of both wheat and cotton did not decline during the initial 4 years of irrigation with different quality water combinations of

TABLE 12.5 Effect of Different Treatments on Yield and N:P:K Soil Status of Rice–Wheat Cropping System in Reclaimed Sodic Soils After 12 Years

Treatments	Yield (t ha^{-1})		Soil available macro-nutrients (kg ha^{-1})		
	Wheat (2004–05)	Rice (2005)	N	P	K
Control	1.97	2.23	91	7.8	160
$N_{120} P_0 K_0$	3.64	3.78	144	6.2	158
$N_{120} P_{26} K_0$	4.25	4.65	168	14.8	155
$N_{120} P_{26} K_{42}$	4.46	4.92	169	15.8	275
$N_{60} P_{13} K_{21}$	3.87	4.08	132	11.4	232
$N_{60} P_{13} K_{21}$ + GM	4.66	6.07	175	18.2	285
$N_{120} P_{26} K_{42}$ +GM	5.02	5.62	201	20.8	301
$N_{60} P_{13} K_{21}$ + FYM	4.22	5.26	176	19.1	291
$N_{120} P_{26} K_{42}$ + FYM	5.17	5.54	198	21.5	298
$N_{180} P_{39} K_{63}$	5.12	5.64	197	24.7	315
CD (P=0.05)	0.49	0.54	9.7	0.98	11.7

2 levels of salinity (EC_{iw} 2 and 4 dS m^{-1}), 2 RSCs (5 and 10 meq L^{-1}) and 3 SARs (10, 20 and 30) although the wheat yields reduced significantly at higher levels of RSC and SAR in sodic water (RSC 10 meq L^{-1} and SAR 30) during 5th year especially when EC_{iw} was 4 dS m^{-1}. The use of gypsum, pyrites and FYM in conjunction with irrigation water having EC 4 dS m^{-1} and SAR 10 resulted in significantly higher yield of rice and wheat [17].

12.4.2 CROP RESIDUES

Over the years, press mud is being used by brick kiln industry and FYM is not available in adequate quantities, as a result researchers are experimenting to find alternative sources of organic products that are available on-site. As an example, large quantities of combine harvested rice and wheat straw are now-a-days being generated at farms. One ton of rice residues contain approximately 6.1 kg N, 0.8 kg P and 11.4 kg K, while one ton of wheat residue contains 5.1 kg N, 1.2 kg P and 10.5 kg K. All the rice and a part of wheat residue is burnt causing immense loss of

TABLE 12.6 Effect of Gypsum With and Without Organic Manures on Yield of Rice and Wheat and Soil Properties

Treatments	Mean yield (t ha⁻¹)		Soil pH	OC (%)	Available nutrients (Kg ha⁻¹)		
	Rice (1994–03)	Wheat (1994–04)			N	P	K
$N_0P_0K_0$	2.60	1.76	8.50	0.32	104	10.2	215
$N_{120}P_{26}K_{42}$	4.9	3.69	8.52	0.35	140	22.6	285
$N_{120}P_{26}K_{42}$ + FYM	5.29	4.16	8.38	0.43	164	22.9	299
$N_{120}P_{26}K_{42}$ + gypsum	5.23	4.10	8.18	0.37	145	19.0	297
$N_{120}P_{26}K_{42}$ + Press mud	5.31	4.46	8.29	0.42	160	24.2	298
$N_{120}P_{26}K_{42}$ + FYM + gypsum	5.35	4.22	8.28	0.42	156	24.5	300
$N_{120}P_{26}K_{42}$ + PM + gypsum	5.41	4.52	8.28	0.40	160	24.0	297
CD at 5 %	0.42	0.34	0.08	0.60	8.9	2.1	20.5

OC = Organic carbon.

nutrients besides causing environmental pollution in the form of toxins and green house gases. On the other hand, if the residues are returned to the soil, it will provide much needed organic matter, energy and nutrients to the soil. The results of a study show that incorporation of wheat residue 50 days prior to rice transplanting either alone or with green manuring or with SPM with recommended dose of NP improved rice yield as compared to recommended dose of N and P fertilizer alone [84, 86] (Table 12.7). Tillage practices also affect the physico-chemical properties of semi-reclaimed sodic or sodic water irrigated soils. No-tillage (NT) practice was found to increase organic carbon and infiltration rate of sodic water irrigated soil in comparison to conventional tillage practices. NT practice also reduced soil pH and SAR. The NT either alone or with residual effect of gypsum or SPM or FYM has been found as an effective option to sustain higher yields of wheat with sodic water irrigation in a rice-wheat system. Besides, it saved 7.22 cm of irrigation water and reduced energy requirement as three disking and planking operations could be avoided [85].

TABLE 12.7 Effect of Crop Residue Management on Yield (Mean of 3 years) and Soil Properties of Semi-Reclaimed Sodic Soil Using Poor Quality Water

Treatments	Grain yield (t ha^{-1})		pH	OC (%)	Available nutrients (kg ha^{-1})		
	Rice	Wheat			N	P	K
$N_0 P_0$	1.05	0.87	9.05	0.28	142	26.8	201
$N_{90} P_{19.5}$ (75 % NP)	2.69	2.31	8.98	0.26	143	25.4	198
$N_{120} P_{26}$ (100 % NP)	3.49	2.94	8.97	0.27	144	27.9	193
100 % NP + Wheat residue burnt	3.72	2.99	8.78	0.32	158	29.3	220
100 % NP + Incorporated wheat residue	4.34	3.19	8.65	0.33	157	29.4	223
100 % NP + Incorporated wheat residue + GM	4.45	3.35	8.70	0.33	157	29.0	220
100 % NP + Incorporated wheat residue + SPM	4.41	3.34	8.79	0.34	156	29.2	219
CD (0.05)	0.58	0.45	0.101	0.011	5.65	1.69	9.40

12.5   SUMMARY

Plants/crops growing in sodic soils or soils irrigated with sodic water are exposed to a number of unfavorable conditions like moisture stress, poor soil physical conditions, nutrients availability and imbalance resulting in their deficiencies and elemental toxicity, etc. Amongst these, inadequate availability of nutrients to plants in sodic soils or soils irrigated with sodic waters have been one of the major reasons for poor plant growth and low crop yields. The relations between salt stress and mineral nutrition of plants are quite complex as activity of nutrient elements is altered because of excess of potentially toxic ions and antagonistic effects on uptake of each other and pH induced changes in their solubility and availability. Salt affected soils have inadequate levels of some nutrients like N, whereas availability of others such as Zn is low. To obtain optimum benefits from applied nutrients, the plant nutrients must be applied in right quantities, at an appropriate time and place using a proper source and in a right combination so as to eliminate/minimize the adverse effect of sodic environment. As such, the chapter includes number of strategies in respect of application of N, P, K, Zn, Mn, Mo so as to improve the soil nutritional status to get good plant growth, higher yields and better nutrient use efficiencies. It also emphasizes on the role of integrated nutrient management including the crop residues for nutrient management under the sodic environment. On the basis of discussions made in this chapter, it emerges that there is a need for systematic studies that deals with physiological aspects and interactions under multiple nutrient deficiencies commonly occurring under sodic soils or sodic water irrigated lands. The thrust areas requiring focus are:

- Develop practices to improve nutrient use efficiency;
- Minimizing the dependence on inorganic fertilizer, as the raw materials for their production are finite;
- Sustaining the crop productivity in post-reclamation phase in relation to soils and water quality vis-à-vis organic matter dynamics and carbon sequestration;
- Identifying varieties having higher nutrient use efficiency as well as tolerance to salt with the long-term objective of multiple stress tolerance;

- Generating resource inventories of salt affected soils and poor quality waters, their nature and properties that influence nutrients availability and nutrients dynamics.

KEYWORDS

- alkali soils
- amendments
- boron
- calcium carbonate
- calcium sulfate
- copper
- crop residues
- DTPA extractable Zn
- ESP
- fertility
- green manuring
- gypsum
- integrated nutrient management
- iron
- leaching
- management (soils)
- Manganese
- molybdenum
- nitrogen
- nitrogen fixation
- Olsen's P
- organic amendments
- phosphorus
- physico-chemical properties
- potassium
- reclamation

- **rice–wheat cropping sequence**
- **saline soils**
- **semi-reclaimed sodic soils**
- *Sesbania*
- **Sodic soils**
- **split application**
- **submergence**
- **volatilization losses**
- **zinc**

REFERENCES

1. Afridi, M. M. R. K., Qadar, A., & Dwivedi, R. S. (1988). Effect of salinity and sodicity on growth and ionic composition of *Oryza sativa* and *Diplachne fusca*. *Journal Indian Botanical Society*, 67, 167–182.
2. Agarwal, R. R., & Yadav, J. S. P. (1954). Diagnostic techniques for the saline and alkali soils of the Indian Gangetic alluvium in Uttar Pradesh. *Journal Soil Sci.*, 5, 330–306.
3. Agarwal, R. R., Yadav, J. S. P., & Gupta, R. N. (1979). *Saline and Alkali Soils of India*. ICAR, New Delhi.
4. Bains, S. S., & Pandey, S. L. (1968). Comparative performance of some *rabi* (winter) season crops in saline and high ground water table conditions. *Proc. Symp. Water Mgmt. Udaipur*, 261–267.
5. Bajwa, M. S., Josan, A. S., & Chaudhary, O. P. (1993). Effect of frequency of sodic and saline-sodic irrigation and gypsum on the build-up of sodium in soil and crop yields. *Irrigation Science,* 13, 21–26.
6. Bandyopadhya, A. K. 1974. Seasonal variation of boron in saline sodic soils. *Ann. Arid and Semi-Arid Zones*, 13, 125–128.
7. Bhardwaj, K. K. R. (1975). Survival and symbiotic characteristics of *Rhizobium* in saline-alkali soils. *Plant and Soil*, 43, 377–385.
8. Bhardwaj, K. K. R., & Abrol, I. P. (1979). Nitrogen management in alkali s*oils. Proc. Natn. Symp. Nitrogen Assimilation and Crop Productivity*, Hisar, India, 83–96.
9. Bhumbla, D. R., Chhabra, R., & Abrol, I. P. (1980). Distribution of boron in alkali soils. *Ann. Report.* Central Soil Salinity Research Institute, Karnal, India.
10. Chaudhary, O. P., Josan, A. S., Bajwa, M. S., & Kapur, M. L. (2004). Effect of sustained sodic and saline-sodic irrigation and application of gypsum and FYM on yield and quality of sugarcane. *Field Crop Research,* 87, 103–116.

11. Chhabra, R. (1985). Crop responses to phosphorus and potassium fertilization of a sodic soil. *Agronomy Journal,* 77, 699–702.
12. Chhabra, R., Abrol, I. P., & Singh, M. V. (1981). Dynamics of phosphorus during reclamation of sodic soils. *Soil Science,* 132, 319–324.
13. Chhabra, R., Ringoet, A., & Lamberts, D. (1976). Kinetics and interaction of chloride and phosphate absorption by intact tomato plants (*Lycopersicom esculentum Mill*) from a dilute nutrient solution. *Z. Pfianzenphysiol,* 78, 253–261.
14. Dahiya, S. S., & Singh, M. (1976). Effect of salinity alkalinity and iron application on the availability of iron, manganese, phosphorus and sodium in pea (*Pisum sativum* L.) *Crop, Plant and Soil,* 44, 697–702.
15. Dargan, K. S., & Gaul, B. L. (1974). Optimum nitrogen levels for dwarf varieties of paddy in semi-reclaimed alkali soils. *Fertilizer News,* 19, 24–25.
16. Dhillon, S. K., Sinha, M. K., & Randhawa, N. S. (1975). Chemical equilibria and Q/I relationship of zinc in selected alkali soils of Punjab. *J. Indian Soc. Soil Sci.,* 23, 38–46.
17. Dubey, S. K., & Mondal, R. C. (1994). Effect of amendments and saline irrigation water on soil properties and yields of rice and wheat in a highly sodic soil. *Journal of Agriculture Science (Cambridge),* 122, 351–357.
18. Ghai, S. K., Rao, D. L. N., & Batra, L. (1988). Nitrogen contribution to wetland rice by green manuring with *Sesbania* species in an alkaline soil. *Biol. Fertility Soils,* 6, 22–25.
19. Gupta, I. C., & Chandra, H. (1972). Effect of gypsum in reducing boron hazards of saline waters and irrigated soils. *Ann. Arid zone,* 11, 228–230.
20. Hassan, N. A. K., Drew, J. V., Knudsen, D., & Olson, R. A. (1970a). Influence of soil salinity on production of dry matter and uptake and distribution of nutrients in barley and corn. II. Corn (*Zea mays L.*). *Agronomy Journal,* 62, 46–48.
21. Hassan, N. A. K., Jackson, W. A., Drew, J. V., Knudsen, D., & Olson, R. A. (1970b). Influence of soil salinity on production of dry matter and uptake and distribution of nutrients in barley and corn. 1. Barley (*Hordeum vulgare*). *Agronomy Journal,* 62, 46–48.
22. Izzo, R., Navari-Izzo, F., & Quartacci, M. F. (1991). Growth and mineral absorption in maize seedlings as affected by increasing NaCl concentrations. *Journal Plant Nutrition,* 14, 687–699.
23. Jewitt, T. N. (1942). Loss of ammonia from ammonium sulfate applied to alkaline soils. *Soil Science,* 54, 401–409.
24. Joshi, Y. S., Qadar, A., Bal, A. R., & Rana, R. S. (1980). Sodium/potassium index of wheat seedlings in relation to sodicity tolerance. *Proc. International Symposium on Management of Salt Affected Soils,* Karnal, India, 457–460.
25. Joshi, Y. C., Qadar, A., & Rana, R. S. (1979). Differential sodium and potassium accumulation related to sodicity tolerance in wheat. *Indian Journal Plant Physiology,* 22, 226–230.
26. Kanwar, J. S. (1961). Clay minerals in saline and alkali soils of the Punjab. *J. Indian Soc. Soil Sci.,* 9, 35–40.
27. Kanwar, J. S., Bhumbla, D. R., & Singh, N. T. (1965). Studies on the reclamation of saline and sodic soils in the Punjab. *Indian J. Agric. Sci.,* 35, 43–51.

28. Kanwar, J. S., & Bhumbla, D. R. (1969). Physico chemical characteristics of sodic soils of Punjab and Haryana and their amelioration by the use of gypsum. *Agrokem Talajt.*, 18, 315–320.

29. Kanwar, J. S., & Singh, S. S. (1961). Boron in normal and saline alkali soils of the irrigated areas of the Punjab. *Soil Science*, 92, 207–211.

30. Kumar, A., & Abrol, I. P. (1982). Note on the effect of gypsum levels on the boron content of soils and its uptake by five forage grasses in a highly sodic soil. *Indian J. Agric Sci.*, 52, 615–617.

31. Kumar, D., Swarup, A., & Kumar, V. (1995). Effect of rates and methods of urea – N application and pre-submergence periods on ammonia volatilization losses from rice fields in a sodic soil. *J. Agric. Sci. (Camb.)*, 125, 95–98.

32. Lindsay, W. L. (1972). Inorganic phase equlibria in soils. In: *Micronutrient in Agriculture* (Mortvedt, J. J. Ed.), Soil Science Society America, Madison, Wisconsin, 48–51.

33. Manchanda, H. R., & Sharma, S. K. (1989). Tolerance of chloride and sulfate salinity in chickpea (*Cicer arietinum*). *Journal Agricultural Science (Cambridge)*, 113, 407–410.

34. Marschner, H. (1995). *Mineral Nutrition of Higher Plants.* Academic Press, London.

35. Martin, J. P., & Bingham, F. T. (1954). Effect of various exchangeable cation ratios in soil on growth and chemical composition of avocado seedlings. *Soil Science*, 78, 349–360.

36. Martin, J. P., Buehrer, T. F., & Caster, A. B. 1942. Threshold pH value for nitrification of ammonia in desert soils. *Soil Sci. Soc. Am. Proc.*, 7, 223–228.

37. Meiri, A., Kamburoff, J., & Poljakoff-Mayber, A. (1971). Response of bean plant to sodium chloride and sodium sulfate salinization. *Annals Botany*, 35, 837–847.

38. Milap Chand, Bhumbla, D. R., Randhawa, N. S., & Sinha, M. K. (1980). The effect of gypsum and organic amendments on the availability of zinc to rice and its uptake by this crop grown in a sodic. *Intern. Symp. on Management of salt-affected Soils*, Karnal, India. 348–355.

39. Mishra, S. G., & Pandey, G. (1976). Zinc in saline and alkali soils of Uttar Pradesh. *J. Indian Soc. Soil Sci.*, 24, 336–338.

40. Mor, R. P., & Manchanda, H. R. (1992). Influence of phosphorus on the tolerance of table pea to chloride and sulfate salinity in a sandy soil. *Arid Soil Research Rehabilitation*, 6, 41–52.

41. Nitant, H. C., & Bhumbla, D. R. (1974). Transformation and movement of nitrogen fertilizers in sodic soils. Annual Report, Central Soil Salinity Research Institute, Karnal, India. 24.

42. NRSA and Associates. (1996). *Mapping of Salt Affected Soils in India, 1:250,000 map sheets, Legend.* National Remote Sensing Agency, Hyderabad, India.

43. Olsen, S. R., Cole, C. V., & Watanabe, F. S. (1954). Estimation of available phosphorus in soils by extraction with sodium bicarbonate. USDA Circular No. 939. Washington, D. C., USA: United States Department of Agriculture.

44. Page, A. L., Chang, A. C., & Adriano, D. C. (1990). Deficiencies and toxicities of trace elements. In: *Agricultural Salinity Assessment and Management; Manuals and Reports on Engineering Practices* (Tanji, K. K. Ed.) American Society of Civil Engineers, New York, 71, 138–160.

45. Pasricha, N. S., & Ponnamperuma, F. N. (1976) Influence of salt and alkali on ionic equilibria in submerged soils. *Soil Science Society of American. Proceeding,* 40, 374–376.

46. Perez-Alfocea, F., Balibrea, M. E., Santa Cruz, A., & Estan, M. T. (1996). Agronomical and physiological characterization of salinity tolerance in a commercial tomato hybrid. *Plant Soil,* 180, 251–257.

47. Qadar, A. (1991). Differential sodium accumulation in shoots of rice genotypes in relation to their sodicity tolerance. *Indian Journal Agricultural Sciences,* 61, 40–42.

48. Qadar, A. (2002). Selecting rice genotypes tolerant to zinc deficiency and sodicity stresses. 1. Differences in zinc, iron, manganese, Copper, phosphorus concentrations and phosphorus/zinc ratio in their leaves. *Journal Plant Nutrition,* 25, 457–473.

49. Qadar A., & Zake, M. A. (2007). Selecting rice genotypes tolerant to zinc deficiency and sodicity, differences in concentration of major cations and sodium/potassium ratio in leaves. *Journal of Plant Nutrition,* 30, 2061–2076.

50. Rahman, S., Vance, G. F., & Munn, L. C. (1993). Salinity induced effects on the nutrient status of soil, corn leaves and kernels. *Communication Soil Science Plant Analysis,* 24, 2251–2269.

51. Rao. D. L. N., & Batra, L. (1983). Ammonia volatilization from applied nitrogen in alkali soils. *Plant Soil,* 20, 219–228.

52. Shahi, H. N., Khind, C. S., & Gill, P. S. (1976). Iron chlorosis in rice (*Oriza sativa* L.). *Plant Soil,* 44, 231–232.

53. Sharma, B. M., & Yadav, J. S. P. (1986) Leaching losses of iron and manganese during reclamation of alkali soil. *Soil Science,* 142, 149–52.

54. Sharma, D. R., & Minhas, P. S. (1998) Effect of irrigation with sodic waters of varying EC, RSC and SAR/adj. SAR on soil properties and yield of cotton – wheat. *Journal of the Indian Society of Soil Science,* 46, 116–119.

55. Shukla, U. S., Mittal, S. B., & Gupta, R. K. (1980). Zinc application in some soils as affected by exchangeable cations. *Soil Science,* 129, 366–370.

56. Shukla, U. C., & Mukhi, A. K. (1985). Ameliorative role of zinc on maize (*Zea mays* L.) growth under salt-affected soil conditions. *Plant Soil,* 87, 423–432.

57. Singh, C. S., Lakshmi Kumar, M., Biswas, A., & Subba Rao, N. S. (1973). Effect of carbonate and biocarbonate of sodium on growth of rhizobia and nodulation in lucerne (*Medicago sativa* L.). *Indian J. Microbiology,* 13, 125–128.

58. Singh, M. V., & Abrol, I. P. (1985). Solubility and absorption of zinc in alkali soil. *Soil Science,* 140, 406–411.

59. Singh, M. V., Chhabra, R., & Abrol, I. P. (1984). Factors Affecting DTPA Extractable Zn in Sodic Soils. *Soil Science* 136, pp. 359–366.

60. Singh, M. V., Chhabra, R., & Abrol, I. P. (1987). Interaction between application of gypsum and zinc sulfate on the yield and chemical composition of rice grown in an alkali soil. *Journal of Agricultural Science (Cambridge),* 108, 275–279.

61. Singh, S. B., Chhabra R., & Abrol, I. P. (1980). Effect of soil sodicity on the yield and chemical composition of cowpea *(Vigna unguiculata)* grown for fodder. *Indian J. Agric. Science,* 50, 852–856.

62. Singh, S. B., Chhabra, R., & Abrol, I. P. (1981). Effect of exchangeable sodium on the yield, chemical composition and oil content of safflower (*Carthamum tinctoriuo* L.) and linseed (*Linum usitatissimum* L.). *Indian Journal of Agricultural Science,* 52, 881–891.

63. Singh, Y., Khind, C. S., & Singh, V. (1991). Efficient management of leguminous green manures in wetland rice. *Advances in Agronomy*, 45, 135–139.
64. Soni, M. L., Swarup, A., & Singh, M. (1996). Influence of rates and methods of manganese application on yield and nutrition of wheat in a reclaimed sodic soil. *Journal of Agricultural. Science (Cambridge)*, 127, 433–439.
65. Subbarao, G. V., Johansen, C., Jana, M. K., & Kumar Rao, J. V. D. K. (1990). Effects of the sodium/calcium ratio in modifying salinity response of pigeon pea (*Cajanus cajan*). *Journal Plant Physiology*, 136, 439–443.
66. Swarup, A. (1980). Effect of submergence and farm yard manure application on yield and nutrition of rice under sodic conditions. *Journal of Indian Society Soil Science*, 28, 532–534.
67. Swarup, A. (1981). Effect of iron and manganese application on the availability of micronutrients to rice in sodic soil. *Plant Soil*, 60, 481–485.
68. Swarup, A. (1986). Effect of gypsum, pyrites, farm yard manure and rice husk on the availability of zinc and phosphorus to rice in submerged sodic soil. *J. Indian Soc. Soil Sci.*, 34, 844–848.
69. Swarup, A. (1988). Effect of *Sesbania bispinosa* decomposition and sodicity on yield of rice and N contribution. *International Rice Research Notes*, 13, 28–29.
70. Swarup. A. (1989). Transformation and availability of iron and manganese in submerged sodic soil in relation to yield and nutrition of rice. *Fertilizer News*, 34, 21–31.
71. Swarup, A. (1991a). Effect of gypsum, green manure, farm yard manure and zinc fertilization on the zinc, iron and manganese nutrition of wetland rice on a sodic soil. *J. Indian Soc. Soil Sci.*, 39, 530–536.
72. Swarup, A. (1991b). Long-term effects of green manuring (*Sesbani aaculeata*) on soil properties and sustainability of rice and wheat yield on a sodic soil. *J. Indian Soc. Soil Sci.*, 39, 777–780.
73. Swarup, A. (1994). Chemistry of salt affected soils and fertility management. In: *Salinity Management for Sustainable Agriculture*, Central Soil Salinity Research Institute, Karnal. 18–40.
74. Swarup, A. (2000). Emerging soil fertility management issues for sustainable crop productivity in irrigated system. *Proc. of National Workshop on Long-Term Soil Fertility Management through Integrated Plant Nutrient Supply*, IISS, Bhopal, India.
75. Swarup, A., & Chhillar, R. K. (1986). Build-up and depletion of soil phosphorus and potassium and their uptake by rice and wheat in a long-term field experiment. *Plant Soil*, 91, 161–170.
76. Swarup, A., & Singh, K. N. (1989). Effect of 12-years rice wheat cropping and fertilizer use on soil properties and crop yields in a sodic soil. *Field Crops Research*, 21, 277–287.
77. Swarup, A., & Yaduvanshi, N. P. S. (2000). Effect of integrated nutrient management on soil properties and yields of rice and wheat in alkali soils. *J. Indian Soc. Soil Sci.*, 48, 279–282.
78. Swarup, A., & Yaduvanshi, N. P. S. (2004). Response of rice and wheat to organic and inorganic fertilizers and soil amendment under sodic water irrigated conditions. *IRRI News Letter*, 29, 49–51.
79. Verma, T. S., & Neue, U. (1984). Effect of soil salinity level and zinc application on growth, yield and nutrient composition of rice. *Plant Soil*, 82, 3–14.

80. Yaduvanshi, N. P. S. (2001a). Ammonia volatilization losses from integrated nutrient management in rice fields of alkali soils. *J. Indian Soc. Soil Sci.*, 49, 276–280.

81. Yaduvanshi, N. P. S. (2001b). Effect of five years of rice-wheat cropping and NPK fertilizer use with and without organic and green manures on soil properties and crop yields in a reclamation sodic soil. *J. Indian Soc. Soil Sci.*, 49, 714–719.

82. Yaduvanshi, N. P. S. (2002). Budgeting of P and K for a rice-wheat cropping sequence on a sodic soil. *Tropical Agriculture (Trinidad)*, 79, 211–216.

83. Yaduvanshi, N. P. S. (2003). Substitution of inorganic fertilizers by organic manures and the effect on soil fertility in a rice-wheat rotation on reclaimed sodic soil in India. *Journal of Agriculture Science (Cambridge)*, 140, 161–169.

84. Yaduvanshi, N. P. S., & Sharma, D. R. (2007). Use of wheat residue and manures to enhance nutrient availability and rice-wheat yields in sodic soil under sodic water irrigation. *J. Indian Soc. Soil Sci.*, 55, 330–334.

85. Yaduvanshi, N. P. S., & Sharma, D. R. (2008). Tillage and residual organic manures/chemical amendment effects on soil organic matter and yield of wheat under sodic water irrigation. *Soil and Tillage Research*, 98, 11–16.

86. Yaduvanshi, N. P. S., & Sharma, D. R. (2010). Effect of organic sources with and without chemical fertilizers on productivity of rice (*Oryza sativa*) and wheat (*Triticum aestivum*) sequence and plant nutrients balance in a reclaimed sodic soil. *Indian J. of Agricultural Sciences*, 80, 482–486.

87. Yaduvanshi N. P. S., Sharma, D. R., & Swarup, A. (2013). Impact of Integrated nutrient management on soil properties and yield of rice and wheat in a long-term experiment on a reclaimed sodic soil. *J. Indian Soc. Soil Sci.*, 61, 188–194.

88. Yaduvanshi, N. P. S., & Swarup, A. (2005). Effect of continuous use of sodic irrigation water with and without gypsum, farm yard manure, press mud and fertilizer on soil properties and yields of rice and wheat in a long-term experiment. *Nutrient Cycle in Agroecosystems*, 73, 111–118.

89. Yaduvanshi, N. P. S., & Swarup, A. (2006). Effect of long-term fertilization and manuring on potassium balance and non-exchangeable K release in a reclaimed sodic soil. *J. Indian Soc. Soil Sci.*, 54, 203–207.

INDEX

Milton Keynes UK
Ingram Content Group UK Ltd.
UKHW022043141024
449569UK00022B/794